Serono Symposia, USA
Norwell, Massachusetts

FERTILIZATION IN MAMMALS
Edited by Barry D. Bavister, Jim Cummins,
and Eduardo R.S. Roldan

*FOLLICLE STIMULATING HORMONE: Regulation of Secretion
and Molecular Mechanisms of Action*
Edited by Mary Hunzicker-Dunn and Neena B. Schwartz

GAMETE PHYSIOLOGY
Edited by Ricardo H. Asch, Jose P. Balmaceda,
and Ian Johnston

*GLYCOPROTEIN HORMONES: Structure, Synthesis, and Biologic
Function*
Edited by William W. Chin and Irving Boime

GROWTH FACTORS IN REPRODUCTION
Edited by David W. Schomberg

*THE MENOPAUSE: Biological and Clinical Consequences of Ovarian
Failure: Evaluation and Management*
Edited by Stanley G. Korenman

MODES OF ACTION OF GnRH AND GnRH ANALOGS
Edited by William F. Crowley, Jr., and P. Michael Conn

MOLECULAR BASIS OF REPRODUCTIVE ENDOCRINOLOGY
Edited by Peter C.K. Leung, Aaron J.W. Hsueh, and
Henry G. Friesen

NEUROENDOCRINE REGULATION OF REPRODUCTION
Edited by Samuel S.C. Yen and Wylie W. Vale

*SIGNALING MECHANISMS AND GENE EXPRESSION
IN THE OVARY*
Edited by Geula Gibori

UTERINE CONTRACTILITY: Mechanisms of Control
Edited by Robert E. Garfield

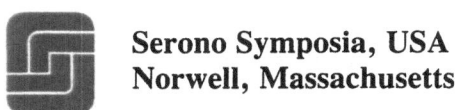

Serono Symposia, USA
Norwell, Massachusetts

Peter C.K. Leung   Aaron J.W. Hsueh
Henry G. Friesen   Editors

Molecular Basis of Reproductive Endocrinology

With 77 Figures

Springer-Verlag
New York Berlin Heidelberg London Paris
Tokyo Hong Kong Barcelona Budapest

Peter C.K. Leung, Ph.D.
Research Division
Department of Obstetrics and
  Gynecology
University of British Columbia
Grace Hospital
Vancouver, British Columbia
Canada V6H 3V5

Aaron J.W. Hsueh, Ph.D.
Division of Reproductive Biology
Department of Gynecology and
  Obstetrics
Stanford University Medical Center
Stanford, CA
USA 94305-5317

Henry G. Friesen, M.D.
Department of Physiology
University of Manitoba
Winnipeg, Manitoba
Canada R3E OW3

Proceedings of the Symposium on the Molecular Basis of Reproductive Endocrinology, sponsored by Serono Symposia, USA, held July 25 to 26, 1991, in Vancouver, British Columbia.

For information on previous volumes, please contact Serono Symposia, USA.

Library of Congress Cataloging-in-Publication Data
Molecular basis of reproductive endocrinology / Peter C.K. Leung,
  Aaron J.W. Hsueh, Henry G. Friesen, editors.
    p. cm.
  "Proceedings of the Symposium on the Molecular Basis of
Reproductive Endocrinology, sponsored by Serono Symposia, USA, held
July 25 to 26, 1991, in Vancouver, British Columbia"—T.p. verso.
  Includes bibliographical references and indexes.

  ISBN-13: 978-1-4613-9262-0   e-ISBN-13: 978-1-4613-9260-6
  DOI: 10.1007/978-1-4613-9260-6

  1. Human reproduction—Endocrine aspects—Congresses.
2. Molecular endocrinology—Congresses. I. Leung, P. C. K.
II. Hsueh, Aaron J. W. III. Friesen, Henry G. IV. Serono Symposia,
USA. V. Symposium on the Molecular Basis of Reproductive
Endocrinology (1991: Vancouver, B.C.)
  [DNLM: 1. Endocrine Glands—physiology—congresses.
2. Reproduction—physiology—congresses. WQ205 M7175 1991]
QP252.M65 1992
612.6—dc20
DNLM/DLC   92-2304

Printed on acid-free paper.

Production coordinated by Technical Texts and managed by Francine Sikorski; manufacturing supervised by Vincent Scelta.
Typeset by Best-set Typesetter Ltd., Hong Kong

9 8 7 6 5 4 3 2 1

SYMPOSIUM ON THE MOLECULAR BASIS OF REPRODUCTIVE ENDOCRINOLOGY

Scientific Committee

Peter C.K. Leung, Ph.D., Chairman
University of British Columbia

Aaron J.W. Hsueh, Ph.D.
Stanford University Medical Center

Henry G. Friesen, M.D.
University of Manitoba

Organizing Secretaries

Bruce K. Burnett, Ph.D.
L. Lisa Kern, Ph.D.
Serono Symposia, USA
100 Longwater Circle
Norwell, Massachusetts

Preface

Recent advances in molecular biology have provided new dimensions in the study of the reproductive system. There has been major progress in our understanding of the molecular mechanisms of hormone action in the past few years. The symposium on "Molecular Basis of Reproductive Endocrinology" was organized to highlight new research findings on the regulation of the hypothalamic-pituitary-gonadal axis. The emphasis of the symposium was on physiological questions answered by the molecular biology approach. Studies on the functional relevance of gonadotropin releasing hormone and LH and FSH gene expression were presented, together with research on the molecular biology of ovarian and testicular steroidogenic enzymes and protein hormones. Also, several novel aspects of hormone gene expression in placental tissues were reviewed.

The symposium was held July 25 to 26, 1991, immediately prior to the 24th Annual Meeting of the Society for the Study of Reproduction, on the campus of the University of British Columbia in Vancouver. Serono Symposia, USA generously financed and coordinated the meeting. We are indebted to Dr. Bruce K. Burnett and Dr. L. Lisa Kern for their professional assistance in the organization of the symposium. We would also like to thank Drs. Victor Gomel, Basil Ho Yuen, and John Challis, who served as session moderators. Most of all, we truly appreciate the efforts of all the invited speakers, poster presenters, and discussants in making this a memorable event as the largest one-day meeting of the Serono Symposia USA, series.

PETER C.K. LEUNG
AARON J.W. HSUEH
HENRY G. FRIESEN

Contents

Part III.   Placenta and Fetus

Part IV.   Poster Presentation Manuscripts

Contributors

YUMIKO ABE, Department of Obstetrics and Gynecology, Gunma University School of Medicine, Maebashi, Japan.

J.P. ADELMAN, Vollum Institute, Oregon Health Sciences University, Portland, Oregon, USA.

SHIGEO AKIRA, Department of Obstetrics and Gynecology, Nippon Medical School, Tokyo, Japan.

O.O. ANAKWE, Division of Reproductive Biology, Department of Obstetrics and Gynecology, University of Pennsylvania School of Medicine, Philadelphia, Pennsylvania, USA.

TSUTOMU ARAKI, Department of Obstetrics and Gynecology, Nippon Medical School, Tokyo, Japan.

GBOLAGADE O. BABALOLA, Department of Obstetrics and Gynecology, University of Pennsylvania, Philadelphia, Pennsylvania, USA.

CAROL A. BAGNELL, Department of Animal Sciences, Rutgers University, New Brunswick, New Jersey, USA.

JANICE M. BAHR, Department of Animal Sciences, Animal Genetics Laboratory, University of Illinois, Urbana, Illinois, USA.

P.A. BAIN, Departments of Obstetrics/Gynecology, the Cell and Molecular Biology Program, and the Reproductive Sciences Program, University of Michigan, Women's Hospital, Ann Arbor, Michigan, USA.

C.T. BOND, Vollum Institute, Oregon Health Sciences University, Portland, Oregon, USA.

TIM D. BRADEN, Department of Pharmacology, University of Iowa College of Medicine, Iowa City, Iowa, USA.

NATHALIE BRETON, Medical Research Council Group in Molecular Endocrinology, CHUL Research Center and Laval University, Quebec, Canada.

GILLIAN D. BRYANT-GREENWOOD, Department of Anatomy and Reproductive Biology, University of Hawaii, Honolulu, Hawaii, USA.

JOHN C. CARLSON, Department of Biology, University of Waterloo, Waterloo, Ontario, Canada.

WILLIAM Y. CHANG, Department of Veterinary Physiology and Pharmacology, College of Veterinary Medicine, The Ohio State University, Columbus, Ohio, USA.

CHERYL L. CLARK, Department of Veterinary Physiology and Pharmacology, College of Veterinary Medicine, Iowa State University, Ames, Iowa, USA.

T. CLARKE, Departments of Obstetrics/Gynecology and Biological Chemistry, University of Michigan, Women's Hospital, Ann Arbor, Michigan, USA.

JEFFREY W. CLEMENS, Department of Cell Biology, Baylor College of Medicine, Houston, Texas, USA.

P. MICHAEL CONN, Department of Pharmacology, University of Iowa College of Medicine, Iowa City, Iowa, USA.

MARCO CONTI, The Laboratories for Reproductive Biology, Department of Pediatrics, University of North Carolina at Chapel Hill, Chapel Hill, North Carolina, USA.

SERDAR COSKUN, Department of Veterinary Physiology and Pharmacology, College of Veterinary Medicine, The Ohio State University, Columbus, Ohio, USA.

JACQUES COUËT, Medical Research Council Group in Molecular Endocrinology, CHUL Research Center and Laval University, Quebec, Canada.

CHRISTOS COUTIFARIS, Department of Obstetrics and Gynecology, University of Pennsylvania, Philadelphia, Pennsylvania, USA.

PATRICK COUTURE, Medical Research Council Group in Molecular Endocrinology, CHUL Research Center and Laval University, Quebec, Canada.

DENIS J. CRANKSHAW, Department of Obstetrics and Gynecology, Faculty of Health Sciences, McMaster University, Hamilton, Ontario, Canada.

W. DAVID CURRIE, Natural Sciences and Engineering Research Council of Canada, Department of Obstetrics and Gynecology, University of British Columbia, Vancouver, British Columbia, Canada.

YVAN DE LAUNOIT, Medical Research Council Group in Molecular Endocrinology, CHUL Research Center and Laval University, Quebec, Canada.

MASAKI DOI, Department of Obstetrics and Gynecology, Gunma University School of Medicine, Maebashi, Japan.

MARY LYNN DUCKWORTH, Department of Physiology, University of Manitoba, Winnipeg, Manitoba, Canada.

MARTINE DUMONT, Medical Research Council Group in Molecular Endocrinology, CHUL Research Center and Laval University, Quebec, Canada.

ERIC DUPONT, Medical Research Council Group in Molecular Endocrinology, CHUL Research Center and Laval University, Quebec, Canada.

EDWARD M. EDDY, Gamete Biology Section, Laboratory of Reproductive and Developmental Toxicology, National Institute of Environmental Health Sciences, National Institutes of Health, Research Triangle Park, North Carolina, USA.

R. FERNALD, Department of Biology, Stanford University, Stanford, California, USA.

SUSAN L. FITZPATRICK, Department of Cell Biology, Baylor College of Medicine, Houston, Texas, USA.

R. FRANCIS, Department of Biology, Stanford University, Stanford, California, USA.

HENRY G. FRIESEN, Department of Physiology, University of Manitoba, Winnipeg, Manitoba, Canada.

MASAKI FUKUDA, Department of Obstetrics and Gynecology, Gunma University School of Medicine, Maebashi, Japan.

MANABU FUKUMOTO, Department of Pathology, Faculty of Medicine, Kyoto University, Kyoto, Japan.

S.H. HAMMOND, Departments of Obstetrics/Gynecology, Biological Chemistry, and the Reproductive Sciences Program, University of Michigan, Women's Hospital, Ann Arbor, Michigan, USA.

YOSHIHISA HASEGAWA, Department of Obstetrics and Gynecology, Gunma University School of Medicine, Maebashi, Japan.

P.J. HORNSBY, Department of Biochemistry and Molecular Biology, Medical College of Georgia, Augusta, Georgia, USA.

BASIL HO-YUEN, Division of Reproductive Endocrinology and Infertility, Department of Obstetrics and Gynecology, University of British Columbia, Vancouver, British Columbia, Canada.

AARON J.W. HSUEH, Division of Reproductive Biology, Department of Gynecology and Obstetrics, Stanford University Medical Center, Stanford, California, USA.

MASAO IGARASHI, Department of Obstetrics and Gynecology, Gunma University School of Medicine, Maebashi, Japan.

JANE A. JACKSON, Department of Animal Sciences, Animal Genetics Laboratory, University of Illinois, Urbana, Illinois, USA.

XIAO-CHI JIA, Division of Reproductive Biology, Department of Gynecology and Obstetrics, Stanford University Medical Center, Stanford, California, USA.

CATHERINE S.-L. JIN, The Laboratories for Reproductive Biology, Department of Pediatrics, University of North Carolina at Chapel Hill, Chapel Hill, North Carolina, USA.

LEE-CHUAN KAO, Department of Obstetrics and Gynecology, University of Pennsylvania, Philadelphia, Pennsylvania, USA.

HILDEGARD KOHLHAUF ALBERTIN, Department of Animal Science, Swiss Federal Institute of Technology, Zürich, Switzerland.

GREGORY S. KOPF, Department of Obstetrics and Gynecology, University of Pennsylvania, Philadelphia, Pennsylvania, USA.

RICHARD C. KURTEN, Department of Cell Biology, Baylor College of Medicine, Houston, Texas, USA.

CLAUDE LABRIE, Medical Research Council Group in Molecular Endocrinology, CHUL Research Center and Laval University, Quebec, Canada.

FERNAND LABRIE, Medical Research Council Group in Molecular Endocrinology, CHUL Research Center and Laval University, Quebec, Canada.

YVES LACHANCE, Medical Research Council Group in Molecular Endocrinology, CHUL Research Center and Laval University, Quebec, Canada.

PHILIP S. LAPOLT, Division of Reproductive Biology, Department of Gynecology and Obstetrics, Stanford University Medical Center, Stanford, California, USA.

FLORENCE LEDWITZ-RIGBY, Department of Biological Sciences, Northern Illinois University, Dekalb, Illinois, USA; and Department of Obstetrics and Gynecology, University of British Columbia, Vancouver, British Columbia, Canada.

DIANA L. LEFEBVRE, Laboratory of Molecular Endocrinology, Royal Victoria Hospital, Montreal, Quebec, Canada.

PETER C.K. LEUNG, Research Division, Department of Obstetrics and Gynecology, University of British Columbia, Grace Hospital, Vancouver, British Columbia, Canada.

S.W. LIN, Department of Obstetrics and Gynecology, University of British Columbia, Vancouver, British Columbia, Canada.

YOUNG C. LIN, Department of Veterinary Physiology and Pharmacology, College of Veterinary Medicine, The Ohio State University, Columbus, Ohio, USA.

VAN LUU-THE, Medical Research Council Group in Molecular Endocrinology, CHUL Research Center and Laval University, Quebec, Canada.

V.B. Mahesh, Department of Physiology and Endocrinology, Medical College of Georgia, Augusta, Georgia, USA.

Claude Martel, Medical Research Council Group in Molecular Endocrinology, CHUL Research Center and Laval University, Quebec, Canada.

Kaoru Miyamoto, Department of Obstetrics and Gynecology, Gunma University School of Medicine, Maebashi, Japan.

Kohji Miyazaki, Department of Obstetrics and Gynecology, Kumamoto University Medical School, Kumamoto, Japan.

Matthew K. Nickerson, Department of Biological Sciences, Northern Illinois University, Dekalb, Illinois, USA.

John H. Nilson, Department of Pharmacology, School of Medicine, Case Western Reserve University, Cleveland, Ohio, USA.

Deborah A. O'Brien, Gamete Biology Section, Laboratory of Reproductive and Developmental Toxicology, National Institute of Environmental Health Sciences, National Institutes of Health, Research Triangle Park, North Carolina, USA.

Kathleen Ohleth, Department of Animal Sciences, Rutgers University, New Brunswick, New Jersey, USA.

Hiroshi Ohmura, Department of Veterinary Physiology and Pharmacology, College of Veterinary Medicine, The Ohio State University, Columbus, Ohio, USA.

Hitoshi Okamura, Department of Obstetrics and Gynecology, Kumamoto University Medical School, Kumamoto, Japan.

A.H. Payne, Departments of Obstetrics/Gynecology, Biological Chemistry, and the Reproductive Sciences Program, University of Michigan, Women's Hospital, Ann Arbor, Michigan, USA.

Georges Pelletier, Medical Research Council Group in Molecular Endocrinology, CHUL Research Center and Laval University, Quebec, Canada.

Vladimir Pliška, Department of Animal Science, Swiss Federal Institute of Technology, Zürich, Switzerland.

NI QUAN, Department of Physiology, University of Manitoba, Winnipeg, Manitoba, Canada.

K. RAJKUMAR, Department of Obstetrics and Gynecology, University of Saskatchewan, Saskatoon, Saskatchewan, Canada.

I.M. RAO, Department of Physiology and Endocrinology, Medical College of Georgia, Augusta, Georgia, USA.

ERIC RHÉAUME, Medical Research Council Group in Molecular Endocrinology, CHUL Research Center and Laval University, Quebec, Canada.

STÉPHANE RICHARD, Laboratory of Molecular Endocrinology, Royal Victoria Hospital, Montreal, Quebec, Canada.

JOANNE S. RICHARDS, Department of Cell Biology, Baylor College of Medicine, Houston, Texas, USA.

MAY C. ROBERTSON, Department of Physiology, University of Manitoba, Winnipeg, Manitoba, Canada.

SHIN-ICHI ROKUKAWA, Department of Obstetrics and Gynecology, Gunma University School of Medicine, Maebashi, Japan.

MASAAKI SAWADA, Department of Biology, University of Waterloo, Waterloo, Ontario, Canada.

INGO SCHROEDTER, Department of Physiology, University of Manitoba, Winnipeg, Manitoba, Canada.

ULRIKE SESTER, Department of Biology, University of Waterloo, Waterloo, Ontario, Canada.

L. SHA, Departments of Obstetrics/Gynecology, University of Michigan, Women's Hospital, Ann Arbor, Michigan, USA.

N. SHERWOOD, Department of Biology, University of Victoria, Victoria, Canada.

JACQUES SIMARD, Medical Research Council Group in Molecular Endocrinology, CHUL Research Center and Laval University, Quebec, Canada.

JEAN SIROIS, Department of Cell Biology, Baylor College of Medicine, Houston, Texas, USA.

GILLIAN L. STEELE, Department of Obstetrics and Gynecology, University of British Columbia, Grace Hospital, Vancouver, British Columbia, Canada.

JEROME F. STRAUSS III, Department of Obstetrics and Gynecology, University of Pennsylvania, Philadelphia, Pennsylvania, USA.

TSUTOMU SUZUKI, Department of Obstetrics and Gynecology, Gunma University School of Medicine, Maebashi, Japan.

JOHANNES V. SWINNEN, The Laboratories for Reproductive Biology, Department of Pediatrics, University of North Carolina at Chapel Hill, Chapel Hill, North Carolina, USA.

HIRONORI TASHIRO, Department of Obstetrics and Gynecology, Kumamoto University Medical School, Kumamoto, Japan.

MICHAEL J. TAYLOR, Department of Veterinary Physiology and Pharmacology, College of Veterinary Medicine, Iowa State University, Ames, Iowa, USA.

CLAUDE TRUDEL, Medical Research Council Group in Molecular Endocrinology, CHUL Research Center and Laval University, Quebec, Canada.

KIYOSHI TSUCHIYA, Department of Obstetrics and Gynecology, Gunma University School of Medicine, Maebashi, Japan.

MEHMET UZUMCU, Department of Veterinary Physiology and Pharmacology, College of Veterinary Medicine, The Ohio State University, Columbus, Ohio, USA.

KIMMO K. VIHKO, Division of Reproductive Biology, Department of Gynecology and Obstetrics, Stanford University Medical Center, Stanford, California, USA.

JEAN-CLAUDE VUILLE, Department of Physiology, University of Manitoba, Winnipeg, Manitoba, Canada.

JEFF E. WELCH, Gamete Biology Section, Laboratory of Reproductive and Developmental Toxicology, National Institute of Environmental Health Sciences, National Institutes of Health, Research Triangle Park, North Carolina, USA.

WINONA L. WONG, Department of Cell Biology, Baylor College of Medicine, Houston, Texas, USA.

XIU MEI WU, Department of Biology, University of Waterloo, Waterloo, Ontario, Canada.

MASAAKI YAMAGUCHI, Department of Obstetrics and Gynecology, Gunma University School of Medicine, Maebashi, Japan.

M. YOO, Department of Biology, Keimyung University, Dalsuh-gu, Taegu, Korea.

G.L. YOUNGBLOOD, Departments of Obstetrics/Gynecology and Biological Chemistry, University of Michigan, Women's Hospital, Ann Arbor, Michigan, USA.

R. ZHAI, Department of Obstetrics and Gynecology, University of Saskatchewan, Saskatoon, Saskatchewan, Canada.

HUI-FEN ZHAO, Medical Research Council Group in Molecular Endocrinology, CHUL Research Center and Laval University, Quebec, Canada.

HANS H. ZINGG, Laboratory of Molecular Endocrinology, Royal Victoria Hospital, Montreal, Quebec, Canada.

Part I

Hypothalamus and Pituitary

Part 1

1

Social Control of Reproduction: Molecular Studies of GnRH in the African Cichlid *Haplochromis burtoni*

C.T. Bond, R. Fernald, R. Francis, N. Sherwood, and J.P. Adelman

Evolution of the GnRH Decapeptide

During 500 million years of evolution, the primary structure of GnRH has been remarkably conserved. Recently, peptide sequence analysis of GnRH-like immunoreactive material from lamprey brain has shown that 5 of 10 amino acid residues are identical between agnathan and mammalian GnRH. In addition, the pyro-glu amino terminal structure and amidation of the carboxy terminus of the decapeptide are conserved (1). To date, the amino acid sequence has been determined and bioactivity demonstrated for 5 separate forms of the GnRH decapeptide (2). Two GnRH decapeptide sequences have been found in the chicken (3, 4). Immunologic and chromatographic evidence indicate that there may be 2 decapeptides in other vertebrates, including some species of reptiles and teleost fish. However, amino acid sequences for these putative second GnRH forms have not yet been determined (5).

Nonmammalian Model for the Study of GnRH

In the teleost *Haplochromis burtoni*, an African cichlid species, sexual development in males is regulated by social interactions (6, 7). A normal population of *H. burtoni* includes two distinct male types: those that are territorial and those that are not. Territorial males are brightly colored and sexually mature and are called "machos" because of the aggressive behavior they display in establishing and defending territories for feeding and breeding. In contrast, nonterritorial males, called "wimps," are cryptically colored, sexually immature, and do not reproduce. The social

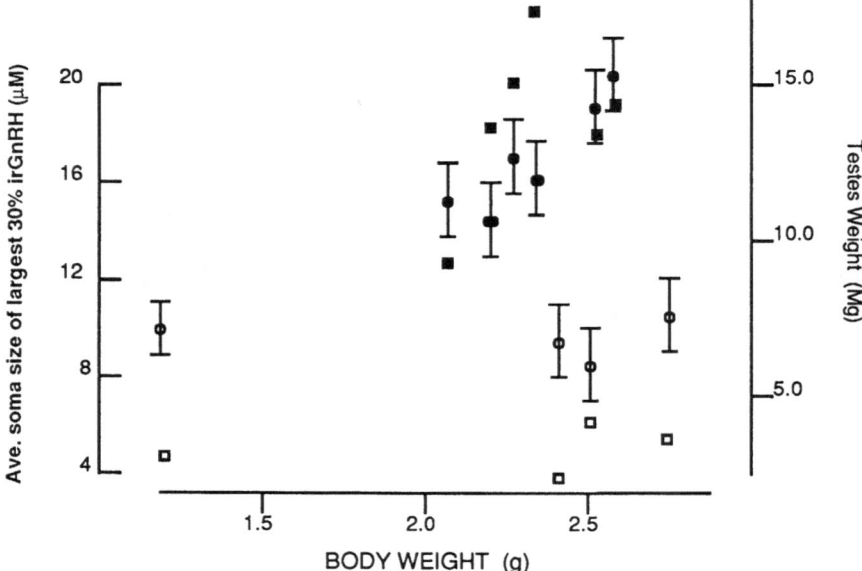

FIGURE 1.1. Preoptic irGnRH cell soma sizes for control (early maturing) (solid circles) and experimental (maturation-supressed) (open circles) animals presented as a function of body weight, illustrating that in control animals the cells are approximately 50% larger. Testes weights for the same control (solid boxes) and experimental (open boxes) animals, are shown, also as a function of body weight. Adapted with permission from Davis and Fernald (8).

dominance of machos over wimps is correlated with the size of forebrain magnocellular neurons that express GnRH immunoreactivity. Concomitant with the emergence of aggressive behavior and establishment of territory is an increase in the size of those neurons that specifically express GnRH (8) (Fig. 1.1). The most significant influence on the development of reproductive capability in wimp males is the proximity of larger, threatening conspecifics.

The attainment of territorial dominance, breeding capability, and enlargement of GnRH neurons are reversible conditions; in the presence of a larger, more dominant male, a previously macho male loses its bright coloration, the testes regress, and the GnRH magnocellular neurons shrink to normal size. This amazing plasticity is probably mediated by diverse sensory inputs and may be reflected in the level of expression of the GnRH gene in these neurons. As a first step in understanding the cellular and molecular basis of this aspect of teleost reproduction, we have cloned the cDNA encoding *H. burtoni* preproGnRH (21).

Results

Prohormone Structural Conservation Between Teleosts and Mammals

Oligonucleotide pools encompassing all possible coding sequences of the first 8 amino acids of the known teleost decapeptide sequence (9) were used as radiolabeled probes on a cDNA library constructed from poly(A)$^+$ RNA isolated from the brains of macho male *H. burtoni*. Hybridization specificity was obtained by washing library screen filters in 3-M tetramethylammonium chloride (10). Sequence analysis of the cDNA clone thus isolated shows remarkable conservation of structure relative to the mammalian prohormones (11). This structural conservation is like that seen for other polyprotein precursors that have been characterized from diverse species (12) (Fig. 1.2).

The teleost preproGnRH contains a signal peptide of 23 amino acids immediately preceding the decapeptide sequence. A dibasic proteolytic cleavage site follows the decapeptide and amide-donating glycine, providing a substrate for maturation of the decapeptide and separation from the 54-amino acid-associated C-terminal peptide.

No Homology of Teleost GnRH-Associated Peptide to Mammalian GAP

Despite the conservation of overall structure between these evolutionarily distant GnRH prohormones, the amino acid sequences differ considerably. There is significant amino acid homology within the signal sequences; 8 of 23 residues are identical, and a basic residue—lysine in the

	Signal Peptide	GnRH
CICHLID	MEAGSRVIMQVLLLALVVQVTLS	QHWSYGWLPG GKR
MOUSE	MILKLMAGILLLTVCLEGCSS	QHWSYGLRPG GKR
RAT	METIPKLMAAVVLLTVCLEGCSS	QHWSYGLRPG GKR
HUMAN	MKPIQKLLAGLILLTWCVEGCSS	QHWSYGLRPG GKR

	GAP
CICHLID	SVGELEATIRMMGTGGVVSLPDEANAQIQERLRPYNIINDDSSHFDRKKRFPNN
MOUSE	NTEHLVESFQEMGKEVDQMAEPQHFECTVHWPRSPLRDLRGALESLIEEEARQKKM
RAT	NTEHLVDSFQEMGKEEDQMAEPQNFECTVHWPRSPLRDLRGALERLIEEEAGQKKM
HUMAN	DAENLIDSFQEIVKEVGQLAETQRFECTTHQPRSPLRDLKGALESLIEEETGQKKI

FIGURE 1.2. Comparison of preprohormone sequences of teleost and mammalian GnRH.

fish, argenine in the mammal—occupies a conserved position. The decapeptide and proteolytic processing site sequences are identical except for the previously known 2-amino acid substitutions. However, the associated C-terminal peptide of the teleost prohormone has no significant homology to mammalian GAP.

Mammalian GAP has been implicated in the control of prolactin release from pituitary lactotrophs (13). Although the physiological properties mediated by prolactin in mammals (lactation and parturition) are not germane in the fish, human GAP has recently been shown to effect prolactin release from the pituitary of the tilapia *Oreochromis mossambicus* (14). Teleosts express two forms of prolactin that are under control of unknown hypothalamic factors (15, 16), and both forms are thought to function in osmoregulation (17, 18).

Examination of all possible reading frames encoded by the teleost GnRH cDNA revealed that a single base insertion 3' of the decapeptide encoding sequence would yield an incomplete open reading frame that contains a significant block of homology to a region of mammalian GAP that is highly conserved among the three known sequences. In light of the prolactin study cited above, and in spite of the fact that this possible island of homology lay within a reading frame that was not contiguous with that of the decapeptide, it was essential to verify the sequence of the GnRH coding sequence isolated from the *H. burtoni* library. To do so, 3 independent reverse transcription reactions were performed on *H. burtoni* brain RNA, followed by multiple PCR reactions employing oligonucleotide primers specific for untranslated regions of the *H. burtoni* GnRH sequence. Nucleotide sequence analysis of PCR reaction products verified the reading frame of the original clone, although the analysis contained occasional transitions from T to C, probably representing PCR mistakes. This sequence was further verified by the subsequent isolation of additional cDNA clones from the original library during low-stringency hybridization experiments described below.

Failure of Low-Stringency Hybridization Studies to Detect a Second GnRH Coding Sequence

Immunologic and chromatographic studies have found evidence that a second decapeptide sequence is expressed in some species of teleost (19); however, exact amino acid sequence data has not been reported. Expression of a second decapeptide might be the result of an earlier gene duplication event, in which case, the coding sequence should bear significant homology to the characterized teleost cDNA. We examined this possibility by a series of hybridization experiments employing randomly primed DNA probes derived from the full-length *H. burtoni* GnRH cDNA applied to *H. burtoni* brain library filters, genomic Southern blots and Northern blots containing poly(A)$^+$ and total RNA from *H. burtoni* macho male brains. None of these experiments yielded any evidence of a related prohormone coding sequence.

The inability to uncover a second GnRH encoding sequence in *H. burtoni* by low-stringency hybridization does not eliminate the possibility that such a gene exists. The alternative GnRH decapeptide might be encoded in an otherwise unrelated DNA milieu, rendering it undetectable by this type of hybridization study. If, indeed, a GnRH-like decapeptide does arise from gene sequences that are completely nonhomologous, it raises questions about the evolutionary relatedness of the decapeptides. Perhaps the decapeptide motif found in GnRH has occurred twice coincidentally, with the second decapeptide encoded by a distinct locus and subserving unrelated functions. There may be circumstantial evidence for this in the distribution of each decapeptide type in the brains of various species (20). Resolution of this question awaits the cloning of 2 GnRH-encoding sequences from a single species.

Multiple GnRH Forms in the Brain of H. burtoni

As discussed above, the exact amino acid sequences of several GnRH motifs have been determined from a range of vertebrate species. Further, specific immunological and chromatographic data strongly suggest the existence of several more GnRHs (5). Within a single species, many lower vertebrates appear to express more than one GnRH, and 2 forms have actually been sequenced from chickens (3, 4). Although the striking reproductive changes in *Haplochromis* correlate with immunohisto-chemically defined changes in GnRH-expressing neurons of the telencephalon, we examined whether multiple GnRHs are expressed in *Haplochromis* brain. Brain extracts from macho males were initially characterized by RIA employing a battery of antibodies that distinguish between the various sequenced forms of GnRH. This analysis indicated several potential GnRHs present in *Haplochromis*. Next, brain extracts were subjected to HPLC, and the elution profiles of immunoreactive species were compared with control profiles of various forms of GnRH synthetic peptides (performed as outlined in reference 1). This analysis demonstrated the presence of 3 distinct forms of GnRH in *Haplochromis*, one of which corresponds to the salmon form that we have proven by molecular cloning. In addition, it is very likely that *Haplochromis* brain contains the chicken II form of GnRH. The third GnRH does not conform to any of the known GnRH motifs and, therefore, may represent a previously unknown version of the decapeptide.

These results are significant for two reasons. First, the inability to detect a cDNA encoding an alternate form of GnRH in our *Haplochromis* cDNA library may initially have been interpreted as a lack of additional GnRH forms expressed in this teleost. However, the HPLC and immuno-logical data strongly support the presence of 2 alternate forms. This finding strengthens our prediction that the nucleic acid sequences encoding the various prohormone forms are very different and calls into question the relatedness of the prohormones themselves, as well as the evolutionary relationship between the genes encoding them. Second, these results call into question the degree to which the form of GnRH that we have isolated by molecular cloning is actually responsible for the immunological and physiological changes in *Haplochromis* as a function of reproductive status. The roles of cGnRHII and the novel decapeptide form in *Haplochromis* remain to be determined. It is imperative that future experiments discriminate between the effects mediated by the various GnRHs. Isolation of molecular clones encoding these prohormones will permit the generation of nucleic acid and immunological probes that will be employed to map the distribution of the different GnRH-expressing neurons and to quantitate changes in the relative RNA and protein levels of the various forms as a function of reproductive status.

Changes in GnRH mRNA Levels During Reproductive Transitions

We have employed the GnRH cDNA described above to examine changes in the levels of this mRNA as a function of male reproductive status. In previous experiments, we have determined that RNA from a single animal can be used to detect the mRNA encoding the GnRH precursor that corresponds to the form we have isolated. In the present experiments, we initially isolated RNA from the endpoints of the transition, macho and wimp, and prepared these samples as a Northern blot. The transferred RNA was stained with methylene blue, and the 18S and 28S ribosomal RNA bands were quantified by densitometry. The blot was then probed under high-stringency conditions; autoradiographic signals were quantified and standardized to the ribosomal bands. The result of this experiment was that no significant differences in steady state GnRH mRNA were apparent. Next, RNA was extracted from brains of animals during the transition from wimp to macho, and vice versa. When these samples were probed and the signals quantified, 3- to 5-fold differences in steady state GnRH mRNA levels were detected during a limited window, 2–4 days after induction of the transition. These results are presently being extended using more detailed time points.

The Northern blot results are important for several reasons. First, they imply that the mRNA for the GnRH form we have characterized as a cDNA is at least partially responsible for reproductive regulation in *Haplochromis*. Second, they support the hypothesis that one of the underlying mechanisms for the changes within GnRH neurons is a change in the steady state mRNA levels. It is likely that these mRNA changes reflect changes in peptide levels, and this is presently being experimentally determined. Third, these results indicate that the transitional period involves an initial phase in which significant changes in GnRH mRNA levels are important, while later stages, including maintenance of the macho or wimp status, do not require elevated (or repressed) levels of GnRH mRNA.

Discussion

Future Studies Employing H. burtoni as a Model System

The dissimilarity of sequence between mammalian GAP and this teleost GAP, combined with studies indicating prolactin-related bioactivity of mammalian GAP in teleost pituitaries, are at present enigmatic. Studies are currently being initiated to examine the biological role of the teleost GAP peptide in the fish. Divergence of associated peptides concurrent with extreme conservation of the principle peptide hormone within a polyprotein precursor is a common thread in evolution (12). Perhaps this format allows evolutionary experimentation within the structure and function of the associated peptides while preserving essential functions mediated by the principle hormone. In this light, it is interesting to note that a frame shift at the appropriate location within the teleost GAP sequence yields a stretch of amino acids with a high degree of homology to a sequence in mammalian GAP that is highly conserved across the three cloned species. Perhaps as prolactin came to serve functions more directly related to reproduction, evolution found a way to tie the regulation of prolactin release to the expression of GnRH.

Haplochromis burtoni, whose reproductive capability is mediated by social factors, offers an exquisite model for the study of regulation of GnRH. The interplay of sensory inputs and development and the combined effect on expression of GnRH can be examined. Also of interest is the role that GnRH might play in mediating the corresponding behaviors of the separate male types. Sexual differences in the regulation of GnRH and subsequent behavioral patterns are also accessible in this model.

References

1. Sherwood NM, Sowers SA, Marshak DR, Fraser BA, Brownstein MJ. Primary structure of gonadotropin-releasing hormone from lamprey brain. J Biol Chem 1986;261:4812–9.
2. King JA, Millar RP. Genealogy of the GnRH family. In: Epple A, Scoues CG, Stetson MH, eds. Progress in comparative endocrinology; vol. 342, Progress in clinical and biological research. Wiley-Liss, 1990:45–59.

3. King JA, Miller RP. Structure of chicken hypothalamic luteinizing hormone-releasing hormone, II. Isolation and characterization. J Biol Chem 1982;257:10729–32.

4. Miyamoto K, Hasegawa Y, Nomura M, Igarashi M, Kangawa K, Matsuo H. Identification of the second gonadotropin-releasing hormone in chicken hypothalamus; evidence that gonadotropin secretion is probably controlled by two distinct gonadotropin-releasing hormones in avian species. Proc Natl Acad Sci USA 1984;81:3874–8.

5. Sherwood N. The GnRH family of peptides. Trends Neurosci 1987; 10:129132.

6. Fernald RD, Hirata N. Field study of *Haplochromis burtoni*: quantitative behavioral observations. Anim Behav 1977;25:964–75.

7. Fraley NB, Fernald RD. Social control of developmental rate in the African cichlid fish *Haplochromis burtoni*. Z Tierpsychol 1982;60:66–82.

8. Davis MR, Fernald RD. Social control of neuronal soma size. J Neurobiol 1990;21:1180–8.

9. Sherwood N, Eiden L, Brownstein M, Spiess J, Rivier J, Vale W. Characterization of a teleost gonadotropin-releasing hormone. Proc Natl Acad Sci USA 1983;80:2794–8.

10. Wood WI, Gitzchier J, Lasky LA, Lawn RM. Base composition independent hybridization in tetramethylammonium chloride: a method for oligo-nucleotide screening of highly complex gene libraries. Proc Natl Acad Sci USA 1985;82:1585–8.

11. Adelman JP, Mason AJ, Hayflick JS, Seeburg PH. Isolation of the gene and hypothalamic cDNA for the common precursor of gonadotropin-releasing hormone and prolactin release-inhibiting factor in human and rat. Proc Natl Acad Sci USA 1986;83:179–83.

12. Sherwood NM, Parker DB. Neuropeptide families: an evolutionary perspective. J Exp Zool 1990;4:53–71.

13. Nikolics K, Mason AJ, Szónyi É, Ramachandran J, Seeburg PH. A prolactin-inhibiting factor within the precursor for human gondatotropin-releasing hormone. Nature 1985;316:511–7.

14. Planas J, Bern HA, Millar RP. Effects of GnRH-associated peptide and its component peptides on prolactin secretion from the tilapia pituitary in vitro. Gen Comp Endocrinol 1990;77:386–96.

15. Nishioka RA, Kelley KM, Bern HA. Control of prolactin and growth hormone secretion in teleost fishes. Zool Sci 1988;5:267–80.

16. Rivas RJ, Nishioka RS, Bern HA. In vitro effect of somatostatin and urotensin II on prolactin and growth hormone secretion in tilapia, *Oreochromis mossambicus*. Gen Comp Endocrinol 1986;63:245–51.

17. Brown PS, Brown SC. Osmoregulatory actions of prolactin and other adenohypophysial hormones. In: Pan PKT, Schreibman M, eds. Vertebrate Endocrinology; vol 2. San Diego: Academic Press:45–84.

18. Loretz CA, Bern HA. Prolactin and osmoregulation in vertebrates. Neuroendocrinology 1982;35:292–304.

19. Sherwood NM. In: Idler DR, Crim LW, Walsh JM, eds. Reproductive physiology of fish. Proc 3rd Int Symposium on Reproductive Physiology of Fish. St. Johns, Newfoundland, August 1987.

20. Katz IA, Millar RP, King JA. Differential regional distribution and release of two forms of gonadotropin-releasing hormone in the chicken brain. Peptides 1990;11:443–50.
21. Bond CT, Francis RC, Fernald RD, Adelman JP. Characterization of complementary DNA encoding the precursor for gonadotropin releasing hormone and its associated peptide from a teleost fish. Mol Endocrinol, 1991;5:931–7.

2

GnRH and Its Mechanism of Action

Tim D. Braden and P. Michael Conn

Gonadotropin releasing hormone (GnRH) plays a central role in regulating the reproductive process. Since isolation of this decapeptide and identification of its structure almost twenty years ago, our understanding of the neural control of reproduction and neuroendocrinology as a whole has experienced tremendous growth. Analogs of GnRH are now being used clinically to treat precocious puberty in children, endometriosis, polycystic ovarian disease, and two of the most prevalent steroid-dependent neoplasia, prostate cancer and breast cancer. In addition, GnRH and its analogs have proven useful in enhancing the reproductive efficiency of animals produced for both food and fiber. Clearly, basic research into the physiology and pharmacology of GnRH can be regarded as particularly successful in light of the relatively short time span from basic studies to practical utility.

In addition to the direct clinical applications of GnRH, the study of this hormone has contributed to our understanding of the mechanisms and pattern of hormone release, as well as the mechanisms by which responsiveness of target glands are regulated. Therefore, studies on GnRH have spanned physiology, pharmacology, endocrinology, reproductive biology, cellular biology, and molecular biology.

Hypothalamic GnRH is synthesized primarily in the arcuate nucleus region and transported to and released from the median eminence into the hypothalamic-hypophyseal portal system. The release of GnRH occurs in a pulsatile fashion that can be regulated by various external signals (i.e., steroid hormones). GnRH has its effects at the pituitary gonadotrope and stimulates the release of the gonadotropins—*luteinizing hormone* (LH) and *follicle stimulating hormone* (FSH)—into the peripheral circulation. The pulsatile nature of GnRH release results in the pulsatile release of LH and FSH. In addition to gonadotropin release,

This work was adapted from The Stevenson Lecture and appeared in Can J Physiol Pharmacol 1991;69:445–58. The work was supported by NIH Grant HD-19899.

GnRH evokes several other cellular responses of gonadotropes, such as desensitization, up- and down-regulation of GnRH receptors, and biosynthesis of gonadotropins and GnRH receptors.

Structure, Binding, and Function Relationships of GnRH and Its Analogs

Structure of the GnRH Gene

The gene encoding for GnRH is a single gene in rat, mouse, and human and is located on the short arm of chromosome 8 in the human (1). Information obtained from cloned cDNAs of hypothalamic and placental origins indicates 4 exons (2, 3). The coding sequence translates for a 92-amino acid precursor protein for GnRH and a 56-amino acid peptide termed *GAP* (GnRH-associated peptide). The first exon consists of a 5'-untranslated region that differs between cDNAs from hypothalamic and placental tissues. The second exon codes for the signal peptide, GnRH, and the first 11-amino acid residues of GAP. The third exon codes for GAP residues 12 through 43. The fourth exon codes for the 13 terminal amino acids residues of GAP and the remaining 3'-untranslated mRNA (reviewed in 4). Identification of the gene for GnRH has led to the isolation and reversal of a defect in a strain of *hypogonadal* (hpg) mice that do not normally progress through puberty. The GnRH gene in hpg mice was found to have a large deletion that omits the third and fourth exons. Using transgenic animals carrying a wild-type GnRH gene and crosses with heterozygous hpg mice, animals that were homozygous for the hpg mutation and yet carriers of the wild-type GnRH transgene were produced. These homozygous hpg mice underwent normal puberty and were fertile (4, 5). Alleviation of the hpg phenotype through gene incorporation suggests the possibility that similar human deficiencies may also be corrected through the use of gene therapy.

Amino Acid Structure

Since publication of the amino acid sequence of native mammalian GnRH (6–8), over 3500 different analogs of GnRH have been synthesized. The linear sequence of natural mammalian GnRH is

$$\text{pyroGlu}^1\text{-His}^2\text{-Trp}^3\text{-Ser}^4\text{-Tyr}^5\text{-Gly}^6\text{-Leu}^7\text{-Arg}^8\text{-Pro}^9\text{-Gly}^{10}\text{amide}$$

GnRH from virtually all mammals has the same structure. Distinct variant forms in the primary structure of GnRH have been identified in the chicken (2 variants [9–11]), salmon (12), and lamprey (13). Generally, the structure of GnRH has been conserved throughout evolution; however, significant variations are observed at positions 7 and 8 (reviewed in

14, 15). Amino acid substitution analysis has led to identification of the structural requirements for binding to the GnRH receptor and activation of the target cell (16). The native GnRH molecule can undergo major conformational changes from a fully extended form to a highly folded form. The formation of a β-type II turn at Gly[6]-Leu[7] results in a configuration that has a high affinity for GnRH receptors (17). This least energy configuration is apparently stabilized by the formation of hydrogen bonds between the pyrrolidone carbonyl residue (position 1) and the glycinamide group (position 10 [18, 19]). Therefore, the close opposition of amino acids in positions 1 and 10 is involved in the binding of GnRH to its receptor (20). Substitution at amino acid 6 with D-amino acids containing bulky hydrophobic side chains constrains the molecule to the receptor-preferred conformation and results in a high binding affinity of both agonists and antagonists with this substitution. A combination of substitutions at Gly[6] and substitution of ethylamide for Gly[10] further enhances the affinity of GnRH analogs for the GnRH receptor (21, 22).

Target Cell Activation

Activation of target cells through the binding of ligand to GnRH receptors is thought to be dependent upon the 3 N-terminal residues of the ligand. Substitution of the first 3 residues of GnRH analogs with hydrophobic D-amino acids results in GnRH antagonists (23). This type of substitution, combined with substitution at position 6 with strong basic amino acids, yields potent antagonists of GnRH, but the combination of these two substitutions has been implicated in the release of histamine (24). Further studies on the requirements for target cell activation using reduced-size GnRH analogs (hexapeptides) have suggested that relationships between residue side chains at positions 3 and 6 can change the activity of reduced-size GnRH analogs from agonists to antagonists (25), with the addition of one methylene group at position 3 changing activity from agonistic to antagonistic. Additionally, subtle changes in residue side chains at position 4 can cause steric hindrances involving position 3 to again change the activity of GnRH analogs.

Common Analogs of GnRH

The half-life of GnRH in vivo is relatively short (<10 min in humans [26]) due to initial degradation of the amino terminal half of the peptide (27). Additionally, GnRH is known to be degraded by a number of enzymes. A pyroglutamate aminopeptidase (cleaves GnRH at pyroGlu[1]-His[2]), a postproline-cleaving enzyme (cleaves GnRH at Pro[9]-Gly[10]amide), and a nonchymotrypsin-like endopeptidase (cleaves GnRH preferentially at Tyr[5]-Gly[6], then at His[2]-Trp[3] [28–31]) have all been isolated from the pituitary gland; however, it is unclear whether these enzymes participate in the physiological inactivation of native GnRH (32). Longer half-lives of GnRH analogs have been achieved with hydrophobic D-amino acid substitutions at position 6 (33). Based upon substitutions to prolong the half-life and improve binding activity of GnRH analogs, several highly potent superagonists of GnRH have become commercially available, such as D-Ser(tBu^6) GnRH ethylamide (Buserelin, Hoechst [34]), D-Leu^6-GnRH ethylamide (Leuprolide, TAP [35]), and 3(2 naphthyl)Ala^6-GnRH (Nafarelin, Syntex [36]). Highly hydrophobic GnRH analogs may have a prolonged biological half-life due to association with binding proteins in serum (37).

Receptors for GnRH

Physical and Chemical Characteristics

The primary site of action of GnRH is the gonadotrope in the anterior pituitary gland; however, binding sites for GnRH have been identified in other tissues. Receptors for GnRH have been observed in the gonads of rats (38, 39) and humans (40) but not in ovine, bovine, or porcine ovaries (41). Additionally, GnRH binding sites have been observed in adrenal glands (42), some cancer tissues (43, 44), and in the central nervous system (45–47). Because of the low concentrations of circulating GnRH and its short half-life, it is unlikely that GnRH released from the hypothalamus occupies a sufficient number of receptors in peripheral tissues to have physiological effects, but local synthesis of GnRH and paracrine effects cannot be excluded. It has been suggested that GnRH binding sites in the hypothalamus may participate in behavioral modification (48, 49). While there is much interest in extrapituitary binding sites for GnRH (50), this chapter concentrates on GnRH receptors on pituitary gonadotropes.

The first step in the mechanism of action of GnRH in stimulating gonadotropin release is the binding of GnRH to its receptor on gonadotropes. Receptors for GnRH are found exclusively in the plasma membrane fraction (51). Receptors for GnRH from rat and bovine pituitary plasma membranes are glycoproteins (52, 53), as evidenced by a loss of agonist and antagonist binding after exposure of receptors to neuraminidase and wheat germ agglutinin. Sialic acid residues on GnRH receptors appear to be required for activation of the receptor, as well as for the appearance of receptors on the cell surface (53). Binding of GnRH to its receptor appears to be dependent upon the presence of both exterior hydrophilic head groups and fatty acid linked to the β-carbon of phospholipids (54). Additionally, 2 carboxylic groups and 2 aromatic amino acids appear to be in the ligand binding portion of the receptor and/or influence the binding of GnRH (55, 56). It appears that the GnRH receptor is linked to a G-protein (57).

Estimates of the size of the GnRH holoreceptor range from 50,000 kd to 700,000 kd depending upon the conditions used to estimate the size. The zwitterionic detergent CHAPS (3-[-3-cholamidopropyl-dimethylammonio]-1-propanesulfonic acid) has proven useful for solubilization of GnRH receptors that retain their ability to bind ligands (58, 59). Solubilization of GnRH receptors with CHAPS followed by nondenaturing sizing gel exclusion has indicated an apparent molecular weight of the GnRH receptor to be 60,000–150,000 (60, 61). Using covalent labeling of the GnRH receptor with a radiolabeled photoaffinity agonist (^{125}I-Tyr^5-azidobenzoyl-D-Lys^6-GnRH) followed by *sodium dodecyl sulfate polyacrylamide gel electrophoresis* (SDS-PAGE) and autoradiography, several laboratories working independently have observed specific radioactive labeling at the apparent molecular weight of 60,000 (62, 63). Radiation inactivation (target size analysis) of the functional size of intact GnRH receptors in plasma membranes indicated an apparent molecular weight of 136,000 for the GnRH holoreceptor (64).

Although the purification of GnRH receptors and generation of polyclonal antibodies against this receptor have recently been reported, the low abundance of receptor and low titer of the antibodies have hampered further characterization of the GnRH receptor. Therefore, it is unclear whether the 60-kd component of the GnRH receptor represents the holoreceptor or simply a ligand binding component. It seems likely that the 60-kd band identified by photoaffinity labeling and Western blotting represents a subunit of the GnRH receptor and that the holoreceptor is made up of this binding subunit and at least one other subunit of similar size that may or may not bind ligand. GnRH receptors have been reported to have been purified to homogeneity by chromatography on wheat germ agglutinin-agarose followed by affinity chromatography using immobilized avidin coupled to biotinylated D-Lys^6-GnRH (65). Analysis of purified radiolabeled GnRH receptors by SDS-PAGE followed by autoradiography indicated 2 bands of activity at 57 kd and 59 kd. Subsequent production of a low-titer polyclonal antibody to GnRH receptors has been reported (59); however, further characterization of GnRH receptors using this antibody has not been published.

Because of the difficulties in purification of GnRH receptors and generation of useful antibodies against the receptor, the molecular biology of the gene for GnRH receptors remains largely unknown. Recently, however, several laboratories have reported expression of GnRH receptors after injection of pituitary mRNA or mRNA from a pituitary cell line into *Xenopus* oocytes (66–68). The specific mRNA encoding the GnRH receptor has an apparent size of >28S, which suggests a length of 6–7 kb (68). These data were obtained from cells whose origin was a pituitary tumor of a transgenic mouse. Although the cell line expresses GnRH binding sites and produces the α-subunit of the glycoprotein hormones, it is unknown whether this GnRH receptor population has the same characteristics as normally expressed GnRH receptors.

Regulation of Number of GnRH Receptors

Changes in the number of receptors for GnRH in the pituitary glands of a number of species have been characterized during many physiological conditions. During the estrous cycle of rats, hamsters, ewes, and cows, the maximum number of GnRH receptors was observed during the proestrous period prior to the preovulatory surge of LH (69–74). After the preovulatory surge of LH, the number of GnRH receptors decreases and may require several days to achieve proestrous levels. Following removal of the gonads, significant increases in the number of GnRH receptors have been observed (72). In contrast, during pregnancy and lactation, the number of GnRH receptors is fewer than those observed during the estrous cycle (69, 72). These observations clearly demonstrate regulation of GnRH receptors in vivo.

Numerous treatments in vitro can alter the number of receptors for GnRH in pituitary cell cultures. Treatment of pituitary cell cultures with physiological concentrations of GnRH results in a biphasic response by the cells with respect to GnRH receptor number (75). Initially, a down-regulation of receptors is observed (<4 h posttreatment) followed by an increase in the number of GnRH receptors (9 h posttreatment). The initial down-regulation of receptors for GnRH is temporally associated with the desensitization of gonadotropes to GnRH, although, clearly, other mechanisms such as uncoupling of receptors from second-messenger systems contribute to desensitization. Homologous down-regulation of GnRH receptors appears to be independent of extracellular calcium, while up-regulation of GnRH receptors is dependent upon extracellular calcium and requires protein synthesis (75, 76).

Up-regulation of GnRH receptors by homologous hormone can be mimicked by treatment of pituitary cells with analogs of adenosine 3',5'-monophosphate, as well as by nonspecific depolarization of pituitary cells with KCl (76) and A23187 (75, 77). While up-regulation of GnRH receptors indicates the ability of gonadotropes to respond to various external signals with an increased number of plasma membrane receptors, this increased receptor number does not increase the sensitivity of gonadotropes to GnRH when LH release is measured as the cellular response (78). Gonadotropes can respond with near-maximal LH release when only 20% of the available GnRH receptors are occupied in vitro (79); and when 50% of the receptors are blocked with a GnRH antagonist, ewes can still respond fully to subsequent GnRH administration with LH release (80). As mentioned earlier, these data indicate that there are "spare" GnRH receptors when LH release is the sole parameter measured; however, it is unknown if the spare receptor conclusion is valid for the other functions of gonadotropes in response to GnRH (i.e., FSH release, receptor synthesis, up-regulation, down-regulation, and gonadotropin biosynthesis).

In addition to regulation of GnRH receptor by homologous hormone, the number of GnRH receptors can be regulated by other hormones, including steroids and protein products from the gonad. As indicated earlier, removal of the gonads can increase the number of GnRH receptors in vivo when hypothalamic-pituitary connections are intact. In the absence of hypothalamic input, 17β-estradiol can increase the number of GnRH receptors (81, 82). Using ovine pituitary gonadotrope cell cultures, Laws et al. (83, 84) have shown that estradiol can increase and that progesterone can decrease the number of receptors for GnRH.

Protein products of the gonads have also been shown to influence the number of GnRH receptors. Wang et al. (85) have shown a decreased number of GnRH receptors when rat pituitary cell cultures were treated with inhibin. This group subsequently showed that inhibin was able to block GnRH-stimulated up-regulation of GnRH receptors (86). The effects of inhibin on the basal number of GnRH receptors were shown to be independent of biosynthesis of GnRH receptors (87), but the ability of inhibin to block up-regulation of GnRH receptors was at least partially due to the ability of inhibin to antagonize GnRH-stimulated synthesis of GnRH receptors (87). In direct contrast, Laws et al. (83) observed that treatment of ovine pituitary cell cultures with inhibin significantly increased the number of GnRH receptors. Given these two distinctly separate observations regarding the effects of inhibin on GnRH receptor populations in two different species, it is clear that other species will have to be examined before defining the general role of inhibin in regulating the number of pituitary receptors for GnRH.

Fate and Replacement of Occupied GnRH Receptors

The presence of receptors on the cell surface is due to a combination of processes that either contribute to the plasma membrane population (synthesis, recycling, and unmasking of receptors) or remove receptors from the cell surface (internalization, degradation, and inactivation). After binding of the agonist, GnRH receptors form patches in coated pits and are internalized. It appears as if GnRH receptors can undergo micro-aggregation as part of the target cell activation process. When a GnRH antagonist was allowed to occupy GnRH receptors, no cellular response was observed. However, when antagonist-occupied GnRH receptors are dimerized and bound to an antagonist antibody, the antagonist begins to function as an agonist, indicating that microaggregation of GnRH receptors may be a mechanism involved in target cell activation (88, 89).

After aggregation, GnRH receptors are internalized and become associated with lysosomes, suggesting a degradation pathway, and/or the Golgi complex and LH granules, suggesting a recycling pathway (90–93).

Evidence for the recycling of GnRH receptors has been presented by Schvartz and Hazum (94) after observing the apparent reappearance of GnRH receptors on the cell surface after internalization. The authors covalently attached a GnRH agonist to GnRH receptors, allowed the complex to be internalized, and subsequently evaluated the susceptability of this complex to extracellular trypsin treatment. After initial internalization of the agonist-receptor complex, trypsin treatment caused the appearance of a characteristic GnRH receptor fragment indicative of the return of GnRH receptors to the cell surface. Administration of lysosomotropic agents (chloroquine and methylamine) and monensin increased the apparent rate of recycling of GnRH receptors, presumably by reducing the degradation of the agonist receptor complex in lysosomes. It should be noted that these studies observed the recycling of covalently linked agonist-receptor complexes that may or may not be routed similarly to normal agonist-receptor complexes. Finally, it appears to be likely that agonist-occupied GnRH receptors are routed through the cell differently than antagonist-occupied receptors (95).

Receptors for GnRH, therefore, can be replaced in plasma membranes by recycling and biosynthesis of new GnRH receptors. Using the density-shift technique, the time required for synthesis of one-half of the population of GnRH receptors in rat pituitary cell cultures is 24–28 h (96). Because of the relatively slow basal synthesis rate, GnRH receptors are likely degraded and new receptors synthesized as part of general membrane turnover. Treatment of cells with GnRH stimulates the synthesis of GnRH receptors and reduces the half-time of synthesis to 12 h (97). This stimulation by GnRH of receptor synthesis appears to be independent of extracellular calcium.

Thus, it appears that GnRH receptors follow a common internalization pathway after agonist binding and that the appearance of receptors on the cell surface is at least due to the processes of recycling and hormone-sensitive biosynthesis.

Cellular Mechanisms of GnRH Action

The primary physiological response to GnRH binding by receptors on gonadotropes is the release of gonadotropins. However, there are other cellular responses evoked by GnRH, including down- and up-regulation of GnRH receptors, desensitization of gonadotropes, gonadotropin biosynthesis, and GnRH receptor biosynthesis. It appears as if the effects of GnRH are mediated through G-protein-linked mechanisms (98, 99). As several pathways exist for information flow within the gonadotrope, it is likely that one or more second-messenger systems may be utilized by the gonadotrope to perform these varied functions in response to GnRH.

LH Release

Studies on the stimulation of LH secretion in response to GnRH first implicated *cyclic adenosine 3',5'-monophosphate* (cAMP) as the second messenger (100–102). However, the role of cAMP was questioned as several studies could not show significant involvement of cAMP in GnRH-stimulated LH release (103–106). A comprehensive study by Conn et al. (107) demonstrated that LH release in vitro could be uncoupled from cAMP. It is well accepted now that cAMP is not a second messenger in GnRH-stimulated LH release (reviewed in 108–110).

Several lines of evidence indicate that calcium functions as a second messenger in acute release of LH in response to GnRH. When extracellular calcium is omitted or chelated, GnRH-stimulated LH release is inhibited (111–117). When calcium is introduced back into media, the release of LH occurs normally (118). As binding of GnRH to its membrane receptors is independent of extracellular calcium (119), an intracellular action of calcium is implicated. Increases in intracellular calcium of gonadotropes by the administration of ionophores, liposomes loaded with calcium, or KCl depolarization stimulate the release of LH with similar efficacy to GnRH (115), and this effect is not mimicked by other cations (Mg^{++} and Na^+). Therefore, agents that are thought to provoke an increase in intracellular calcium of gonadotropes also cause LH release.

A second line of evidence for calcium as a second messenger in GnRH-stimulated LH release is the observation that administration of GnRH causes a measurable increase in intracellular calcium. Initially, a transmembrane flux of calcium in response to GnRH was observed (111, 120). Subsequent studies have utilized probes that fluoresce in the presence of calcium, Quin2 and Fura2, to show that GnRH stimulates a transient increase in intracellular calcium levels (121–123) associated with gonadotropin release. This increase in intracellular calcium is not provoked by occupancy of the receptor alone, as the stimulatory actions of GnRH on intracellular calcium levels are not observed after treatment of pituitary cells with a GnRH antagonist.

Further supporting evidence indicating a second-messenger role of calcium in GnRH-stimulated LH release is that modulators of calcium channel function can alter gonadotropin release. Treatment of pituitary cells with such calcium channel blocking agents as verapamil and methoxyverapamil can block stimulated LH release (124). These agents also block gonadotropin release in vivo (125, 126).

Finally, treatment of pituitary cells with the calcium channel agonist maitotoxin can stimulate the release of LH (127). Calcium ion channels associated with the GnRH receptor appear to be receptor-operated-type channels, as depolarization of the gonadotrope does not occur after stimulation by GnRH (128), although depolarization of gonadotropes with KCl does cause LH release. Based upon studies using calcium channel antagonists, GnRH-sensitive calcium ion channels also possess some of the characteristics of voltage-sensitive calcium channels (124, 129). Taken together, these data provide strong evidence that calcium is the second messenger that mediates the acute release of LH in response to GnRH.

Two primary biochemical pathways have been identified in transducing changes in intracellular calcium levels and altered cellular response. The intracellular "calcium receptors," calmodulin and *protein kinase C* (PKC), have been identified in gonadotropes, and both appear to respond to activation of gonadotropes by GnRH.

Since its identification, PKC has been implicated in numerous systems for mediation of agonist-induced cellular responses. This enzyme activity is dependent upon calcium and phospholipids (reviewed in 130). Several observations indicate that PKC is involved in GnRH-stimulated functions of the gonadotrope. First, GnRH and GnRH agonists cause the redistribution of PKC activity from the cytosolic to a particulate fraction of the pituitary both in vivo and in vitro (131–133). Generation of the activators of PKC, calcium and phospholipid, is achieved by phosphoinositide phosphorylation and hydrolysis (reviewed in 134, 135). These phosphorylations and subsequent hydrolysis by phospholipase C-type reactions result in the formation of *diacylglycerols* (DAG) and inositol 1,4,5-trisphosphate (IP$_3$ [136]). Diacylglycerol can function to activate PKC directly. IP$_3$ causes the release of calcium from intracellular nonmitochondrial stores (137) and may regulate plasma membrane calcium channels as well (138). GnRH stimulates the turnover of the polyphosphoinositide cycle (139, 140), suggesting stimulation of the phospholipase C-type reaction and generation of the activators of PKC. Additionally, as described above, if intracellular calcium levels are increased in gonadotropes, LH release is stimulated. Gonadotropes can respond to administration of synthetic DAG with enhanced LH release (141, 142). The stimulation of LH release by administration of protein kinase activators appears independent of extracellular calcium, as evidenced by chelation of extracellular calcium, antagonism of calcium channels, or coculture with calmodulin inhibitors (141, 143), and, in fact, is synergistically enhanced in the presence of calcium ionophores (142). Finally, administration of phorbol esters (i.e., *phorbol myristate acetate* [PMA]) that can directly activate PKC also causes LH release. These data strongly implicated a role for PKC in GnRH-stimulated LH release.

There are, however, a number of observations that question the involvement of PKC. First, PKC activators do not stimulate LH release to the same extent as does GnRH (141, 142, 144). Additionally, administration of PKC activators (PMA) and GnRH results in an additive stimulation of LH release, suggesting independent mechanisms of action (140). The most definitive studies to indicate a dissociation between PKC activity and GnRH-stimulated LH release have been obtained through the use of cells that have been depleted of PKC activity. Because of the lack of useful specific inhibitors of PKC, protocols were developed to "down-regulate" PKC activity in order to study cellular responses in the absence of PKC activity. After exposure of pituitary cells to high doses of PMA (relative to those sufficient to stimulate LH release) for several hours, PKC was depleted from these cells, as evidenced by (1) no measurable PKC phosphorylating activity, (2) no PKC activity as measured by the ability of PMA to stimulate LH release, (3) no demonstrable binding sites for phorbol esters (PKC is an intracellular "receptor" for phorbol esters), and (4) the absence of any immunologically detectable PKC (145). Using cells depleted of PKC activity, McArdle et al. (146) showed that LH release in response to GnRH or calcium ionophore was intact. The absolute levels of LH release in PKC-depleted cells were less than in PKC-intact cells due to PMA treatment for depletion of PKC, which caused LH release during pretreatment. Consequently, PKC-depleted cells contained less LH than control cells. When results are corrected for cellular content of LH, it is clear that the absence of PKC activity does not affect GnRH-stimulated LH release. These and other observations (144, 147, 148) provide strong evidence that PKC does not mediate the effects of GnRH to stimulate acute LH release.

An alternative calcium-sensitive system is the calmodulin pathway. Calmodulin is a ubiquitous calcium binding protein that upon activation by calcium can regulate the activity of many regulatory enzymes, such as adenylate cyclase, phosphorylase kinase, myosine light chain kinase, calcineurin, and phosphodiesterase (149). Additionally, several cytoskeletal proteins, likely involved in the cellular secretion mechanism, can be regulated by calmodulin (150). Administration of GnRH to ovariectomized rats causes redistribution of calmodulin, as measured by radioimmunoassay, from the cytosolic to the plasma membrane fraction of pituitary tissue (151). Moreover, calmodulin associates with patches of agonist-occupied GnRH receptors on gonadotropes (152).

In addition to these observations of physical associations of calmodulin and GnRH receptors, biochemical evidence also supports a role for calmodulin in the action of GnRH. Administration of agents that inhibit the activity of calmodulin (e.g., pimozide) inhibits GnRH-stimulated LH release (153, 154). Notably, these calmodulin inhibitors show similar actions in gonadotropes, even though they are from different classes of calmodulin inhibitors. These calmodulin inhibitors can block LH release stimulated by calcium ionophores (142), suggesting that calmodulin is required for GnRH- and ionophore-stimulated LH release. Five major calmodulin binding components in gonadotropes have been suggested through utilization of a calmodulin gel overlayer assay that evaluates calcium-dependent calmodulin binding to cellular products after SDS-PAGE (155). Subsequent work towards identifying these calmodulin components has suggested the presence of spectrin, caldesmon, and calcineurin as three of the calmodulin binding proteins present in the pituitary gonadotrope (156). These data provide strong evidence for a role of calmodulin in GnRH-stimulated LH release, and one calmodulin binding protein, caldesmon, has been implicated in GnRH-stimulated LH release (157).

Biosynthesis of LH

In addition to stimulating the release of LH, GnRH is also required for biosynthesis of LH. LH is composed of two subunits, α and β. The α-subunit is common for LH, FSH, *thyroid stimulating hormone* (TSH), and *human chorionic gonadotropin* (hCG). The β-subunit differs among these glycoprotein hormones and confers biological specificity (158). With the complex nature of LH, there are several specific processes related to biosynthesis that are regulated, including mRNA production for both subunits, translation, subunit assembly, and glycosylation. GnRH stimulates and is required for normal production of mRNA for the β-subunit of LH (159–161). In pituitary cell cultures, PKC appears to be required for GnRH-stimulated β-subunit mRNA production (159). There are conflicting reports whether GnRH stimulates translation of the α- and β-subunits. Starzec et al. (162, 163) have shown that GnRH stimulates translation of both the α- and β-subunits of LH; however, this was not observed by others (164, 165). Moreover, Starzec et al. (163) have suggested that GnRH-stimulated translation of LH can be mimicked by cAMP analogs and phorbol esters in a nonadditive manner indicative of a similar mechanism of action. Lastly, glycosylation of LH is regulated by GnRH (164, 165). Glycosylation of LH can be stimulated by phorbol esters and DAG (166). Additionally, D600 (a calcium entry blocker) and pimozide (calmodulin inactivator) can prevent GnRH-stimulated glycosylation of LH (164). These data indicate the calcium requirement for LH glycosylation and suggest mediation by calmodulin and/or PKC.

The difficulty in describing a single mechanism for GnRH-stimulated biosynthesis of LH is clear. However, it appears as if LH biosynthesis, in part, is regulated in a similar manner to GnRH-stimulated release of LH as calcium and calmodulin are required for glycosylation; yet the stimulatory action of GnRH on the production and translation of individual subunits and their mRNAs is different from LH release, as PKC may mediate mRNA production and translation.

Desensitization

Gonadotropes, like many hormone-responsive cell types, can become refractory to specific hormones. *Desensitization* is the condition in which cellular responsiveness to a subsequent stimulation by the same hormone is decreased (167–170). Desensitization of the pituitary gland to GnRH and its analogs has been exploited for its clinical value. Long-term exposure to GnRH analogs can result in virtually the complete cessation of LH and FSH release as well as subsequent gonadal steroid secretion. This "chemical castration" has proven useful for treatment of individuals with steroid-dependent neoplasia, as well as for controlling ovarian follicular development in patients for in vitro fertilization. The loss of responsiveness of gonadotropes to GnRH after prior exposure can be dissociated from the mechanism of LH release. As indicated above, GnRH-stimulated LH release is dependent upon extracellular calcium. When extracellular calcium is chelated during exposure to GnRH, gonadotropes still become desensitized (171, 172).

Additionally, desensitization to GnRH can neither be provoked by calcium ionophores nor blocked by the addition of calcium channel antagonists (171). Desensitization of gonadotropes to GnRH can be achieved after a brief exposure to GnRH (20 min) and lasts for at least 12 h after a single exposure (172). Occupancy of the GnRH receptor alone is insufficient to cause desensitization, as the administration of antagonists of GnRH does not induce desensitization. Microaggregation of GnRH receptors, dimerization by antibody-antagonist-receptor complexes, does appear to be sufficient stimulus for desensitization (173); however, internalization of occupied GnRH receptors is not required (174). Desensitization of gonadotropes to GnRH is also associated with a loss of responsiveness to maitotoxin, which activates calcium ion channels, indicating that loss of activation of calcium ion channels by GnRH may contribute to desensitization (127, 175). Desensitization of gonadotropes has also been shown to be affected by changes in membrane fluidity (175, 176).

Finally, desensitization occurs normally in gonadotropes depleted of PKC and is not induced by activators of PKC, suggesting a lack of involvement of PKC in mediating desensitization (177). Therefore, it appears that after binding of GnRH to its receptors, the receptors undergo microaggregation that stimulates LH release and desensitization. Subsequent known effects of GnRH (i.e., calcium entry) are required for LH release, but are not required for desensitization. It has been suggested that down-regulation of GnRH receptors participates in cellular desensitization. Clearly, down-regulation of receptors could reduce the sensitivity of gonadotropes to GnRH in the short term; however, there also appears to be an uncoupling of the receptor-effector system that contributes to desensitization.

Regulation of GnRH Receptor Populations

As indicated above, GnRH can stimulate many alterations in the numbers of its own receptors, including down- and up-regulation, unmasking of receptors, and biosynthesis of receptors. Although these varied effects are all the result of GnRH stimulation, it appears as though these alterations are independently regulated.

Down-regulation of GnRH receptors occurs within the first 3–4 h of GnRH treatment. GnRH antagonists do not induce down-regulation, indicating that this effect is not due simply to occupancy of the GnRH receptor. Down-regulation appears to be independent of calcium flux caused by GnRH, as chelation of extracellular calcium does not interfere with down-regulation of GnRH receptors (75). Moreover, down-regulation is not evoked by agents that raise intracellular calcium (75). In gonadotropes exposed to high concentrations of phorbol esters to activate PKC, there is decreased binding of GnRH agonist, which suggests a role in down-regulation (177); however, treatment of gonadotropes that are depleted of PKC activity with GnRH is followed by normal down-regulation of GnRH receptors (177). These data suggest that down-regulation of GnRH receptors utilizes a different intracellular second-messenger system than does GnRH-stimulated release of LH.

Up-regulation of GnRH receptors in response to homologous hormone occurs several hours after down-regulation. Again, occupancy of the receptor alone is not responsible for up-regulation, as GnRH antagonists cannot substitute for GnRH agonists in causing this effect. In direct contrast to down-regulation, up-regulation of GnRH receptors is dependent upon extracellular calcium and can be stimulated by agents that elevate intracellular calcium levels (75–77). Additionally, up-regulation of GnRH receptors requires protein synthesis as well as microtubule function (77). Depolarization of gonadotropes by KCl administration can stimulate up-regulation (76), and these effects can be observed after treatment of gonadotropes with analogs of cAMP (76).

Recent evidence suggests that there are at least two mechanisms that contribute to up-regulation of GnRH receptors. Treatment of gonadotropes with GnRH significantly stimulates the synthesis rate of GnRH receptors (97). This stimulatory effect of GnRH is found to occur normally, even when extracellular calcium is chelated and cannot be caused by the administration of calcium ionophore. Therefore, GnRH-stimulated synthesis of its own receptors is independent of extracellular calcium, but up-regulation of GnRH receptors is dependent on extracellular calcium. Presumably, stimulation of the synthesis of GnRH receptors contributes to up-regulation but is independent of calcium, suggesting more than one mechanism of up-regulation.

Treatment of cells with phorbol esters can lead to increases in receptor number; however, this appears to be an unmasking of GnRH receptors already present on gonadotropes and requires the simultaneous administration of the phorbol ester and GnRH agonist (178). As the effect of phorbol ester to unmask GnRH receptors occurs within 20 min of treatment, the observed effects of phorbol esters are likely not similar to the up-regulation induced by GnRH. Moreover, these "phorbol ester unmasked" GnRH receptors appear to be selectively uncoupled to phosphoinositide metabolism (63) and have a similar rate of synthesis to GnRH receptors normally present on gonadotropes (96). These observations indicate that the unmasking of GnRH receptors by phorbol esters and GnRH-stimulated up-regulation are probably two separate events. It is clear from the many varied observations involving the up-regulation of GnRH receptors that up-regulation is a complex process and is likely mediated, in part, by several intracellular mechanisms that we are just beginning to identify.

Summary and Conclusions

Hypothalamic GnRH controls the release of gonadotropins from the pituitary gland. The effect of GnRH to stimulate gonadotropin release is used to improve reproductive function and efficiency. Long-term administration of GnRH or its analogs can result in the reduction of LH and FSH release, with a concomitant decrease in sex steroid production. This characteristic forms the basis for one of the primary clinical uses of GnRH analogs: treatment of steroid-dependent neoplasia. In addition to effects on gonadotropin release, GnRH also regulates gonadotropin biosynthesis, up- and down-regulation of GnRH receptors, desensitization, GnRH receptor biosynthesis, and gonadotropin subunit mRNA production. These varied effects of GnRH appear to be mediated by the interaction of different intracellular second-messenger systems. Thus, GnRH actions illustrate the ability of a single hormone utilizing a single receptor type to activate several different intracellular mechanisms within a single cell type.

References

1. Yang-Feng TL, Seeburg PH, Francke U. Human luteinizing hormone-releasing hormone gene (LHRH) is located on short arm of chromosome 8 (region 8p11.2–p21). Somatic Cell Mol Genet 1986;12:95–100.
2. Seeburg PH, Adelson JP. Characterization of cDNA for precursor of human luteinizing hormone releasing hormone. Nature 1984;311:666–8.
3. Adelman JP, Mason AJ, Hayflick JS, Seeburg PH. Isolation of the gene and hypothalamic cDNA for the common precursor of gonadotropin-releasing hormone and prolactin-inhibiting factor in human and rat. Proc Natl Acad Sci USA 1986;83:177–83.

4. Seeburg PH, Mason TJ, Stewart TA, Nikolics K. The mammalian GnRH gene and its pivotal role in reproduction. Recent Prog Horm Res 1987; 43:69–91.

5. Mason AJ, Hayflick JS, Zoeller T, et al. Truncating the gonadotropin releasing hormone gene is responsible for hypogonadism in the hpg mouse. Science 1986;234:1366–71.

6. Schally AV, Arimura A, Baba Y, et al. Isolation and properties of the FSH and LH-releasing hormone. Biochem Biophys Res Commun 1971;43:393–9.

7. Matsuo J, Baba Y, Nair RMG, Arimura A, Schally AV. Structure of the porcine LH- and FSH-releasing hormone, I. The proposed amino acid sequence. Biochem Biophys Res Commun 1971;43:1334–9.

8. Burgus R, Butcher M, Amoss M, et al. Primary structure of the ovine hypothalamic luteinizing hormone-releasing factor (LRF). Proc Natl Acad Sci USA 1972;69:278–82.

9. King JA, Millar RP. Structure of chicken hypothalamic luteinizing hormone-releasing hormone, I. Structural determination on partially purified material. J Biol Chem 1982;257:10722–8.

10. King JA, Millar R. Structure of chicken hypothalamic luteinizing hormone-releasing hormone, II. Isolation and characterization. J Biol Chem 1982; 257:10729–32.

11. Miyamoto K, Hasegawa Y, Nomura M, Igarashi M, Kangawa K, Matsuo H. Identification of the second gonadotropin-releasing hormone in chicken hypothalamus: evidence that gonadotropin secretion is probably controlled by two distinct gonadotropin-releasing hormones in avian species. Proc Natl Acad Sci USA 1984;81:3874–8.

12. Sherwood NM, Eiden L, Brownstein M, Spiess J, Rivier J, Vale WW. Characterization of a teleost gonadotropin-releasing hormone. Proc Natl Acad Sci USA 1983;80:2794–8.

13. Sherwood NM, Sower SA, Marshak DR, Fraser BA, Brownstein MJ. Primary structure of gonadotropin-releasing hormone from lamprey brain. J Biol Chem 1986;261:4812.

14. Sherwood NM. Evolution of a neuropeptide family: gonadotropin-releasing hormone. Am Zool 1986;26:1041–54.

15. Millar RP, King JA. Evolution of gonadotropin-releasing hormone: multiple usage of a peptide. NIPS 1988;3:49–53.

16. Karten MJ, Rivier JE. Gonadotropin-releasing hormone analog design. Structure-function studies toward the development of agonists and antagonists: rationale and perspective. Endocr Rev 1986;7:44–66.

17. Monahan MW, Amoss MS, Anderson HA, Vale W. Synthetic analogs of the hypothalamic luteinizing hormone factor with measured agonist or antagonist properties. Biochemistry 1973;12:4616–20.

18. Coy DH, Seprodi J, Vilchez-Martinez JA, Pedroza E, Gardner J, Schally AV. 1979. Structure function studies and prediction of conformational requirements for LH-RH. In: Collin R, Barbeau A, Ducharme JR, Rockefort JG, eds. Central nervous system effects of hypothalamic hormones and other peptides. New York: Raven Press, 1979;317–23.

19. Nikolics K, Coy DH, Vilchez-Martinez JA, Coy EJ, Schally AV. Synthesis and biological activity of position 1 analogs of LH-RH. Int J Pept Protein Res 1977;9:57–62.

20. Momany FA. Conformational energy analysis of the molecule, luteinizing hormone-releasing hormone, 1. Native decapeptide. J Am Chem Soc 1975;98:2990–6.
21. Fujino M, Kobayashi S, Obayashi M, et al. Structure-activity relationships in the C-terminal part of luteinizing hormone releasing hormone (LH-RH). Biochem Biophys Res Commun 1972;49:863–9.
22. Fujino M, Fukuda T, Shinagawa S, Kobayashi S, Yamazaki I, Nakayama R. Synthetic analogs of luteinizing hormone releasing hormone (LH-RH) substituted in position 6 and 10. Biochem Biophys Res Commun 1974; 60:406–13.
23. Conn PM, Rogers DC, Seay S, et al. Receptor-effector coupling in the pituitary gonadotrope. In: McKerns KW, Naor Z, eds. Biochemical endocrinology. New York: Plenum Press, 1984:153–73.
24. Hook WA, Karten M, Siraganian RP. Histamine release by structural analogs of LHRH [Abstract 5336]. Fed Proc 1985;44:1323.
25. Haviv F, Palabrica CA, Bush EN, et al. Active reduced size hexapeptide analogues of luteinizing hormone-releasing hormone. J Med Chem 1989; 32:2340–4.
26. Bennett HPJ, McMartin C. Peptide hormones and their analogues: distribution, clearance from the circulation, and inactivation in vivo. Pharmacol Rev 1979;30:247–92.
27. Griffiths EC, Kelly AJ. Mechanism of inactivation of hypothalamic regulatory hormones. Mol Cell Endocrinol 1979;14:3–17.
28. Horsthemke B, Bauer K. Substrate specificity of an adenohypophyseal endopeptidase capable of hydrolyzing luteinizing hormone-releasing hormone: preferential cleavage of peptide bonds involving the carboxyl terminus of hydrophobic and basic amino acids. Biochemistry 1982;21:1033–6.
29. Tate SS. Purification and properties of a bovine brain thyrotropin-releasing-factor deamidase, a post proline cleaving enzyme of limited specificity. Eur J Biochem 1981;118:17–23.
30. Knisatschek H, Baur K. Characterization of "thyroliberin deaminating enzyme" as a post-proline-cleaving enzyme. J Biol Chem 1979;254:10936–43.
31. Horsthemke B, Bauer K. Characterization of a nonchymotrypsin-like endopeptidase from anterior pituitary that hydrolyzes luteinizing hormone-releasing hormone at the tyrosyl-glycine and histidyl-tryptophan bonds. Biochemistry 1981;19:2867–73.
32. Handelsman DJ, Swerdloff RS. Pharmacokinetics of gonadotropin-releasing hormone and its analogs. Endocr Rev 1986;7:95–105.
33. Coy DH, Vilchez-Martinez JA, Coy EJ, Schally AV. Analogs of luteinizing hormone-releasing hormone with increased biological activity produced by D-amino acid substitutions in position 6. J Med Chem 1976;19:423–5.
34. Coy DH, Coy EJ, Schally AV, Vilchez-Martinez JA, Hirotsu Y, Arimura A. Synthetic and biological properties of [D-Ala-6-DES-Gly-NH$_2$-10]-LH-RH ethylamide, a peptide with greatly enhanced LH- and FSH-releasing activity. Biochem Biophys Res Commun 1974;57:335–40.
35. Vilchez-Martinez JA, Coy DH, Arimura A, Coy EJ, Hirotsu Y, Schally AV. Synthesis and biological properties of [Leu-6]-LH-RH and [D-Leu-6,

DES Gly NH$_2$-10]-LH-RH ethylamide. Biochem Biophys Res Commun 1974;59:1226–32.
36. Nestor JJ, Ho TL, Tahilramani R, McRae GI, Vickery BH. Long acting LHRH agonists and antagonists. In: Labri F, Belanger A, Dupont A, eds. LHRH and its analogues, basic and clinical aspects. Excerpta Medica, 1984:24–35.
37. Danforth DR, Gordon K, Leal JA, Williams RF, Hodgen GD. Extended presence of antide (Nal-Lys GnRH antagonist) in circulation: prolonged duration of gonadotropin inhibition may derive from antide binding to serum proteins. J Clin Endocrinol Metab 1990;70:554–6.
38. Clayton RN, Harwood JP, Catt KJ. Gonadotropin-releasing hormone analogue binds to luteal cells and inhibits progesterone production. Nature 1979;282:90–2.
39. Jones PBC, Conn PM, Marian J, Hsueh AJW. Binding of gonadotropin-releasing hormone agonist to rat ovarian granulosa cells. Life Sci 1980; 27:2125–32.
40. Latouche J, Crumeyrolle-Arias M, Jordon D, et al. GnRH receptors in human granulosa cells: anatomical localization and characterization by autoradiographic study. Endocrinology 1989;125:1739–41.
41. Brown JL, Reeves JJ. Absence of specific luteinizing hormone releasing hormone receptors in ovine, bovine or porcine ovaries. Biol Reprod 1983;29:1179–82.
42. Eidne KA, Hendricks DT, Millar RP. Demonstration of a 60K molecular weight luteinizing hormone-releasing hormone receptor in solubilized adrenal membrane by a ligand-immunoblotting technique. Endocrinology 1985;116:1792–5.
43. Eidne KA, Flanagan CA, Millar RP. Gonadotropin releasing hormone binding sites in human breast carcinoma. Science 1985;229:989–91.
44. Fekete M, Zalanti A, Schally AV. Presence of membrane binding sites for [D-Trp^6]-luteinizing hormone-releasing hormone in experimental pancreatic cancer. Cancer Lett 1989;45:87–91.
45. Reubi JC, Palcios JM, Maurer R. Specific luteinizing-hormone-releasing hormone receptor binding sites in hippocampus and pituitary: an autoradiographical study. Neuroscience 1987;21:847–56.
46. Jennes L, Dalati B, Conn PM. Distribution of gonadotropin releasing hormone agonist binding sites in the rat central nervous system. Brain Res 1988;452:156–64.
47. Jennes L, Janovick J, Braden T, Conn PM. Gonadotropin releasing hormone binding sites in rat hippocampus: different structure/binding relationships compared to the anterior pituitary. Mol Cell Neurosci 1990;1:121–7.
48. Moss RL, McCann S. Induction of mating behavior in rats by luteinizing hormone-releasing hormone. Science 1973;181:177–9.
49. Pfaff DW. Luteinizing hormone-releasing factor potentiates lordosis behavior in hypophysectomized ovariectomized female rats. Science 1973;182:1148–9.
50. Hsueh AJW, Jones PB. Extrapituitary actions of gonadotropin-releasing hormone. Endocr Rev 1981;2:437–61.

51. Marian J, Conn PM. Subcellular localization of the receptor for gonadotropin-releasing hormone in pituitary and ovarian tissue. Endocrinology 1983;112:104–12.

52. Hazum E. GnRH-receptor of rat pituitary is a glycoprotein: differential effect of neuroaminidase and lectins on agonists and antagonists binding. Mol Cell Endocrinol 1982;26:217–22.

53. Schvartz I, Hazum E. Tunicamycin and neuraminidase effects on luteinizing hormone (LH)-releasing hormone binding and LH release from rat pituitary cells in culture. Endocrinology 1985;116:2341–6.

54. Hazum E, Garritsen A, Keinan D. Role of lipids on gonadotropin releasing hormone agonist and antagonist binding to rat pituitary. Biochem Biophys Res Commun 1982;105:8–13.

55. Keinan D, Hazum E. Mapping of gonadotropin releasing hormone receptor binding site. Biochemistry 1985;24:7728–32.

56. Hazum E. Binding properties of solubilized gonadotropin releasing hormone receptor: role of carboxylic groups. Biochemistry 1987;26:7011–4.

57. Perrin MH, Haas Y, Porter J, Rivier J, Vale W. The gonadotropin-releasing hormone pituitary receptor interacts with a guanosine triphosphate binding protein: differential effects of guanyl nucleotides on agonist and antagonist binding. Endocrinology 1989;124:798–804.

58. Perrin MM, Haas Y, Rivier JE, Vale WW. Solubilization of the gonadotropin-releasing hormone receptor from bovine pituitary plasma membranes. Endocrinology 1983;112:1538–40.

59. Hazum E, Schvartz I, Popliker M. Production and characterization of antibodies to gonadotropin-releasing hormone receptors. J Biol Chem 1987;262:531–4.

60. Iwashita M, Hirota J, Izumi S-I, Chen H-C, Catt KJ. Solubilization and characterization of the rat pituitary gonadotrophin-releasing hormone receptor. J Mol Endocrinol 1988;1:187–96.

61. Ogier SA, Mitchell R, Fink G. Solubilization of a large molecular weight form of the rat LHRH receptor. J Endocrinol 1987;115:151–9.

62. Hazum E. Photoaffinity labeling of luteinizing hormone receptor of rat pituitary membrane preparations. Endocrinology 1981;109:1281–3.

63. Huckle WR, Hawes BE, Conn PM. Protein kinase C-mediated gonadotropin releasing hormone sequestration is associated with uncoupling of phosphoinositide turnover hydrolysis. J Biol Chem 1989;264:8619–26.

64. Conn PM, Venter JC. Radiation inactivation (target size analysis of the gonadotropin-releasing hormone receptor: evidence for a high molecular weight complex). Endocrinology 1985;116:1324–6.

65. Hazum E, Schvartz I, Waksman Y, Keinan D. Solubilization and purification of rat pituitary gonadotropin-releasing hormone receptor. J Biol Chem 1986;261:13043–8.

66. Eidne KA, McNiven AI, Taylor PL, et al. Functional expression of rat pituitary gonadotropin-releasing hormone receptors in *Xenopus* oocytes. J Mol Endocrinol 1988;1:R9–12.

67. Yoshida S, Plant S, Taylor PL, Eidne KA. Chloride channels mediate the response to gonadotropin-releasing hormone (GnRH) in *Xenopus* oocytes injected with rat anterior pituitary mRNA. Mol Endocrinol 1989;3:1953–60.

68. Sealfon SC, Gillo B, Mundomattom S, et al. Gonadotropin releasing hormone receptor expression in *Xenopus* oocytes. Mol Endocrinol 1990;4:119–24.
69. Clayton RN, Solano AR, Garcia-Vila A, Dufau ML, Catt KJ. Regulation of pituitary receptors for gonadotropin releasing hormone during the rat estrous cycle. Endocrinology 1980;107:699–706.
70. Savoy-Moore RT, Schwartz NB, Duncan JA, Marshall JC. Pituitary gonadotropin-releasing hormone receptors during the rat estrous cycle. Science 1980;209:942–4.
71. Adams TE, Spies HG. Binding characteristics of gonadotropin-releasing hormone receptors throughout the estrous cycle of the hamster. Endocrinology 1981;108:2245–53.
72. Marian J, Cooper RL, Conn PM. Regulation of the rat pituitary gonadotropin-releasing hormone receptor. Mol Pharmacol 1981;19:339–405.
73. Crowder ME, Nett TM. Pituitary content of gonadotropins and receptors for gonadotropin-releasing hormone (GnRH) and hypothalamic content of GnRH during the periovulatory period of the ewe. Endocrinology 1984; 114:234–9.
74. Nett TM, Cermak D, Braden T, Manns J, Niswender GD. Pituitary receptors for GnRH and estradiol, and pituitary content of gonadotropins in beef cows, I. Changes during the estrous cycle. Dom Anim Endocrinol 1987;4:123–32.
75. Conn PM, Rogers DC, Seay SG. Biphasic regulation of the gonadotropin-releasing hormone receptor by receptor microaggregation and intracellular Ca^{2+} levels. Mol Pharmacol 1984;25:51–5.
76. Young LS, Naik SI, Clayton RN. Adenosine 3',5'-monophosphate derivatives increase gonadotropin-releasing hormone receptors in cultured pituitary cells. Endocrinology 1984;114:2114–22.
77. Young LS, Naik SI, Clayton RN. Pituitary gonadotrophin-releasing hormone receptor up-regulation in vitro: dependence on calcium and microtubule function. J Endocrinol 1985;107:49–56.
78. Young LS, Naik SI, Clayton RN. Increased gonadotrophin releasing hormone receptors on pituitary gonadotrophs: effect on subsequent LH secretion. Mol Cell Endocrinol 1985;41:69–78.
79. Naor Z, Clayton RN, Catt KJ. Characterization of gonadotropin-releasing hormone receptors in cultured rat pituitary cells. Endocrinology 1980; 107:1144–52.
80. Wise ME, Nieman D, Stewart J, Nett TM. Effect of number of receptors for gonadotropin-releasing hormone on the release of luteinizing hormone. Biol Reprod 1984;31:1007–13.
81. Clarke IJ, Cummins JT, Crowder ME, Nett TM. Pituitary receptors for gonadotropin-releasing hormone in relation to changes in pituitary and plasma gonadotropins in ovariectomized hypothalamo/pituitary-disconnected ewes, II. A marked rise in receptor number during the acute feedback effects of estradiol. Biol Reprod 1988;39:349–54.
82. Gregg DW, Nett TM. Direct effects of estradiol-17β on the number of gonadotropin-releasing hormone receptors in the ovine pituitary. Biol Reprod 1989;40:288–93.

83. Laws SC, Beggs MJ, Webster JC, Miller WL. Inhibin increases and progesterone decreases receptors for gonadotropin-releasing hormone in ovine pituitary culture. Endocrinology 1990;127:373–80.
84. Laws SC, Webster JC, Miller WL. Estradiol alters the effectiveness of gonadotropin-releasing hormone (GnRH) in ovine pituitary cultures: GnRH receptors versus responsiveness to GnRH. Endocrinology 1990;127:381–6.
85. Wang QF, Farnworth PG, Findlay JK, Burger HG. Effect of 31K bovine inhibin on the specific binding of gonadotropin-releasing hormone to rat anterior pituitary cells in culture. Endocrinology 1988;123:2161–6.
86. Wang QF, Farnworth PG, Findlay JK, Burger HG. Inhibitory effect of pure 31-kilodalton bovine inhibin on gonadotropin-releasing hormone (GnRH)-induced up-regulation of GnRH binding sites in cultured rat anterior pituitary cells. Endocrinology 1989;124:363–8.
87. Braden TD, Farnworth PG, Burger HG, Conn PM. Regulation of the synthetic rate of gonadotropin-releasing hormone receptors in rat pituitary cell cultures by inhibin. Endocrinology 1990;127:2387–92.
88. Conn PM, Rogers DC, Stewart JM, Niedal J, Sheffield T. Conversion of a gonadotropin-releasing hormone antagonist to an agonist. Nature 1982; 296:653–5.
89. Conn PM. Ligand dimerization: a technique for assessing receptor-receptor interactions. Methods Enzymol 1983;103:49–58.
90. Pelletier G, Dube D, Guy J, Sequin C, Lefebvre FA. Binding and internalization of a luteinizing hormone-releasing hormone agonist by rat gonadotrophic cells, a radiographic study. Endocrinology 1982;111:1068–76.
91. Duello TM, Nett TM, Farquhar MG. Fate of a gonadotropin-releasing hormone agonist internalized by rat pituitary gonadotrophs. Endocrinology 1983;112:1–10.
92. Jennes L, Stumpf WE, Conn PM. Internalization pathways of electron opaque gonadotropin-releasing hormone derivatives bound by cultured gonadotropes. Endocrinology 1983;113:1683–9.
93. Hazum E, Conn PM. Molecular mechanism of gonadotropin-releasing hormone (GnRH) action, I. The GnRH receptor. Endocr Rev 1988;9:379–85.
94. Schvartz I, Hazum E. Internalization and recycling of receptor-bound gonadotropin-releasing hormone agonist in pituitary gonadotropes. J Biol Chem 1987;262:17046–50.
95. Jennes L, Coy D, Conn PM. Receptor-mediated uptake of GnRH agonist and antagonists by cultured gonadotropes: evidence for differential intracellular routing. Peptides 1986;7:459–63.
96. Braden TD, Hawes BE, Conn PM. Synthesis of gonadotropin-releasing hormone receptors by gonadotrope cell cultures: both preexisting receptors and those unmasked by protein kinase-C activators show a similar synthetic rate. Endocrinology 1989;125:1623–9.
97. Braden TD, Conn PM. Altered rate of synthesis of gonadotropin-releasing hormone receptors: effects of homologous hormone appear independent of extracellular calcium. Endocrinology 1990;126:2577–82.
98. Andrews WV, Staley DD, Huckle WR, Conn PM. Stimulation of luteinizing hormone (LH) release and phospholipid breakdown by guanosine tri-

phosphate in permeabilized pituitary gonadotropes: antagonist action suggests association of a G-protein and gonadotropin-releasing hormone receptor. Endocrinology 1986;119:2537–46.

99. Waters SB, Hawes BE, Conn PM. Stimulation of luteinizing hormone release by sodium fluoride is independent of protein kinase-C activity and unaffected by desensitization to gonadotropin-releasing hormone. Endocrinology 1990;126:2583–91.

100. Borgeat P, Chavaney G, Dupont A, Labrie F, Arimura A, Schally AV. Stimulation of adenosine 3′,5′-cyclic monophosphate accumulation in anterior pituitary gland in vitro by synthetic luteinizing hormone-releasing hormone. Proc Natl Acad Sci USA 1972;69:2677–81.

101. Menon KMJ, Cuanaga KP, Azhar S. GnRH action in rat anterior pituitary gland: regulation of protein, glycoprotein and LH synthesis. Acta Endocrinol (Copenh) 1977;86:473–88.

102. Adams TE, Wagner TOF, Sawyer HR, Nett TM. GnRH interactions with anterior pituitary, II. Cyclic AMP as an intracellular mediator in the GnRH activated gonadotroph. Biol Reprod 1979;21:735–47.

103. Naor Z, Koch Y, Chobsieng P, Zor U. Pituitary cyclic AMP production and mechanisms of luteinizing hormone release. FEBS Lett 1975;58: 318–21.

104. Naor Z, Zor U, Meidan R, Koch Y. Sex differences in pituitary cyclic AMP response to gonadotropin-releasing hormone. Am J Physiol 1978;235:E37–41.

105. Ratner A, Wilson MC, Srivastave L, Peake GT. Dissociation between LH release and pituitary cyclic nucleotide accumulation in response to synthetic LH releasing hormone in vivo. Neuroendocrinology 1976;20:35–42.

106. Rigler GL, Peake GT, Ratner A. Effect of luteinizing hormone releasing hormone on accumulation of pituitary cyclic AMP and GMP in vitro. J Endocrinol 1978;76:367–8.

107. Conn PM, Morrell DV, Dufau ML, Catt KJ. Gonadotropin-releasing hormone action in cultured pituicytes: independence of luteinizing hormone release and adenosine 3′,5′-monophosphate production. Endocrinology 1979; 104:448–53.

108. Conn PM, Huckle WR, Andrews WV, McArdle CA. The molecular mechanism of action of gonadotropin releasing hormone (GnRH) in the pituitary. Recent Prog Horm Res 1987;43:29–68.

109. Conn PM, McArdle CA, Andrews WV, Huckle WR. The molecular basis of gonadotropin releasing hormone (GnRH) action in the pituitary gonadotrope. Biol Reprod 1987;36:17–35.

110. Huckle WR, Conn PM. Molecular mechanism of gonadotropin releasing hormone action, II. The effector system. Endocr Rev 1988;9:387–95.

111. Hopkins CR, Walker AM. Calcium as a second messenger in the stimulation of luteinizing hormone secretion. Mol Cell Endocrinol 1978;12:189–208.

112. Wakabayashi K, Kamberi IA, McCann SM. In vitro response of the rat pituitary to gonadotropin releasing factors and to ions. Endocrinology 1969;85:1046–56.

113. Samli MH, Geshwind II. Some effects of energy-transfer inhibitors and of Ca^{2+}-free and K^+-enhanced media on the release of LH from the rat pituitary gland in vitro. Endocrinology 1968;82:225–31.

114. Adams TE, Nett TM. Interactions of GnRH with the pituitary, III. Role of divalent cations, microtubules and microfilaments in the GnRH activated gonadotroph. Biol Reprod 1979;21:1073–86.
115. Marian J, Conn PM. Gonadotropin releasing hormone stimulation of cultured pituitary cells requires calcium. Mol Pharmacol; 16:196–201.
116. Stern JE, Conn PM. Perifusion of rat pituitaries: requirements of optimal GnRH-stimulated LH release. Am J Physiol 1981;240:E504–9.
117. Conn PM, Rogers DC, Sandhu FS. Alteration of the intracellular calcium level stimulated gonadotropin release from cultured rat anterior pituitary cells. Endocrinology 1979;105:1122–7.
118. Bates MD, Conn PM. Calcium mobilization in the pituitary gonadotrope: relative roles of intra- and extracellular sources. Endocrinology 1984; 115:1380–5.
119. Marian J, Conn PM. The calcium requirement of GnRH-stimulated LH release is not mediated through specific action on the receptor. Life Sci 1980;27:87–92.
120. Conn PM, Kilpatrick D, Kirshner N. Ionophoretic Ca^{2+} mobilization in rat gonadotropes and bovine adrenomedullary cells. Cell Calcium 1980;1:129–33.
121. Clapper D, Conn PM. Gonadotropin-releasing hormone stimulation of pituitary gonadotrope cells produces an increase in intracellular calcium. Biol Reprod 1985;32:269–78.
122. Chang JP, McCoy EE, Graeter J, Tasaka K, Catt KJ. Participation of voltage-dependent calcium channels in the action of gonadotropin-releasing hormone. J Biol Chem 1986;261:9105–8.
123. Leong DA. Spatial mapping of cytosolic calcium oscillations in single gonadotropes [Abstract 7]. Proc 71st annu meet of the Endocr Soc. Seattle, WA, 1989:24.
124. Conn PM, Rogers DC, Seay SG. Structure-function relationships of calcium ion channel antagonists at the pituitary gonadotrope. Endocrinology 1983; 113:1592–5.
125. Barbino A, DeMarinis L. Calcium antagonists and hormone release, II. Effects of verapamil on basal, gonadotropin-releasing hormone, and thryotropin-releasing hormone induced pituitary hormone release in normal subjects. J Clin Endocrinol Metab 1980;51:749–53.
126. Veldhuis JD, Borges JLC, Drake CR, Rogal AD, Kaiser DL, Thorner MO. Divergent influences of the structurally dissimilar calcium entry blockers diltiazem and verapamil on thyrotropin- and gonadotropin-releasing hormone-stimulated anterior pituitary hormone secretion in man. J Clin Endocrinol Metab 1985;60:144–9.
127. Conn PM, Staley DD, Yasumoto T, Huckle WR, Janovick J. Homologous desensitization with gonadotropin-releasing hormone (GnRH) also diminishes gonadotrope responsiveness to maitotoxin: a role for the GnRH receptor regulated calcium ion channel in mediation of cellular desensitization. Mol Endocrinol 1987;1:154–9.
128. Mason WT, Waring DW. Electrophysiological recording from gonadotrophs; evidence for Ca^{2+} channels mediated by gonadotropin-releasing hormone. Neuroendocrinology 1985;41:258–68.

129. Stojilkovic SS, Izumi S, Catt KJ. Participation of voltage-sensitive calcium channels in pituitary hormone release. J Biol Chem 1988;263:13054–61.
130. Nishizuka Y. Studies and perspectives of protein kinase C. Science 1986;233:305–12.
131. Naor Z, Zer J, Zakut H, Hermon J. Characterization of pituitary calcium-activated, phospholipid-dependent protein kinase: redistribution by gonadotropin releasing hormone. Proc Natl Acad Sci USA 1985;82:8203–7.
132. Hirota K, Hirota T, Aguilera G, Catt KJ. Hormone induced redistribution of calcium-activated phospholipid dependent protein kinase in pituitary gonadotrophs. J Biol Chem 1985;260:3243–6.
133. McArdle CA, Conn PM. Hormone-stimulated redistribution of gonadotrope protein kinase C in vivo, dependence on Ca^{2+} influx. Mol Pharmacol 1986;29:570–6.
134. Nishizuka Y. Turnover of inositol phospholipids and signal transduction. Science 1984;225:1365–70.
135. Rana RS, Hokin LE. Role of phosphoinositides in transmembrane signaling. Physiol Rev 1990;70:115–64.
136. Berridge MJ. Inositol phosphate and diacylglycerol as second messengers. Biochem J 1984;220:345–60.
137. Streb H, Irvine RJ, Berridge MJ, Schulz I. Release of calcium from a nonmitochondrial intracellular store in pancreatic acinar cells by inositol 1,4,5-trisphosphate. Nature 1983;306:67–9.
138. Kuno M, Gardner P. Ion channels activated by inositol 1,4,5-trisphosphate in plasma membranes of T-lymphocytes. Nature 1987;326:301–4.
139. Andrews WV, Conn PM. Gonadotropin-releasing hormone stimulates mass changes in phosphoinositides and diacylglycerol accumulation in purified gonadotrope cell cultures. Endocrinology 1986;118:1148–58.
140. Huckle WR, Conn PM. The relationship between gonadotropin-releasing hormone-stimulated luteinizing hormone release and inositol phosphate production: studies with calcium antagonists and protein kinase C activators. Endocrinology 1987;120:160–9.
141. Conn PM, Ganong BR, Ebeling J, Staley D, Neidel JE, Bell, BM. Diacylglycerols release LH: structure-activity relations reveal a role for protein kinase C. Biochem Biophys Res Commun 1985;126:532–9.
142. Harris CE, Staley D, Conn PM. Diacylglycerols and protein kinase C, potential amplifying mechanism for Ca^{2+} mediated gonadotropin-releasing hormone-stimulated luteinizing hormone release. Mol Pharmacol 1985;27:532–6.
143. Naor Z, Eli Y. Synergistic stimulation of luteinizing hormone (LH) release by protein kinase C activators and Ca^{2+}-ionophore. Biochem Biophys Res Commun 1985;130:848–53.
144. Lewis CE, Richards PSM, Moris JF. Heterogeneity of responses to LH-releasing hormone and phorbol ester among rat gonadotrophs: a study using a reverse haemolytic plaque assay for LH. J Mol Endocrinol 1989;2:55–63.
145. McArdle CA, Conn PM. The use of protein kinase C-depleted cells for investigation of the role of protein kinase C in stimulus-response coupling in the pituitary. Methods Enzymol 1989;168:287–301.

146. McArdle CA, Huckle WR, Conn PM. Phorbol esters reduce gonadotrope responsiveness to protein kinase C activators but not to Ca^{2+}-mobilizing secretagogues. Does protein kinase C mediate gonadotropin-releasing hormone action? J Biol Chem 1987;262:5028–35.

147. Beggs MJ, Miller WL. GnRH-stimulated LH release from ovine gonadotrophs in culture is separate from phorbol ester stimulated LH release. Endocrinology 1989;124:667–74.

148. van der Merwe PA, Millar RP, Davidson JS. Calcium stimulated luteinizing-hormone (lutropin) exocytosis by a mechanism independent of protein kinase C. Biochem J 1990;268:493–8.

149. Means AR, Dedman JR. Calmodulin, an intracellular calcium receptor. Nature 1980;285:73–7.

150. Chafouleas JG, Guerriero V, Means AR. Possible regulatory roles of calmodulin and myosin light chain kinase in secretion. In: Conn PM, ed. Cellular regulation of secretion and release. New York; Academic Press, 1985:445–58.

151. Conn PM, Chafouleas JG, Rogers D, Means AR. Gonadotropin-releasing hormone stimulates calmodulin redistribution in rat pituitary. Nature 1981;292:264–5.

152. Jennes L, Bronson D, Stumpf WE, Conn PM. Evidence for an association between calmodulin and membrane patches containing gonadotropin-releasing hormone-receptor complexes in cultured gonadotropes. Cell Tissue Res 1985;239:311–5.

153. Conn PM, Rogers DC, Sheffield T. Inhibition of gonadotropin-releasing hormone-stimulated luteinizing hormone release by pimozide: evidence for a sight of action after calcium mobilization. Endocrinology 1981;109:1122–6.

154. Conn PM, Bates MD, Rogers DC, Seay SG, Smith WA. 1984a. GnRH-receptor-effector-response coupling in the pituitary gonadotrope: a Ca^{2+}-mediated system. In: Fotherby K, Pal SB, eds. The role of drugs and electrolytes in hormogenesis. Berlin: Walter de Gruyer, 1984:85–103.

155. Wooge CH, Conn PM. Characterization of calmodulin-binding components in the pituitary gonadotrope. Mol Cell Endocrinol 1988;56:41–51.

156. Natarajan K, Ness J, Wooge CH, Janovick J, Conn PM. Specific identification and subcellular localization of three calmodulin-binding proteins in the rat gonadotrope: spectrin, caldesmon, and calcineurin. Biol Reprod 1991;44:43–52.

157. Janovick JA, Natarajan K, Longo F, Conn PM. Caldesmon: a bifunctional (calmodulin and actin) binding protein which regulates stimulated gonadotropin release. Endocrinology 1991.

158. Pierce JC, Parsons TF. Glycoprotein hormones: structure and function. Annu Rev Biochem 1981;50:465–95.

159. Andrews WV, Maurer RA, Conn PM. Stimulation of rat luteinizing hormone-β messenger RNA levels by gonadotropin releasing hormone, apparent role for protein kinase C. J Biol Chem 1988;263:13755–61.

160. Hamernik DL, Nett TM. Gonadotropin-releasing hormone increases the amount of messenger ribonucleic acid for gonadotropins in ovariectomized ewes after hypothalamic-pituitary disconnection. Endocrinology 1988;122: 959–66.

161. Lalloz MRA, Detta A, Clayton RN. Gonadotropin-releasing hormone is required for enhanced luteinizing hormone subunit gene expression in vivo. Endocrinology 1988;122:1681–8.

162. Starzec A, Counis R, Jutisz M. Gonadotropin-releasing hormone stimulates the synthesis of the polypeptide chains of luteinizing hormone. Endocrinology 1986;119:561–5.

163. Starzec A, Jutisz M, Counis R. Cyclic adenosine monophosphate and phorbol ester, like gonadotropin-releasing hormone, stimulate the biosynthesis of luteinizing hormone polypeptide chains in a nonadditive manner. Mol Endocrinol 1989;3:618–24.

164. Liu T-C, Jackson GL. Synthesis and release of luteinizing hormone in vitro by rat anterior pituitary cells: effects of gallopamil hydrochloride (D600) and pimozide. Endocrinology 1985;117:1608–14.

165. Vogel DL, Magner JA, Sherins RJ, Weintraub BD. Biosynthesis, glycosylation, and secretion of rat luteinizing hormone α- and β-subunits: differential effects of orchidectomy and gonadotropin-releasing hormone. Endocrinology 1986;119:202–13.

166. Liu T-C, Jackson GL. Stimulation by phorbol ester and diacylglycerol of luteinizing hormone glycosylation and release by rat anterior pituitary cells. Endocrinology 1987;121:1589–95.

167. deKoning JA, van Dietan MJ, van Rees GP. Refractoriness of the pituitary gland after continuous exposure to luteinizing hormone releasing hormone. J Endocrinol 1978;79:311–8.

168. Smith MA, Vale W. Desensitization to gonadotropin-releasing hormone observed in superfused pituitary cells on cytodex beads. Endocrinology 1981;108:752–9.

169. Badger TM, Loughlin JS, Nadaff PG. The luteinizing hormone-releasing hormone (LHRH)-desensitized rat pituitary: luteinizing hormone-responsiveness to LHRH in vitro. Endocrinology 1983;112:793–9.

170. Keri G, Nikolics K, Teplan I, Molnar J. Desensitization of luteinizing hormone release in cultured pituitary cells by gonadotropin-releasing hormone. Mol Cell Endocrinol 1983;30:109–20.

171. Smith WA, Conn PM. GnRH-mediated desensitization of the pituitary gonadotrope is not calcium dependent. Endocrinology 1983;112:408–10.

172. Jinnah HA, Conn PM. Gonadotropin-releasing hormone-mediated desensitization of cultured rat anterior pituitary cells can be uncoupled from luteinizing hormone release. Endocrinology 1986;118:2599–604.

173. Smith WA, Conn PM. Microaggregation of the gonadotropin-releasing hormone-receptor: relation to gonadotrope desensitization. Endocrinology 1984;114:553–9.

174. Gorospe WC, Conn PM. Agents that decrease gonadotropin-releasing hormone (GnRH) receptor internalization do not inhibit GnRH-mediated gonadotrope desensitization. Endocrinology 1987;120:222–9.

175. Gorospe WC, Conn PM. Restoration of the LH secretory response in desensitized gonadotropes. Mol Cell Endocrinol 1988;59:101–10.

176. Gorospe WC, Conn PM. Membrane fluidity regulated development of gonadotrope desensitization to GnRH. Mol Cell Endocrinol 1987;53:131–40.

177. McArdle CA, Gorospe WC, Huckle WR, Conn PM. Homologous down-regulation of gonadotropin-releasing hormone receptors and desensitization of gonadotropes: lack of dependence on protein kinase C. Mol Endocrinol 1987;1:420–9.
178. Huckle WR, McArdle CA, Conn PM. Differential sensitivity of gonadotropin-releasing hormone receptors to activators of protein kinase C: a marker for receptor activators. J Biol Chem 1988;263:3296–302.
179. Huckle WR, Conn PM. The role of protein kinase C in pituitary gonadotropin releasing hormone action. In: Lakowski JM, Perez-Polo JR, Rassin DK, eds. Neural control of reproductive function. New York: Alan Liss, 1988:441–6.

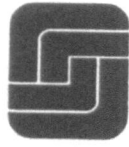

3

Molecular Mechanisms Underlying Placenta- and Pituitary-Specific Expression of the Glycoprotein Hormone α-Subunit Gene

John H. Nilson

All vertebrates synthesize three different pituitary glycoprotein hormones: *luteinizing hormone* (lutropin, LH), *follicle stimulating hormone* (follitropin, FSH), and *thyroid stimulating hormone* (thyrotropin, TSH). *Chorionic gonadotropin* (CG) is structurally related to LH, but is synthesized in the placenta of only primates and equids. All these glycoprotein hormones are heterodimers composed of a noncovalently associated α- and β-subunit. The α-subunit is encoded by a single gene in all mammals examined to date. Thus, within a species, all glycoprotein hormones contain the same α-subunit. In contrast, each glycoprotein hormone contains a unique β-subunit encoded by a discrete β-subunit gene. The β-subunit, then, is responsible for the distinctive biological character of each glycoprotein hormone (reviewed in references 1, 2).

Synthesis of the full spectrum of glycoprotein hormones in primates and equids demands expression of the single-copy α-subunit gene in both pituitary and placenta. Recently, the major regulatory elements required for placenta-specific expression of the α-subunit gene have been identified and characterized (3–9). Accessory elements that further enhance this property have also been described (10, 11). In contrast, regulatory elements responsible for pituitary-specific expression have proven more difficult to study. Nevertheless, recent experiments with transgenic mice are beginning to provide intriguing clues regarding both location and character of the *cis*-acting elements and *trans*-acting factors required for pituitary-specific expression (12, 13, and unpublished data).

The intent of this review is to summarize major findings that have led my colleagues and me to postulate that distinct combinatorial arrays of regulatory elements are responsible for expression of the α-subunit gene in pituitary and placenta. Much of the insight regarding these mechanisms

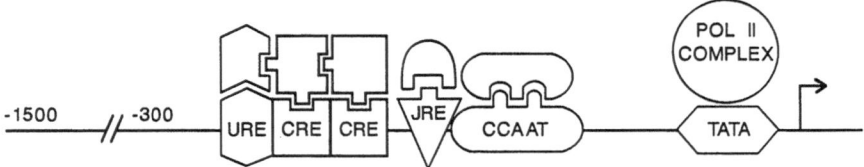

FIGURE 3.1. Regulatory elements in the proximal region of the human α-subunit promoter. The line represents the proximal 1500 bp of 5′-flanking sequence of the human α-subunit gene. The arrow indicates the transcription initiation site. The geometric shapes superimposed on the line represent discrete *cis*-acting elements, whereas those above the line represent *trans*-acting factors that bind specifically to each element. Cooperative interactions are represented by lock and key configurations. (URE = upstream regulatory element; CRE = cAMP response element; JRE = junctional response element; CCAAT = CCAAT box element; TATA = consensus TATA box.)

has come from comparative analysis of the promoter-regulatory regions of the human and bovine α-subunit genes. The human promoter-regulatory region confers both pituitary- and placenta-specific expression, whereas the bovine promoter-regulatory region is active only in pituitary. Characterization of these two promoter-regulatory regions has entailed the use of a variety of experimental techniques, including gene transfection studies (3, 12), DNA-protein binding assays (14, 15), PCR cloning (9), and transgenic animals (12, 13).

Required Elements for Placenta-Specific Expression of the Human α-Subunit Gene

At least two DNA sequence elements are required for placenta-specific expression of the human α-subunit gene (3–8). These elements are located in the 5′-flanking region between nucleotides −180 and −100 relative to the start site of transcription. The element between −180 and −146 is referred to as the *upstream regulatory element* (URE) (Fig. 3.1) and appears to bind a protein unique to choriocarcinoma cells and, presumably, placenta (3–5). The URE may be subdivided into at least two regions, each of which binds a distinct protein (8, 16).

Alone, the URE has no effect on transcription (3, 4). URE activity is completely dependent on the adjacent sequence element located between −146 and −110. This 36-bp sequence is composed of two 18-bp direct repeats containing a conserved palindrome TGACGTCA. Each repeat is designated as a *cAMP response element* (CRE) because a single 18-bp element can confer cAMP responsiveness to either the minimal α-subunit promoter (-100 to $+44$) or a heterologous promoter (3, 4, 6) (Fig. 3.1). Furthermore, each CRE binds a ubiquitous 43-kd nuclear phosphoprotein (*cAMP response element binding protein* [CREB]) (18–20). Changing virtually any base within the core palindrome disrupts binding of CREB (14).

Contribution to Promoter Activity in Placenta of a Complex Array of Tightly Packed Regulatory Elements Contained in the Human α-Subunit Promoter

In a recent study (11), we used a library of oligodeoxyribonucleotides, 20 bp long, to construct 21 "block-replacement" (21) vectors that collectively contain 10-bp transversion mutations at nonoverlapping intervals throughout approximately 200-bp of 5'-flanking sequence of the human α-subunit gene. These vectors were analyzed by transfection, while each discrete region was further characterized by gel mobility shift (22) and cross competition assays (14), methylation interference analysis (23), and UV crosslinking studies (24). Such an approach led to identification of two additional *cis*-acting elements that bind distinct proteins (Fig. 3.1). One of the elements contains a canonical CCAAT box that binds a ubiquitous factor distinct from the previously characterized CCAAT binding factors CTF/NF1 (25), C/EBP (26), CP1 (27), and NF-Y (28). We designated this factor *α-subunit CCAAT binding factor* (αCBF) (11) and suggest that it may contribute to the placenta-specific property of the human α-subunit gene. The other element is referred to as the *junctional regulatory element* (JRE) because of its close juxtaposition between the CRE and CCAAT box elements (10) (Fig. 3.1). The JRE also binds a ubiquitous protein, but this protein augments activity of the α-subunit promoter only in choriocarcinoma cells, suggesting that it may act by forming a higher-order complex with a placenta-specific factor (10). It remains to be determined whether either of these elements may interact synergistically with the CRE to provide maximal transcriptional activity.

Lack of a Functional CRE in the Promoter Proximal Region of the α-Subunit Gene from Nonprimates

We have used PCR (29, 30) to survey the promoter proximal-regulatory regions from several different mammals to determine if there is a correlation between the presence of functional CRE and placental expression of the α-subunit gene. Sequence analysis of primate genes revealed at least one copy of a perfectly conserved 18-bp CRE (Fig. 3.2). The

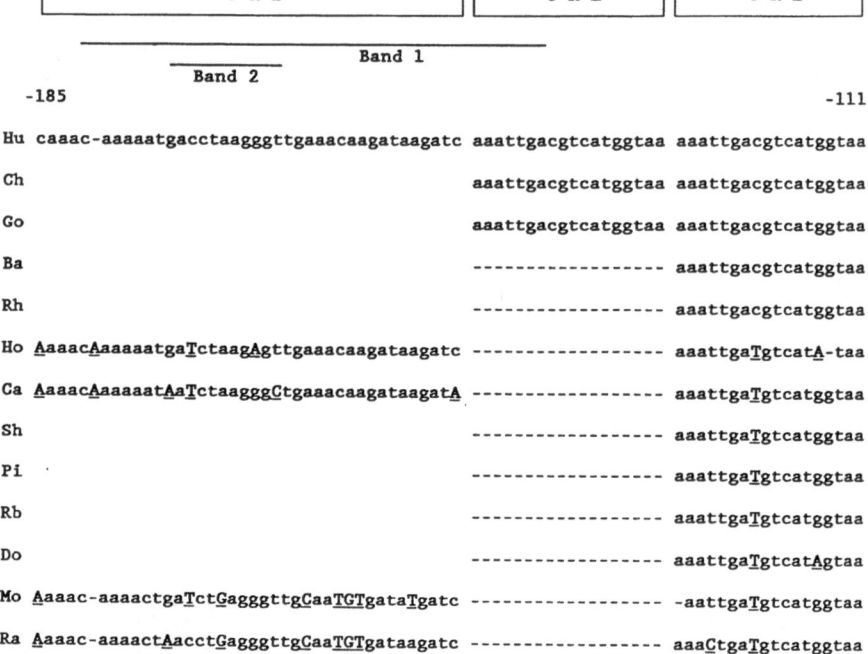

FIGURE 3.2. Comparative analysis of the proximal regions of mammalian α-subunit promoters. PCR was used to amplify a discrete segment from the 5′-flanking region of the α-subunit gene from a number of mammals. This amplified region is comparable to the region extending between −185 and −111 of the human α-subunit gene. PCR and sequence analysis were performed as described in reference 9. Positions of the URE and tandem CRE in the human α-subunit flanking region are shown by the boxed diagram at the top of the figure. Dotted lines represent 5′-flanking regions that contain only a single CRE. URE sequence determination was confined to a limited number of mammals. Nucleotides different from those of the human URE and CRE are capitalized and underlined. Band 1 and Band 2 refer to the binding sites of two URE-specific *trans*-acting factors as described in reference 8. (Hu = human; Ch = chimpanzee; Go = gorilla; Ba = baboon; Rh = rhesus monkey; Ho = horse; Ca = cattle; Sh = sheep; Pi = pig; Rb = rabbit; Do = donkey; Mo = mouse; Ra = rat.)

α-subunit genes of humans and higher primates (gorilla and pygmy chimpanzee) contain tandem CREs. In contrast, α-subunit genes of Old World monkeys (baboon and rhesus monkey) contain a single CRE. Since bona fide CG is synthesized in placentas of baboons (31, 32) and rhesus monkeys (33), these data suggest that a single CRE, acting in conjunction with a URE, is sufficient to direct placenta-specific expression of the α-subunit gene. Furthermore, the presence of a single CRE in the α-subunit gene of lower primates suggests that acquisition of tandem CREs was a recent evolutionary event.

All other nonprimates, including horses, contain a CRE-like sequence with a conserved C-to-T transition at the fourth position of the human α-TGA$\underline{C}$GTCA sequence (9, 12, 34, 35) (Fig. 3.2). This transition disrupts the perfect palindrome and renders the bovine and, presumably, the other nonprimate homologues incapable of binding CREB (12). Thus, there is a correlation between the presence of a functional CRE and placenta-specific expression of the α-subunit gene. Furthermore, the bovine α-subunit promoter is inactive after transfection into human choriocarcinoma cells. Activity, however, can be rescued through a single nucleotide change that converts the bovine CRE-like sequence to an authentic CRE. This underscores the critical role that CREB plays in conferring placenta-specific expression to the α-subunit promoter. Moreover, lack of a CREB binding site provides at least one explanation for inactivity of the α-subunit promoter in placenta of most nonprimates. Because the CRE-like sequences of nonprimate α-subunit genes cannot bind CREB and thus confer responsiveness to cAMP in placental transfection systems (12), we refer to the sequence element as TGAT (or T in Fig. 3.3) rather than CRE. This abbreviation underscores the C/T transition that differentiates the CRE-like element of nonprimates (**TGA$\underline{T}$GTCA**) from the CRE core primates (**TGA$\underline{C}$GTCA**).

Conservation of URE in Mammalian α-Subunit Genes

Earlier studies indicated that the promoter proximal region of the bovine α-subunit gene contained a functional homologue to the human URE (9, 12) (Fig. 3.2). Furthermore, URE-binding activity appears to be conserved in placentas from mammals that fail to express their α-subunit gene. For example, we have shown recently that a chimeric transgene containing the human α-subunit promoter-regulatory region expresses in mouse pituitary as well as in mouse placenta (12). Promoter regions of the α-subunit genes from horse, rat, and mouse also contain sequence homologues to the human URE at the appropriate location (immediately adjacent to the TGAT element; Fig. 3.2). Although there are modest nucleotide changes, there is no single conserved change as there is with the C/T transition that distinguishes the primate CRE (**TGA$\underline{C}$GTCA**) from the nonprimate TGAT element (**TGA$\underline{T}$GTCA**). This suggests the possibility that URE homologues in nonprimate α-subunit genes may be functionally significant (see below).

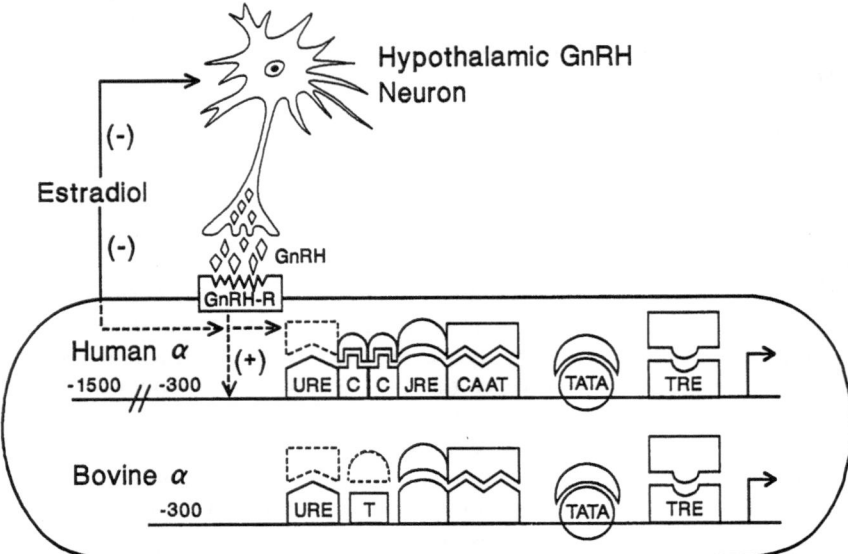

FIGURE 3.3. A model of regulatory elements involved in hormonal responsiveness and pituitary-specific expression of the human and bovine α-subunit genes. The diagrammed human or bovine α-subunit promoters direct expression of the CAT transgene specifically to mouse pituitary gonadotropes, depicted by the ellipse. Chronic treatment with estradiol suppresses transcription of both transgenes, as indicated by the (−) symbol. The solid arrows indicate that this effect could occur by suppressing secretion of gonadotropin releasing hormone (GnRH) from hypothalamic neurons or at the level of the pituitary. Exogenous administration of GnRH reverses the suppressive effect of estradiol, as indicated by the (+) symbol, suggesting that the two hormonal pathways converge at a common site. The dotted arrows indicate one of several possible sites of convergence—an estrogen receptor complex may interfere with transduction of the GnRH signal through the GnRH receptor (GnRH-R) by affecting the binding of a yet-to-be-described *trans*-acting factor or a known factor, such as the URE binding protein. As in Figure 3.1, geometric shapes represent *cis*-acting elements and their cognate *trans*-acting factors. The *trans*-acting factors with dotted lines represent pituitary proteins that bind either the URE or T; these proteins appear different from the placental proteins that bind to these same elements. The presence of a JRE and a CAAT box have not been confirmed for the bovine promoter; thus, these elements are depicted without an abbreviation to underscore this fact. (See Fig. 3.1 key; TRE = thyroid hormone response element.)

Gonadotrope-Specific Elements Within the Promoter Proximal-Regulatory Region of Both the Human and Bovine α-Subunit Gene

We have established lines of transgenic mice containing chimeric constructs composed of the *chloramphenicol acetyltransferase* (CAT) gene linked to the promoter-regulatory region of either the human (-1500 to $+45$) or bovine (-313 to $+48$) α-subunit gene (12). Both transgenes are expressed in pituitary but not in heart, lung, liver, kidney, pancreas, and spleen. The human α-CAT transgene also expresses to a much lower extent in brain for reasons that remain unexplained. As indicated in Figure 3.3, the primary difference between these two constructs, besides the length of their 5'-flanking region, is that the bovine flanking region lacks a CREB-binding CRE. Thus, the binding of CREB cannot be essential for pituitary-specific expression of the bovine α-subunit gene; presumably, this is true for the human α-subunit transgene as well. Instead, there must be other regulatory elements located within these promoter regions that confer this property. For the bovine α-subunit promoter, this element(s) must reside within the proximal 300 bp of 5'-flanking sequence since this is the extent of 5'-flanking sequence present in the transgene. While the human α-subunit chimeric transgene contains a longer 5'-flanking region (1500 bp), we suggest on the basis of nucleotide homology that a pituitary-specific element(s) also resides within the proximal 300 bp of 5'-flanking sequence.

Additional studies with transgenic mice (13) indicated that activity of both the human and bovine α-CAT transgenes were suppressed when castrated female mice were treated chronically with estradiol. As indicated in Figure 3.3, estrogen could be exerting a direct effect by promoting the binding of estrogen receptor to a high-affinity binding site located within the promoter proximal regions of the human and bovine α-subunit genes. Subsequent analysis revealed that the promoter-regulatory regions contained in both CAT transgenes lacked a high-affinity binding site for estrogen receptor (13). Thus, if estrogen is acting directly at the level of the pituitary to suppress transcription of the α-subunit, an indirect mechanism must be involved. Alternatively, estrogen could act by inhibiting the secretion of GnRH from the hypothalamus, which in turn would lead to reduction in transgene activity. This assumes that the α-CAT transgenes are expressed in gonadotropes and that their promoter-regulatory regions contain a response element that mediates the effect of GnRH.

Although the steroid replacement paradigm in transgenic mice cannot be used to distinguish between direct versus indirect actions of estrogen, we felt it could be used to address whether flanking regions from either promoter contained a regulatory element that mediates responsiveness to

GnRH. Thus, we used the castration-estrogen replacement approach to suppress endogenous secretion of GnRH and then superimposed pulses of exogenous GnRH to determine whether the negative effect of estradiol could be reversed. These studies have recently been completed (manuscript submitted) and indicate that pulsatile administration of GnRH (every other hour for 7 continuous days) to ovariectomized mice chronically treated with estradiol reverses the negative effect of the steroid on either the human or bovine α-CAT transgene. This response to GnRH provides clear evidence that the regulatory regions of both transgenes contain a GnRH response element. Moreover, expression of both transgenes must be directed to gonadotropes, as these are the only cell types in pituitary with receptors for GnRH (1, 2).

Importance of Conservation of CRE/TGAT and URE Elements as Possible Clues for Understanding Mechanisms Underlying Pituitary- and Gonadotrope-Specific Expression of the α-Subunit Gene

While the GnRH experiments with transgenic mice provide evidence that the promoter proximal-regulatory regions of the human and bovine α-subunit genes contain regulatory elements that confer hormonal regulation and gonadotrope-specific expression, the location of this element(s) and the identity of its cognate DNA-binding protein(s) remain unknown. There are, however, important clues that emerge through comparative analysis of the 5'-flanking regions of the α-subunit gene from a number of diverse mammals. For example, there is a striking conservation of the URE (Fig. 3.2), even in nonprimates that fail to express their α-subunit gene in placenta. In addition, the CRE has been strongly conserved in α-subunit promoter of primates, whereas the TGAT sequence (CRE-like) has been conserved similarly in nonprimates. The conservation of these elements suggests that they may be functionally important and participate in pituitary-specific expression. Participation of the URE in pituitary-specific expression may seem paradoxical, as this element has been reported to bind a protein unique to placenta (4). However, preliminary data (not shown) indicate that there is URE-binding activity in whole-cell extracts prepared from pituitary and that this activity is different than that detected in placenta. Similarly, although there are no proteins in human choriocarcinoma cells that bind to the nonprimate TGAT sequence, there is a strong binding activity present in pituitary. Interestingly, the protein(s) in pituitary that binds to the nonprimate TGAT sequence can also bind the primate CRE. Thus, in primates, the URE and CRE may confer both pituitary- and placenta-specific patterns of expression to the α-subunit promoter, but do so by binding different sets of proteins unique to each tissue. In nonprimates, we envision that the URE and TGAT elements collaborate to confer pituitary-specific expression to the α-subunit promoter. Clearly, additional experiments that incorporate both the transgenic mouse model and in vitro DNA-protein binding assays will be required to test this hypothesis.

Acknowledgments. I gratefully acknowledge the following colleagues who are responsible for designing, performing, and interpreting the experiments cited herein: Bogi Anderson, Joseph A. Bokar, Colin M. Clay, Joan N. Clay, Todd A. Farmerie, Robert A. Fenstermaker, Debora L. Hamernik, Giulia Catignani Kennedy, Ruth A. Keri, Robin H. Pittman, and Michael W. Wolfe. This work was supported by NIH Grants DK-28559, Cancer Research Grant P-30 CA-43730, and the Ohio Edison Biotechnology Center.

References

1. Gharib SD, Wierman ME, Shupnik MA, Chin WW. Molecular biology of the pituitary gonadotropins. Endocr Rev 1990;11:177–99.
2. Chin WW. Organization and expression of the glycoprotein hormone genes. In: Imura H, ed. The pituitary gland. New York: Raven Press, 1984:103–25.
3. Silver BJ, Bokar J, Virgin JB, Vallen E, Milsted A, Nilson JH. Cyclic AMP regulation of the human glycoprotein hormone α-gene is mediated by an 18 base pair element. Proc Natl Acad Sci USA 1987;84:2198–202.
4. Delegeane AM, Ferland LH, Mellon PL. Tissue-specific enhancer of the human glycoprotein hormone α-subunit gene: dependence on cyclic AMP-inducible elements. Mol Cell Biol 1987;7:3994–4002.
5. Jameson JL, Jaffe RC, Deutsch PJ, Albanese C, Habener JF. The gonadotropin α gene contains multiple protein binding domains that interact to modulate basal and cAMP-responsive transcription. J Biol Chem 1988;263:9879–86.
6. Deutsch PJ, Jameson JL, Habener JF. Cyclic AMP responsiveness of human gonadotropin α gene transcription is directed by a repeated 18-bp enhancer. J Biol Chem 1987;262:12169–74.
7. Deutsch PJ, Hoeffler JP, Jameson JL, Habener JF. Cyclic AMP and phorbol ester-stimulated transcription is directed by a repeated 18-bp enhancer. Proc Natl Acad Sci USA 1988;85:7922–6.
8. Jameson JL, Powers AC, Gallagher GD, Habener JF. Enhancer and promoter element interactions dictate cyclic adenosine monophosphate mediated and cell-specific expression of the glycoprotein hormone α-gene. Mol Endocrinol 1989;3:763–72.
9. Fenstermaker RA, Farmerie TA, Clay CM, Hamernik DL, Nilson JH. Different combinations of regulatory elements may account for expression of the glycoprotein hormone α-subunit gene in primate and horse placenta. Mol Endocrinol 1990;4:1480–7.

10. Andersen B, Kennedy GC, Nilson JH. A *cis*-acting element located between the CREs and CCAAT box augments cell-specific expression of the glycoprotein hormone α subunit gene. J Biol Chem 1990;265:21874–80.
11. Kennedy GC, Andersen B, Hamernik DL, Nilson JH. The human α subunit glycoprotein hormone gene utilizes a unique CCAAT binding factor. J Biol Chem 1990;265:6279–85.
12. Bokar JA, Keri RA, Farmerie TA, et al. Expression of the glycoprotein hormone α-subunit gene in the placenta requires a functional cyclic AMP response element, whereas a different *cis*-acting element mediates pituitary-specific expression. Mol Cell Biol 1989;9:5113–22.
13. Keri RA, Andersen B, Kennedy GC, et al. Estradiol inhibits transcription of the human glycoprotein hormone α-subunit gene despite the absence of a high affinity binding site for estrogen receptor. Mol Endocrinol 1991;5:725–33.
14. Bokar JA, Roesler WJ, Vandenbark GR, Kaetzel DM, Hanson RW, Nilson JH. Characterization of the cAMP responsive elements from the genes for the α subunit of chorionic gonadotropin and P-enolpyruvate carboxykinase: conserved features of nuclear protein binding between tissues and species. J Biol Chem 1988;263:19740–7.
15. Andersen B, Kennedy GC, Hamernik DL, Bokar JA, Bohinski R, Nilson JH. Amplification of the transcriptional signal mediated by the tandem cAMP response elements of the glycoprotein hormone alpha-subunit gene occurs through several distinct mechanisms. Mol Endocrinol 1990;4:573–82.
16. Steger DJ, Altschmied J, Buscher M, Mellon, PL. Evolution of placenta-specific gene expression: comparison of the equine and human gonadotropin α-subunit genes. Mol Endocrinol 1991;5:243–55.
17. Wurzel JM, Curabla LM, Gurr JA, Goldschmide AM, Kourides IA. The luteotropic activity of rat placenta is not due to a chorionic gonadotropin. Endocrinology 1983;113:1854–7.
18. Hoeffler JP, Meyer TE, Yun Y, Jameson JL, Habener JF. Cyclic AMP-responsive DNA-binding protein: structure based on a cloned placental cDNA. Science 1988;242:1430–3.
19. Yamamoto KK, Gonzalez GA, Biggs WH, Montminy MR. Phosphorylation-induced binding and transcriptional efficacy of nuclear CREB. Nature 1988;334:494–8.
20. Gonzalez GA, Yamamoto KK, Fischer WH, Karr D, Menzel P, Biggs II W, et al. A cluster of phosphorylation sites on the cyclic AMP-regulated nuclear factor CREB predicted by its sequence. Nature 1989;337:749–52.
21. Karlsson O, Edlund T, Moss JB, Rutter WJ, Walker MD. A mutational analysis of the insulin transcription control region: expression in beta cells is dependent on two related sequences within the enhancer. Proc Natl Acad Sci 1987;84:8819–23.
22. Fried M, Crothers DM. Equilibria and kinetics of lac repressor-operator interactions by polyacrylamide gel electrophoresis. Nucleic Acids Res 1981; 9:6505–25.
23. Baldwin AS. Methylation interference assay for analysis of DNA-protein interactions. In: Ausubel FM, Brent R, Kingston RE, et al., eds. Current protocols in molecular biology; vol. 2. New York: John Wiley and Sons, 1987: unit 12.3.

24. Chodosh LA. UV crosslinking of proteins to nucleic acids. In: Ausubel FM, Brent R, Kingston RE, et al., eds. Current protocols in molecular biology; vol. 2. New York: John Wiley and Sons, 1987: unit 12.5.
25. Santoro C, Mermod N, Andrews PC, Tjian R. A family of human CCAAT-box-binding proteins active in transcription and DNA replication: cloning and expression of multiple DNAs. Nature 1988;334:218–24.
26. Johnson PF, Landschulz WH, Graves BJ, McKnight SL. Identification of a rat liver nuclear protein that binds to the enhancer core element of three animal viruses. Genes Dev 1987;1:133–46.
27. Chodosh LA, Baldwin AS, Carthew RW, Sharp PA. Human CCAAT-binding proteins have heterologous subunits that are functionally interchangeable. Cell 1988;53:11–24.
28. Koch W, Benoist C, Mathis D. Anatomy of a new B-cell-specific enhancer. Mol Cell Biol 1989;9:303–11.
29. Higuchi R. Simple and rapid preparation of samples for PCR. In: Erlich HA, ed. PCR technology: principles and applications for DNA amplification. New York: Stockton Press, 1989:31–8.
30. Saiki RK, Gelfand DH, Stoffel S, et al. Primer-directed enzymatic amplification of DNA with a thermostable DNA polymerase. Science 1988; 239:487–91.
31. Crawford RJ, Tregear GW, Niall HD. The nucleotide sequences of baboon chorionic gonadotropin β-subunit genes have diverged from the human. Gene 1986;46:161–9.
32. Tullner WW, Gray CW. Chorionic gonadotropin excretion during pregnancy in a gorilla. Proc Soc Exp Biol Med 1968;128:954–6.
33. Neill JD, Knobil E. On the nature of the initial luteotropic stimulus of pregnancy in the rhesus monkey. Endocrinology 1972;90:34–8.
34. Nilson JH, Bokar JA, Clay CM, et al. Different combinations of regulatory elements may explain why placenta-specific expression of the glycoprotein hormone α-subunit gene occurs only in primates and horses. Biol Reprod 1990;44:231–7.
35. Nilson JH, Bokar JA, Keri RA, et al. CRE-binding proteins interact cooperatively to enhance placental-specific expression of the glycoprotein hormone alpha-subunit gene. NY Acad Sci 1989;564:77–85.

4

Role of Inhibins and Activins in Reproduction

MASAO IGARASHI, KAORU MIYAMOTO, YOSHIHISA HASEGAWA,
MASAKI FUKUDA, TSUTOMU SUZUKI, YUMIKO ABE,
MASAAKI YAMAGUCHI, MASAKI DOI, SHIN-ICHI ROKUKAWA,
AND KIYOSHI TSUCHIYA

It is well established that mammalian reproduction is controlled by hypothalamic, pituitary, and gonadal hormones. Until recent years, steroid hormones had been considered the only known gonadal hormones. However, recent progress in endocrinology has begun to clarify the fact that nonsteroidal gonadal factors play an important role in reproduction. Among these nonsteroidal gonadal factors, inhibin and activin are the most important.

The existence of inhibin, a hormone of the testis, was suggested in 1923, 59 years ago. Mottram and Cramer (1) demonstrated that the injection of a water-soluble extract of the testis inhibited the appearance of castration cells in the pituitary. In 1932, McCullagh (2) named this testicular hormone *inhibin*. Since then, many endocrinologists and chemists have attempted to purify and characterize this new testicular hormone. However, the results have been controversial. For this reason, some referred to inhibin as a ghost hormone and believed it might be an artifact. Why has purification of inhibin been so difficult for over 60 years? There might be several reasons, but we think the main reasons are as follows. Assaying low concentrations of inhibin specifically and accurately has proven difficult. The second reason relates to the specific chemical character of inhibin, which exhibits strong interactions with other proteins and gel supports. After several trials and errors, we resolved these difficulties and succeeded in purifying porcine ovarian follicular inhibin in 1984. We presented our data at the Annual Meeting of The Endocrine Society in Baltimore, Maryland, in 1985 and published the data in the same year (3).

A few months before our presentation, Robertson et al. (4) published the results of purification of bovine ovarian inhibin. According to their results, the molecular weight was 58,000 and the N-terminal amino acid sequences were quite different from ours. A few months after our presentation, Ling and Guillemin of the Salk Institute (5) published their results in which the molecular weight and N-terminal amino acid sequences coincided perfectly with our results.

Purification of Inhibin from Porcine and Bovine Follicular Fluid

As mentioned above, inhibin was first detected in testicular extracts in 1923. But, in 1983, de Jong (6) demonstrated the existence of inhibin in ovarian follicular fluid. Ovarian follicular fluid is much more suitable for purification of inhibin than testicular extract because of the higher content of inhibin and the ease of both collection and purification. Thus, we began purification of inhibin from ≈1 L of porcine ovarian follicular fluid aspirated from over 2000 porcine ovaries.

In advance of purification, one of our coworkers, Miyamoto, found that incubation of follicular fluid in 8 M urea solution induced degeneration of most of the contaminating proteins, but induced no change of inhibin. Therefore, most chromatographies in the present study have been carried out in the presence of 8 M urea solutions.

Inhibin activity was assayed by suppression of spontaneous FSH release from the cultured cells of rat anterior pituitary for 3 days. Initially, batchwise chromatographies on a matrix gel Red A and a phenyl-Sepharose column have been carried out for the purification of porcine follicular fluid. On a gel filtration on Sephacryl S 200 in the presence of 8 M urea, after phenyl-Sepharose chromatography, two inhibin-containing fractions were observed. The smaller inhibin-containing fractions from 38 to 43 were desalted and further separated on a DEAE-Sepharose CL6B column in the presence of 8 M urea. Three distinct peaks (1, 2, and 3) with inhibin activity were observed. Peak 2, eliciting the highest bioactivity, was further purified by reverse phase *high-pressure liquid chromatography* (HPLC). Inhibin activity was found in a single and well-separated peak. Analytical SDS-PAGE of the final product from peak 2 under nonreducing conditions gave a single band corresponding to molecular weight 32,000. Under reducing conditions, when inhibin was irreversibly inactivated, 2 polypeptide bands of molecular weight 20,000 and 13,000 were observed, indicating that 32,000 inhibin consists of 2 polypeptide chains linked by disulfide bridges.

The purified 32,000 inhibin preparations inhibited spontaneous FSH release from rat anterior pituitary cells in a dose-dependent manner, but LH release was not affected (3). After completing purification of porcine follicular inhibin, we purified bovine follicular inhibin by using procedures exactly the same as those employed in isolation of porcine inhibin (7). A predominant form of inhibin was 32-kd protein in porcine follicular fluid, but major components of inhibin in bovine follicular fluid were 32-kd and 55-kd proteins.

Control Mechanism of Inhibin Secretion from Granulosa Cells

In order to clarify the control mechanism of inhibin secretion from ovarian granulosa cells, *granulosa cells* (GC) were obtained and cultured from the ovaries of immature female estrogen (*diethylstilbestrol* [DES])-primed rats. Inhibin secretion was stimulated in vitro by addition of FSH (8). The addition of testosterone to the culture system augmented the stimulatory action of FSH upon inhibin secretion. The addition of LH was first ineffective, but after a 2-day addition of FSH into cultured GC, LH became more effective than FSH in stimulation of inhibin secretion. Besides FSH, insulin and platelet extract are also effective alone and synergistic with FSH in secretion of inhibin in vitro. The addition of testosterone also augmented the stimulatory action of insulin and platelet extract upon inhibin secretion in vitro.

In our GC culture system, hydrocortisone alone had no effect on inhibin, estradiol, or progesterone secretion. However, the addition of hydrocortisone made the cells less responsive to FSH stimulation in regard to secretion of both inhibin and estradiol, but more sensitive with respect to progesterone production.

Among 3 inhibitors of cytodifferentiation of GC, a potent protein kinase C activator, TPA, showed the strongest inhibitory effect on inhibin secretion, both alone and in combination with FSH. EGF and GnRH agonist suppressed the FSH action on inhibin production, although administration of EGF or GnRH agonist alone did not induce any change in inhibin secretion (8).

Mode of Action of Inhibin at the Pituitary Level

In order to clarify the mode of action of inhibin at the pituitary level, dispersed pituitary cells were prepared by the method described previously (3). Purified porcine follicular fluid 32-kd inhibin or cycloheximide, an inhibitor of protein synthesis, was added to each well.

During the first 3 days of incubation, purified inhibin suppressed basal FSH secretion but did not suppress basal LH secretion (9). At the same time, the amount of FSH in the cells was also reduced by the inhibin treatment, whereas the amount of LH in the cells was not affected. It is evident, therefore, that the total amount of FSH was severely suppressed by the inhibin treatment. These results indicate that inhibin acts to suppress biosynthesis of FSH. On the other hand, biosynthesis of LH does not seem to be affected by inhibin. In the next experiment, the effects of a protein biosynthesis inhibitor, cycloheximide, on the basal gonadotropin secretion were tested. Surprisingly, cycloheximide, like inhibin, suppressed only FSH but not LH regarding the basal secretion and cell contents. These results indicate that suppression of FSH secretion is a consequence of reduction of FSH biosynthesis. On the other hand, neither inhibin nor cycloheximide could affect biosynthesis of LH since the total amount of LH was not reduced. It is noteworthy that a general inhibitor of protein biosynthesis selectively suppressed FSH biosynthesis and secretion.

On the other hand, the LH-RH-stimulated release of both FSH and LH was suppressed by the treatment with inhibin. This observation clearly indicates that the suppression of LH-RH-stimulated release of LH is one of the intrinsic actions of inhibin on the pituitary cells. It should be emphasized again that cycloheximide also mimicked the inhibin action on both FSH and LH under LH-RH-stimulated conditions.

TPA could stimulate LH or FSH secretion to the same extent as LH-RH in our culture system during the 6-h incubation. The addition of inhibin suppressed the TPA-stimulated release of FSH and LH in a dose-dependent manner. These results indicate that inhibin exerts the action, at least in part, by acting on the C-kinase system in the pituitary gonadotrope (9).

Changes in Serum Inhibin Levels in Mammals

It is well known that LH-RH stimulates both LH and FSH secretion. However, serum FSH levels are not always parallel to serum LH levels. Since inhibin suppresses only FSH secretion but not LH secretion, it is very important and interesting to clarify how inhibin is secreted simultaneously with FSH, LH, and estradiol. A coworker (10–13) generated various polyclonal and monoclonal antibodies against porcine and bovine follicular 32-kd inhibin. Some antibodies among them showed crossreactivity with human follicular fluid inhibin and goat follicular fluid inhibin. Thus, radioimmunoassay of inhibin using these polyclonal antibodies to inhibin became available.

In the rat estrous cycle (11, 12, 15), there are 2 peaks of serum FSH. The first FSH peak occurs in the afternoon of the proestrous day, and the second FSH peak occurs in the morning of the estrous day. We previously demonstrated that the first FSH peak is induced by a preovulatory LH/FSH surge that is controlled by hypothalamic LH-RH, but the second FSH surge, lacking an LH peak, is not dependent on hypothalamic LH-RH. It was demonstrated that the first LH/FSH surge is accompanied by a serum inhibin peak, but the second FSH peak occurred at the lowest value of serum inhibin. During the estrous cycle of pigs (11, 12, 14), the first and second FSH peaks were accompanied by low inhibin levels. At the time of the LH surge, serum inhibin levels were not low. In the follicular phase during the estrous cycle of the cow (11, 12), FSH levels increased, and at that time, inhibin levels remained low. At the time of the LH/FSH surge, inhibin levels were relatively high. In the subsequent luteal phase, serum inhibin levels were high and serum FSH was low.

During the estrous cycle of the goat (11, 12), serum inhibin and estradiol levels increased in the follicular phase in accordance with ovarian follicle maturation. At the time of the LH/FSH surge, serum inhibin levels showed maximum values. In the luteal phase, serum inhibin levels were not so high.

During the normal menstrual cycle in women (12, 16), there was an increase of serum FSH for about 11 days in the follicular phase and an induced increase of estradiol and inhibin, which inhibited FSH secretion. Five days later, the estradiol peak induced an LH/FSH surge. On the same day as the LH/FSH surge, inhibin peaked, then remained at low levels for 4 days. Thereafter, inhibin showed higher levels during the luteal phase. We confirmed a higher concentration of inhibin in the extract of human corpus luteum biopsied during laparotomy. A textbook of reproductive endocrinology reasons that serum FSH remains low in the luteal phase due to the combined feedback action of progesterone and estradiol. However, the true reason is elucidated by these results that demonstrate secretion of inhibin from the corpus luteum.

In summary, in the follicular phase, FSH stimulates the secretion of inhibin and estradiol in accordance with follicle development and maturation. The increase in inhibin blocks the secretion of FSH and the increase in estradiol induces the LH/FSH surge. At that time, serum inhibin levels are high. In the luteal phase, serum inhibin, secreted from the corpus luteum, shows high levels that inhibit secretion of FSH. Higher levels of inhibin secretion in the luteal phase are observed in humans and cattle. In human in vitro fertilization (17), serum inhibin levels seem to be a better monitor for follicle maturation than estradiol. Moreover, Tsuchiya (17) demonstrated in our laboratory that the serum inhibin level can be predictive of prognosis for in vitro fertilization, but the serum estradiol level cannot. The serum inhibin levels in amenorrheic women were significantly lower in hypoestrogenic hypothalamic or pituitary amenorrhea but higher in PCO, compared to the levels in the early follicular phase of normal menstruating women (unpublished observation). In adult men, serum inhibin showed distinct circadian rhythm, with higher levels in daytime and lower levels at night. The highest peak was demonstrated at 9 A.M. (18). During pregnancy (19), 2 peaks of serum inhibin were observed. The first peak of inhibin in the first trimester was secreted from the corpus luteum, and the second peak occurred at the end of the pregnancy, secreted from the placenta (19). At the time of delivery of the newborn, maternal serum inhibin levels were about twice as high as the newborn's. Serum inhibin levels of the male newborn were significantly higher than those of the female newborn on day 1, day 6, and day 30 following birth (unpublished observation). In male puberty, the increase of serum inhibin was earlier than the increase of testosterone. The highest peak of serum inhibin was observed in men from 17 to 29 years old, while the peaks of serum testosterone were observed from 30 to 39 years old (unpublished observation). In the luteal phase of women, injection of hCG induced an increase in inhibin and progesterone, but no change in estradiol, and a decrease in serum FSH and LH (unpublished observation). In men, serum inhibin levels began to decrease from 30 years and decreased gradually until 70 years, while serum FSH showed reverse changes against inhibin. On the other hand, the decrease of serum testosterone levels began from 40 years (18).

History of Research on Activin

In 1986, a dimer of the inhibin β-subunit was discovered simultaneously in porcine ovarian follicular fluid by two separate groups. Vale et al. (20) discovered a homodimer βAβA and named it *FSH releasing protein* (FRP), and Ling et al. (21) discovered a heterodimer βAβB and named it *activin*. At the Serono Symposium in Tokyo (22), a new nomenclature was proposed: activin A, activin AB, and activin B. In 1987, independent of activin studies, Eto et al. (23) isolated a new polypeptide, called *erythroid differentiation factor* (EDF), from conditioned medium of phorbol ester-treated human monocytic leukemia cells. This EDF has the ability to cause differentiation of mouse Friend erythroleukemia cells into normal blood cells. Surprisingly, the amino acid sequences of EDF perfectly coincided with activin A. The University of Tokyo group (24) and our group (25) confirmed that EDF stimulates FSH secretion from the cultured rat pituitary cells in vitro.

Physiologic Action of Activin in Reproduction

In our laboratory (25), the rat pituitary cell culture system demonstrated that activin increased FSH not only in culture medium but also in cell content, while it induced no change in LH release and content. These results clearly indicate that activin stimulates both biosynthesis and release of FSH from the pituitary, but induces no change of LH. When activin and inhibin were added into the same culture medium, actions of activin and inhibin were competitive with each other, and the released dose of FSH depended on their reciprocal doses. If we compared the activity of activin and inhibin at the same molal equivalent, the inhibitory action of inhibin was more dominant than the stimulatory action of activin (26).

In our laboratory, Sugino et al. (27) demonstrated the existence of activin receptor in ovarian GC. TGFβ has a similar chemical structure and biological action to that of activin, but did not bind to this activin receptor. Inhibin also did not bind to activin receptor. The addition of FSH in the culture medium induced an increase in the number of activin receptors. The existence of activin receptors in ovarian GC strongly suggests an autocrine or paracrine action of activin in these cells. Actually, we could demonstrate a series of autocrine and paracrine actions of activin.

First, activin could induce an increase of FSH receptor in the GC (28). Such action of the activin receptor is very specific because the other proteins, such as inhibin, TGFβ, insulin, and EGF, did not increase FSH receptor content. Second, the addition of activin to the GC culture system for 72 h stimulated production of inhibin (29). Third, the addition of activin to the GC culture system stimulated production of estradiol in vitro (unpublished observation). Fourth, activin augmented FSH action to induce LH receptor formation (29). The addition of FSH to the GC culture system for 60 h induced LH receptors. If activin was simultaneously added with FSH, the time of induction of LH receptors was shortened and the number of LH receptors increased. Fifth, activin was demonstrated to augment maturation of the oocyte (30). It has been reported that such growth factors as EGF, TGFβ, IGF-I, and IGF-II stimulate *germinal vesicle breakdown* (GVBD). In 1989, Robertson et al. reported that inhibin inhibited GVBD, but activin neither inhibited nor stimulated GVBD. However, our coworker succeeded in demonstrating the stimulatory action of activin upon GVBD (30).

Research on the in vivo effect of exogenously administered activin has been poorly reported. Doi, in our laboratory, demonstrated that a sub-cutaneous 1–3 day injection of activin A (20 μg twice daily) in immature female rats induced a significant increase in serum FSH, inhibin, and estradiol levels and uterine weight. Besides these changes, FSH receptor content in the ovary was significantly increased by injection of activin A. Six daily injections of 20-μg activin A increased not only uterine weight but also ovarian weight. Histological findings of the ovaries revealed that injection of activin increased the number of developing follicles and the size of the follicles. In the hypophysectomized rat, injection of activin alone did not induce any significant change, but simultaneous injection of PMS and activin induced significant augmentative effects of activin upon the action of FSH. Significant increases in serum inhibin and estradiol levels and uterine and ovarian weight were observed in the activin-combined PMS group, compared to the PMS alone group. From these results, it is clear that activin induces not only an increase in FSH secretion from the pituitary, but also has autocrine or paracrine stimu-

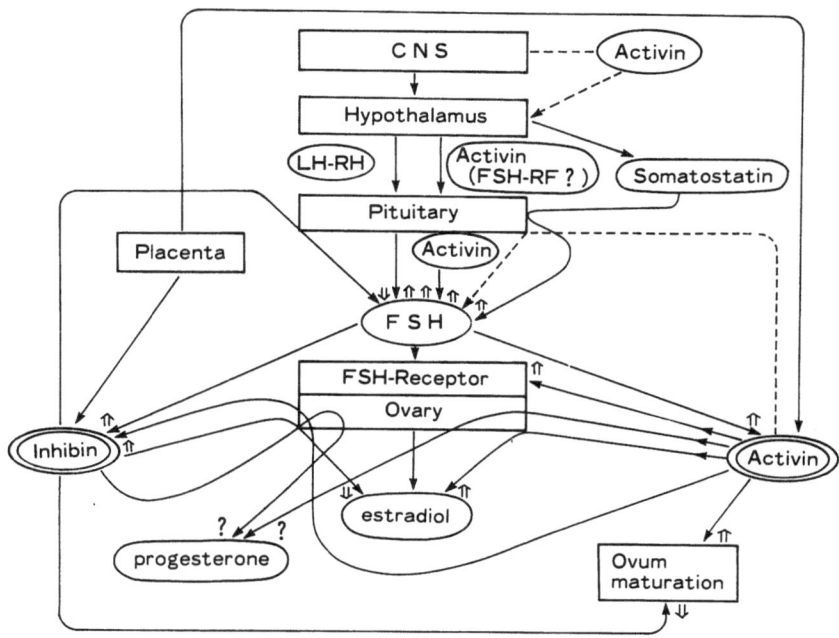

FIGURE 4.1. Mode and site of action of inhibin and activin in reproduction. Black arrows show sites of action of inhibin or activin. White upward arrows show stimulatory effects. White downward arrows show inhibitory effects.

latory action at the ovarian level. Recently, Woodruff et al. (33) reported that an intraovarian injection of 1-μg recombinant activin blocked histological follicular development. It is not clear so far why their results and ours on the paracrine action of activin are so contradictory. Further studies are to be required. The physiologic roles of inhibin and activin in reproduction are illustrated in Figure 4.1.

Site of Production of Inhibin and Activin

In 1986, immunohistochemical localization of inhibin in porcine and bovine ovaries was published from our laboratory (32). A consistent staining of inhibin was demonstrated in the GC and follicular fluid, but not in the theca of the ovaries, using a monoclonal antibody to bovine 32,000 inhibin and polyclonal antiserum to porcine 32,000 inhibin after acetone fixation followed by celloidin embedding. The mRNA studies from Vale et al. (33) revealed the α-subunit to be concentrated in the ovary, testis, pituitary, adrenal, brain, and spinal cord, while the βA subunit is concentrated in the placenta, ovary, bone marrow, brain, and spinal cord. The βB subunit is concentrated in the ovary, pituitary, adrenal, and brain.

Extragonadal Action of Activin

Many reports, including ours, confirmed that activin stimulates synthesis and release of FSH from the pituitary (25–28). This activin-mediated FSH secretion is enhanced by somatostatin (34). In addition to FSH secretion, it is noteworthy that activin reduces not only GH-RH-mediated GH release and TRH-mediated PRL release (35), but also release and biosynthesis of GH in the pituitary (36, 37).

As mentioned earlier, EDF, isolated from the culture medium of human monocyte leukemia cell line, has the same chemical structure as activin A. EDF, or activin A, was demonstrated to exist abundantly in the bone marrow (23). Subsequently, the important role of activin A in erythrodifferentiation, hemoglobin synthesis, potentiation of the effect of erythropoietin on stem cells, and so on, was reported. Thereafter, using EDF, or activin A, it was demonstrated that activin stimulates both glucose production from cultured rat hepatocytes (38) and insulin secretion from rat pancreatic islets (39). In in vivo experiments in the Genentech laboratory (40) and ours (unpublished observation), hypoglycemia was demonstrated in the activin-injected rat for 3 days.

In the central nervous system, activin seems to play important physiologic roles. Schubert reported activin to be a nerve cell survival

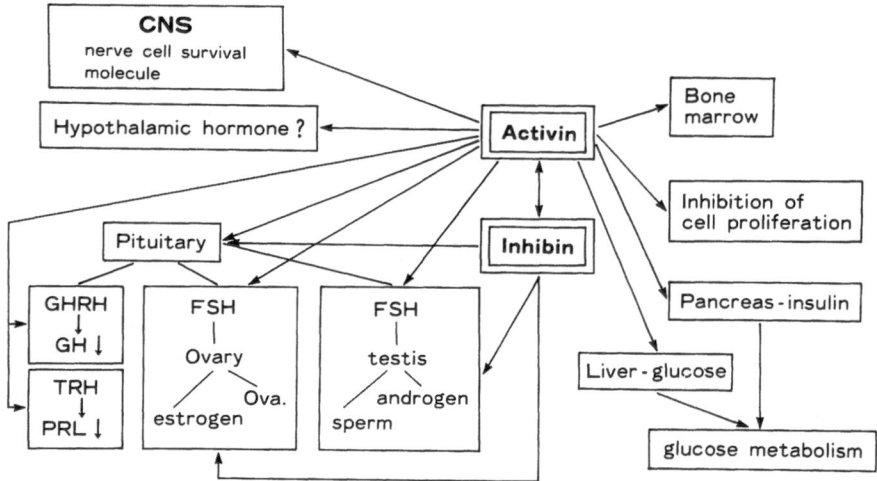

FIGURE 4.2. Site of action of inhibin and activin in the body. Inhibin acts mainly in the pituitary and the gonads, while activin acts in such organs as CNS, bone marrow, pancreas, and liver, as well as pituitary and gonads.

molecule (41). According to another report by Hashimoto et al. (42), activin acts as an inhibitor of neural differentiation of P19 cells.

Immunohistochemical studies from Vale's laboratory (33, 43) showed the inhibin A subunit was positively stained in the ventrolateral medullary reticular formation and in the caudal part of the nucleus of the solitary tract (NTS), of which the most prominent terminal fields are paraventricular (PVN), supraoptic nuclei of the hypothalamus. Vale et al. (43) already demonstrated that infusion of activin into the PVN of rats elevated plasma oxytocin levels. In the adrenal, it is reported that recombinant activin A inhibits proliferation of human fetal adrenal cells in vitro (44). As illustrated in Figure 4.2, activin seems to play a role in the physiologic actions of numerous organs and tissues, but the role of inhibin seems to be concentrated in the pituitary and the gonads.

Conclusions

Inhibins and activin, like growth factors, have various paracrine and autocrine local actions. Recent progress in their purification and determination of molecular structure elucidates important roles of inhibin and activin in reproduction. Still more aspects of their physiologic role might be clarified in the near future.

Acknowledgments. We thank Drs. H. Sugino and T. Nakamura, Frontier Research Program, RIKEN, for their collaboration on activin research, Drs. H. Matsuo and K. Kangawa, Miyazaki Medical College, for their collaboration on determination of amino acid sequences, Drs. Y. Eto and Shibai, the Ajinomoto Central Laboratory, Tokyo, for their supply of activin A, Drs. S. Sasamoto and K. Taya for their supply of inhibin polyclonal antibody, and NIDDK, Dr. A.F. Parlow, National Hormone and Pituitary Program, USA, for their supply of RIA kits of rat and human FSH, LH, and other hormones. This research was supported by grants from the Ministry of Education, government of Japan.

References

1. Mottram JC, Cramer W. On the general effects of exposure to radium on metabolism and tumour growth in the rat and the special effects on testis and pituitary. Q J Exp Physiol 1923;13:209–29.
2. McCullagh DR. Dual endocrine activity of the testis. Science 1932;76:19–20.
3. Miyamoto K, Fukuda M, Nomura M, et al. Isolation of porcine follicular inhibin of 32K daltons. Biochem Biophys Res Commun 1985;129:396–403.
4. Robertson DM, Foulds LM, Levershall L, et al. Isolation of inhibin from bovine follicular fluid. Biochem Biophys Res Commun 1985;126:220–6.

5. Ling N, Ying SY, Ueno N, Esch F, Denoroy L, Guillemin R. Isolation and partial characterization of a Mr 32,000 protein with inhibin activity from porcine follicular fluid. Proc Natl Acad Sci USA 1985;82:7217–21.
6. de Jong FH, Sharpe RM, Evidence for inhibin-like activity in bovine follicular fluid. Nature 1976;263:71–2.
7. Fukuda M, Miyamoto K, Hasegawa Y, et al. Isolation of bovine follicular fluid inhibin of about 32KDa. Mol Cell Endocrinol 1986;44:55–60.
8. Suzuki T, Miyamoto K, Hasegawa Y, et al. Regulation of inhibin production by rat granulosa cells. Mol Cell Endocrinol 1987;54:185–95.
9. Fukuda M, Miyamoto K, Hasegawa Y, Ibuki Y, Igarashi M. Action mechanism of inhibin in vitro-cycloheximide mimics inhibin actions on pituitary cells. Mol Cell Endocrinol 1987;51:41–50.
10. Hasegawa Y, Miyamoto K, Fukuda M, Takahashi Y, Igarashi M. Immunological studies of ovarian inhibin. Endocrinol Jpn 1986;33:645–54.
11. Hasegawa Y, Miyamoto K, Igarashi M, Yanaka T, Sasaki K, Iwamura S. Changes in serum concentrations of inhibin during the estrous cycle of the rat, pig and cow. In: Burger HG, deKretser DM, Findlay JK, Igarashi M, eds. Inhibin-non-steroidal regulation of follicle stimulating hormone secretion. Serono Symposia Publications. New York: Raven Press, 1988;42:119–33.
12. Hasegawa Y. Changes in the serum concentrations of inhibin in mammals. In: Hodgen GD, Rosenwaks Z, Spieler JM, eds. Nonsteroidal gonadal factors; physiological roles and possibilities in contraceptive development. Proc Conrad Int Workshop, Jan 6–8, 1988. Norfolk: Jones Institute Press, 1988: 91–109.
13. Hamada T, Watanabe A, Kokuho T, et al. Radioimmunoassay of inhibin in various mammals. J Endocrinol 1989;122:697–704.
14. Hasegawa Y, Miyamoto K, Iwamura S, Igarashi M. Changes in serum concentrations of inhibin in cycle pigs. J Endocrinol 1988;118:211–9.
15. Hasegawa Y, Miyamoto K, Igarashi M. Changes in serum concentrations of immunoreactive inhibin during the oestrous cycle of the rat. J Endocrinol 1989;121:91–100.
16. Hasegawa Y, Miyamoto K, Kudoh T, et al. Changes in serum concentrations of inhibin during the menstrual cycles of women [Abstract]. Proc 8th Int Cong Endocrinol 1988:138.
17. Tsuchiya K, Hasegawa Y, Seki M, Miyamoto K, Ito M, Igarashi M. Correlation of serum inhibin concentrations with results in an ovarian hyperstimulation program. Fertil Steril 1989;52:1–7.
18. Yamaguchi M, Mizunuma H, Miyamoto K, Hasegawa Y, Ibuki Y, Igarashi M. Immunoreactive inhibin concentration in adult men: presence of a circadian rhythm. J Clin Endocrinol Metab 1991;72:554–9.
19. Abe Y, Hasegawa Y, Miyamoto K, et al. High concentrations of plasma immunoreactive inhibin during normal pregnancy in women. J Clin Endocrinol Metab 1990;71:133–7.
20. Vale W, Rivier J, Vaughan J, et al. Purification and characterization of an FSH releasing protein from porcine ovarian follicular fluid. Nature 1986; 321:776–9.
21. Ling N, Ying SY, Ueno N, et al. Pituitary FSH is released by a heterodimer of the β subunits from the two forms of inhibin. Nature 1986;321:779–82.

22. Burger HG, deKretser DM, Findlay JK, Igarashi M. Inhibin-non-steroidal regulation of follicle stimulating hormone secretion. Serono Symposia Publications. New York: Raven Press, 1982:42.
23. Eto Y, Tsuji T, Takezawa M, Takano S, Yakagawa Y, Shibai H. Purification and characterization of erythroid differentiation factor (EDF) isolated from human leukemia cell THP-1. Biochem Biophys Res Commun 1987;142:1095–103.
24. Kitaoka M, Yamashita N, Eto Y, Shibai H, Ogata E. Stimulation of FSH release by erythroid differentiation factor (EDF). Biochem Biophys Res Commun 1987;146:1382–5.
25. Miyamoto K, Hasegawa Y, Fukuda M, Igarashi M. Structure and function of inhibin and its related peptides. In: Hodgen DG, Rosenwaks Z, Spieler JM, eds. Nonsteroidal gonadal factors; physiological roles and possibilities in contraceptive development. Proc Conrad Int Workshop, Jan 6–8, 1988. Norfolk: Jones Institute Press, 1988:56–73.
26. Hasegawa Y. Production and function of inhibin and its related peptides. Jpn J Anim Reprod 1988;34:1–7.
27. Sugino T, Nakamura T, Hasegawa Y, et al. Identification of a specific receptor for erythroid differentiation factor on follicular granulosa cell. J Biol Chem 1988;26330:15249–52.
28. Hasegawa Y, Miyamoto K, Abe Y, et al. Induction of follicle stimulating hormone receptor by erythroid differentiation factor on rat granulosa cell. Biochem Biophys Res Commun 1988;156:668–74.
29. Sugino H, Nakamura T, Hasegawa Y, et al. Erythroid differentiation factor can modulate follicular granulosa cell functions. Biochem Biophys Res Commun 1988;153:281–8.
30. Itoh M, Igarashi M, Yamada K, et al. Activin A stimulates meiotic maturation of rat oocyte in vitro. Biochem Biophys Res Commun 1990;166:1479–84.
31. Woodruff TK, Lyon RJ, Hasen SE, Rice GC, Mather JP. Inhibin and activin locally regulate rat ovarian folliculogenesis. Endocrinology 1990;127:3196–205.
32. Rokukawa S, Inoue K, Miyamoto K, Kurosumi K, Igarashi M. Immunohistochemical localization of inhibin in porcine and bovine ovaries. Arch Histol Jpn 1986;49:603–11.
33. Vale W, Rivier C, Hsueh A, et al. Chemical and biological characterization of the inhibin family of protein hormones. Recent Prog Horm Res 1988;44:1–34.
34. Kitaoka M, Kojima I, Ogata E. A stimulatory effect of somatostatin: enhancement of activin A-mediated FSH secretion in rat pituitary cell. Biochem Biophys Res Commun 1989;162:958–62.
35. Kitaoka M, Kojima I, Ogata E. Activin-A: a modulator of multiple types of anterior pituitary cells. Biochem Biophys Res Commun 1988;157:48–54.
36. Billestrup N, Gouzalez-Manchon C, Potter E, Vale W. Inhibition of somatotroph growth and growth hormone biosynthesis by activin in vitro. Mol Endocrinol 1990;4:356–62.
37. Bileszikjian LM, Corrigan AZ, Vale W. Activin A modulates growth hormone secretion from cultures of rat anterior pituitary cells. Endocrinology 1990;126:2369–76.

38. Mine T, Kojima I, Ogata E. Stimulation of glucose production by activin-A in isolated rat hepatocytes. Endocrinology 1989;125:586–91.
39. Totsuka Y, Tabuchi M, Kojima I, Shibai H, Ogata E. A novel action of activin A: stimulation of insulin secretion in rat pancreatic islets. Biochem Biophys Res Commun 1988;156:335–9.
40. Schwall R, Schmelzer CH, Matsuyama E, Mason AJ. Multiple actions of recombinant activin-A in vivo. Endocrinology 1989;125:1420–3.
41. Schubert D, Kimura H, La Corbiere M, Vaughan J, Karr D, Fisher WH. Activin is a nerve cell survival molecule. Nature 1990;344:868–70.
42. Hashimoto M, Kondo S, Sakurai T, Eto Y, Shibai H, Muramatsu M. Activin/EDF as an inhibitor of neural differentiation. Biochem Biophys Res Commun 1990;173:193–200.
43. Sawchenko PE, Plotsky PM, Pfeiffer SW, et al. Inhibin β in central neural pathways involved in the control of oxytocin secretion. Nature 1988;334:615–7.
44. Spencer SJ, Rabinovici J, Jaffe RB. Human recombinant activin A inhibits proliferation of human fetal adrenal cells in vitro. J Clin Endocrinol Metab 1990;71:1678–80.

Part II
Ovary and Testis

5

Hormonal Regulation and Tissue-Specific Expression of Steroidogenic Enzymes

A.H. Payne, P.A. Bain, T. Clarke, L. Sha, G.L. Youngblood, S.H. Hammond, M. Yoo, and O.O. Anakwe

The biosynthesis of gonadal and adrenal steroid hormones requires the activities of several cytochrome P450 enzymes (Fig. 5.1). These enzymes, members of a cytochrome P450 multigene family, are heme-containing proteins that function as terminal oxidases in an electron transport chain. The initial rate-limiting step in the production of all steroid hormones is the conversion of the C_{27} steroid, cholesterol, to the C_{21} steroid, pregnenolone, that is catalyzed by the cholesterol side-chain cleavage ($P450_{scc}$) enzyme. This enzyme is located in the inner mitochondrial membrane. The production of C_{19} steroids from C_{21} steroids requires the activities of cytochrome $P450_{17\alpha}$, which is associated with the smooth endoplasmic reticulum. $P450_{17\alpha}$ catalyzes two reactions, the hydroxylation of the C_{21} steroids pregnenolone or progesterone (17α-hydroxylase activity), followed by cleavage of the 2-carbon side chain (C_{17-20} lyase activity), to yield the C_{19} steroids dehydroepiandrosterone or androstenedione, respectively. $P450_{arom}$ is the terminal oxidase in the synthesis of estrone and estradiol from androstenedione and testosterone, respectively. This enzyme is associated with the smooth endoplasmic reticulum and is found in several tissues, including Leydig cells (1), Sertoli cells (2), ovarian granulosa cells (3), and placenta and adipose tissue (4). Two additional P450 enzymes are specific for the biosynthesis of adrenal steroid hormones: $P450_{21}$, which is associated with the smooth endoplasmic reticulum, and $P450_{11\beta}$, which is associated with the inner mitochondrial membrane. The non-P450 enzyme, *3β-hydroxysteroid dehydrogenase/$\Delta^5 \rightarrow \Delta^4$ isomerase* (3βHSD), catalyzes the conversion of the Δ^5-3β-hydroxysteroids, pregnenolone and dehydroepiandrosterone, to the Δ^4-3-ketosteroids, progesterone and androstenedione, respectively (Fig. 5.1). This enzyme is associated with the smooth endoplasmic reticulum and is essential for the biosynthesis of all steroid hormones.

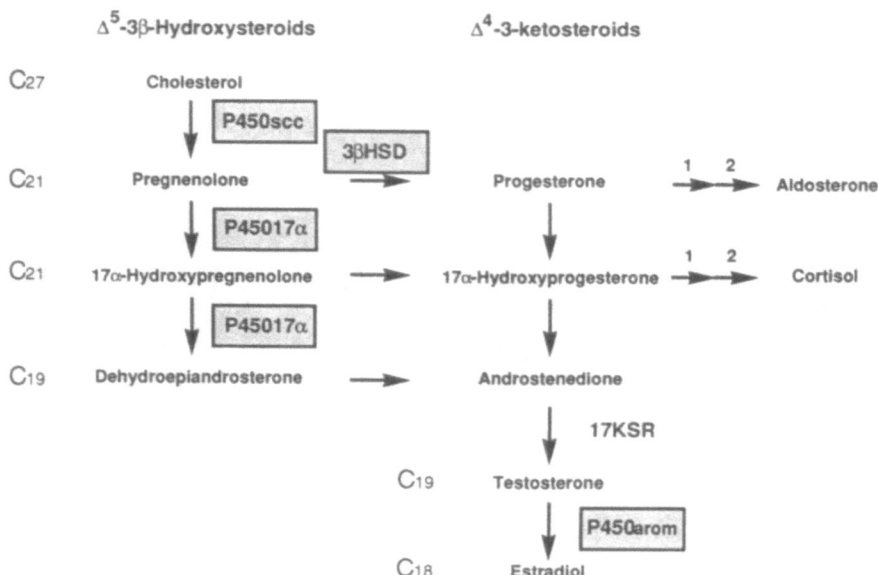

FIGURE 5.1. Steroid biosynthetic pathway in gonads and adrenal glands. (P450$_{scc}$ = cytochrome P450 cholesterol side-chain cleavage; P450$_{17\alpha}$-cytochrome P450$_{17\alpha}$-hydroxylase/C$_{17-20}$ lyase; 3βHSD-3β-hydroxysteroid dehydrogenase/Δ^5 → Δ^4 isomerase; 17KSR-17-ketosteroid reductase; P450$_{arom}$-cytochrome P450 aromatase; 1-cytochrome P450 21-hydroxylase; 2-cytochrome P450 11β-hydroxylase.)

Another non-P450 enzyme involved in the biosynthesis of testosterone and estradiol, *17-ketosteroid reductase* (17KSR), is found in numerous tissues, including the testis and ovary, but not in adrenal glands (5). This chapter focuses on the tissue-specific expression in the mouse of P450$_{scc}$, P450$_{17\alpha}$, and 3βHSD, as well as regulation of expression of these enzymes by *cyclic AMP* (cAMP) and steroid hormones.

Gene Location

P450$_{scc}$ and P450$_{arom}$

We have identified the chromosomal location in the mouse genome for the structural genes, Cyp11a and Cyp19 that encode P450$_{scc}$ and P450$_{arom}$, respectively (6). Using *restriction fragment length variations* (RFLV) and recombinant strains of mice, we established that Cyp11a and Cyp19 are closely linked on mouse chromosome 9. When the strain distribution patterns of the P450$_{scc}$ and P450$_{arom}$ genes were compared to the other markers previously mapped in these recombinant inbred strains of mice, it was found that Cyp11a and Cyp19 are most closely linked to another gene encoding a P450 enzyme, P_1450, Cyp1. The structural genes for P_1450 (CYP1), $P450_{scc}$ (CYP11A1), and $P450_{arom}$ (CYP19) (7) in the human genome have been mapped to chromosome 15. However, the distance between the loci for the human $P450_{scc}$ and other loci has not been determined. The information presented from our laboratory, along with other studies, indicates conservation between homologous human and mouse chromosomal regions and predicts that the human gene that encodes $P450_{scc}$ will be found closely linked to CYP19 on human chromosome 15 (6).

$P450_{17\alpha}$

A mouse Leydig cell $P450_{17\alpha}$ cDNA isolated from a mouse Leydig cell library was used to map the chromosomal location of the structural gene encoding $P450_{17\alpha}$, Cyp17 (7). Using DNA from a mouse intersubspecific

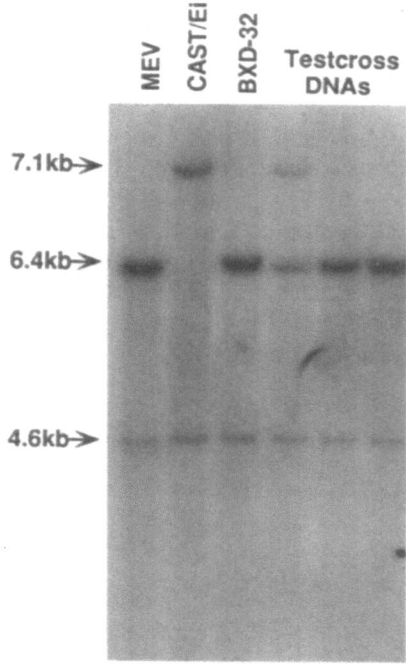

FIGURE 5.2. Autoradiogram of a Southern blot of EcoRI-digested genomic DNA hybridized with a full-length mouse $P450_{17\alpha}$ cDNA. The first 3 lanes show the hybridized fragments of DNA from progenitor strains; the last 3 lanes are representative testcross strains showing the presence or absence of the CAST/Ei allele. Arrows indicate the length of the fragments in kilobases. Reprinted with permission from Youngblood, Sartorius, Taylor, and Payne (7).

testcross, Cyp17 has been localized to mouse chromosome 19, distal to the gene Got1 that encodes glutamate oxaloacetate transaminase-1. Southern analysis of mouse genomic DNA is consistent with the presence of a single gene that encodes for $P450_{17\alpha}$ in the mouse genome (Fig. 5.2) (7). Another cytochrome P450, P4502c (Cyp2c) also is located at the distal end of chromosome 19. The human homologs, CYP17, CYP2C, and GOT1 have been mapped to human chromosome 10, with CYP2C and GOT1 mapped to the distal region. However, the region of chromosome 10 where CYP17 is located has not been identified. The data obtained for the mouse genome predicts that CYP17 will be found in the homologous region of chromosome 10 close to GOT1 (7).

3βHSD

The mouse gene(s) that encode 3βHSD has been mapped to mouse chromosome 3. Using RFLVs as genotypic markers for linkage analysis, the gene(s) for 3βHSD, Hsd3b, was localized to the distal portion of mouse chromosome 3, near Amy1, the gene that encodes for salivary amylase (Bain, Taylor, Payne, unpublished data). The human locus for 3βHSD recently has been mapped by in situ hybridization to chromosome 1p13 (8), in a region of human chromosome 1 that exhibits extensive conservation of gene order and genetic distances with the region of mouse chromosome 3 that includes Amy1 (9, 10). The observation of the close linkage of Hsd3b to Amy1 suggests that Hsd3b is included in this homologous region as well. Multiple restriction fragments are recognized by two different, nonoverlapping 3βHSD Leydig cell cDNA probes (Fig. 5.3). The recognition of these multiple bands indicates the presence of more than one structural gene or pseudogene(s) located on mouse chromosome 3 (compare Figs. 5.2 and 5.3).

Regulation of Expression of $P450_{scc}$, $P450_{17\alpha}$, and 3βHSD in Leydig Cells

Regulation by cAMP

Our laboratory has used mouse Leydig cell cultures to study the regulation of $P450_{scc}$, $P450_{17\alpha}$, and 3βHSD expression. Primary cultures of mouse Leydig cells are preferable to the more frequently used rat Leydig cell culture for studying the regulation of the steroidogenic enzymes and cAMP-stimulated testosterone production since mouse Leydig cell cultures remain responsive to cAMP stimulation for at least 21 days (11 and Payne, unpublished).

Anakwe and Payne (11) investigated the rate of de novo synthesis of $P450_{scc}$ and $P450_{17\alpha}$ in mouse Leydig cell cultures maintained for 15 days.

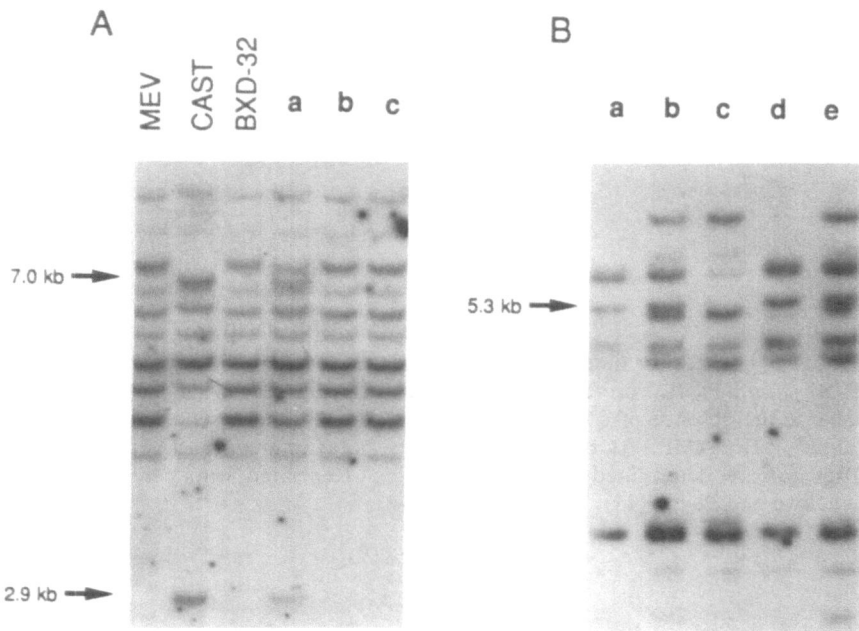

FIGURE 5.3. Autoradiogram of Southern blots of genomic DNA from progenitor and progeny mice used to map 3βHSD to mouse chromosome 3. A: EcoRI-digested genomic DNA hybridized with a 140-bp NcoI-BamHI fragment from the 5′ end of 3βHSD-I, as described in reference 20. This is the same blot as illustrated in Fig. 5.2 after removal of the P450$_{17\alpha}$ probe. B: HindIII digested genomic DNA of (CAST × MEV)F$_2$ mice hybridized with a SacI-BglII 906-bp fragment from the coding region of 3βHSD-I, as described in reference 20. Arrows indicate the length of the fragments of the variant restriction fragments representative of the CAST/Ei allele.

Leydig cells were incubated in a synthetic, serum-free medium containing 0.1% *bovine serum albumin* (BSA) and insulin (500 µg/mL). Cultures were incubated for 7 days in the absence of cAMP followed by 4 days in the presence or absence of 50-µM cAMP. Figure 5.4 illustrates that in the absence of cAMP, the amount of immunoreactive P450$_{17\alpha}$ protein decreases to 50% by day 2 of culture and is essentially undetectable from days 4 through 11. Treatment with cAMP starting on day 7 results in an increase in the amount of P450$_{17\alpha}$. In sharp contrast, the amount of immunoreactive P450$_{scc}$ in untreated cells remains relatively unchanged during the 11 days of culture. Treatment with cAMP starting on day 7 results only in a slight increase in the amount of P450$_{scc}$ after 4 days of treatment. Previous studies by Malaska and Payne (12) demonstrated that

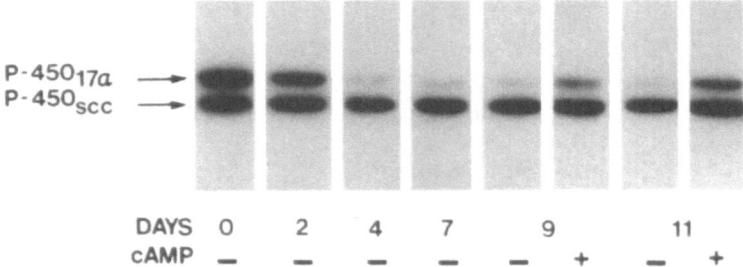

FIGURE 5.4. Effect of time in culture and treatment with cAMP on the amount of $P450_{17\alpha}$ and $P450_{scc}$ protein. Purified mouse Leydig cells were incubated for 7 days in the absence of cAMP, followed by 4 days in the absence or presence of 50-μM 8-Br-cAMP. At the indicated times, total immunoreactive $P450_{17\alpha}$ and $P450_{scc}$ were determined by immunoblotting. Reprinted with permission from Anakwe and Payne (11).

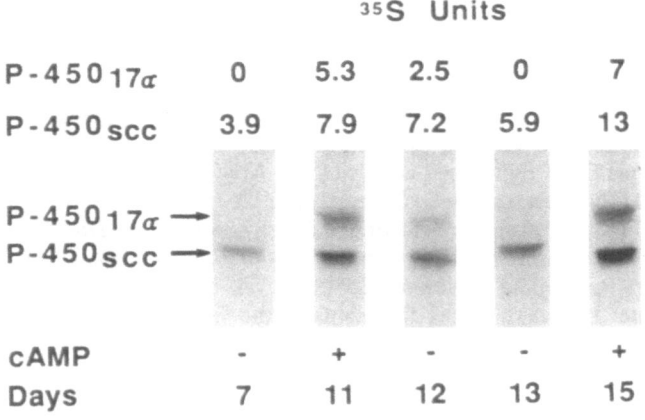

FIGURE 5.5. Effect of cAMP withdrawal and replacement on the rate of de novo synthesis of $P450_{17\alpha}$ and $P450_{scc}$. Mouse Leydig cells were maintained in culture for 7 days without cAMP. From days 7 to 11, cultures were treated with 50-μM 8-Br-cAMP. On day 11, cAMP was removed by washing cells for 2 h in media only. Cells were then incubated in the absence of cAMP from days 11 to 13 followed by incubation in the presence of cAMP from days 13 to 15. At the indicated time, de novo synthesis was determined by incubating cells for 3 h with [^{35}S]methionine as described in reference 11. Reprinted with permission from Anakwe and Payne (11).

cAMP treatment of mouse Leydig cells is necessary for the induction of P450$_{17\alpha}$ enzyme activity. The absolute dependence of P450$_{17\alpha}$ de novo synthesis on cAMP stimulation of Leydig cells is shown in Figure 5.5. De novo synthesis of the P450 enzymes was measured by the incorporation of [^{35}S]methionine into newly synthesized P450$_{scc}$ and P450$_{17\alpha}$. In the absence of cAMP treatment, no incorporation of [^{35}S]methionine into P450$_{17\alpha}$ was observed on day 7 of culture, while newly synthesized P450$_{scc}$ was demonstrable. Treatment with 50-μM cAMP from day 7 to 11 resulted in the detection of newly synthesized P450$_{17\alpha}$ and in a 2-fold increase in the rate of synthesis of P450$_{scc}$. The withdrawal of cAMP on day 11 caused a 50% decrease in the rate of de novo synthesis of P450$_{17\alpha}$ by day 12 and complete cessation of P450$_{17\alpha}$ synthesis by day 13. Replacement of cAMP to these same cultures on day 13 for an additional 2 days restored P450$_{17\alpha}$ synthesis. In contrast to the absolute requirement for cAMP for the de novo synthesis of P450$_{17\alpha}$, cAMP withdrawal had little effect on the rate of de novo synthesis of P450$_{scc}$ (Fig. 5.5).

The role of cAMP for de novo synthesis of the P450 enzymes reflects the effect of cAMP on the expression of P450$_{17\alpha}$ and P450$_{scc}$ mRNA levels in Leydig cells. As was observed with enzyme synthesis, P450$_{scc}$ mRNA is constitutively expressed and increased approximately 2-fold

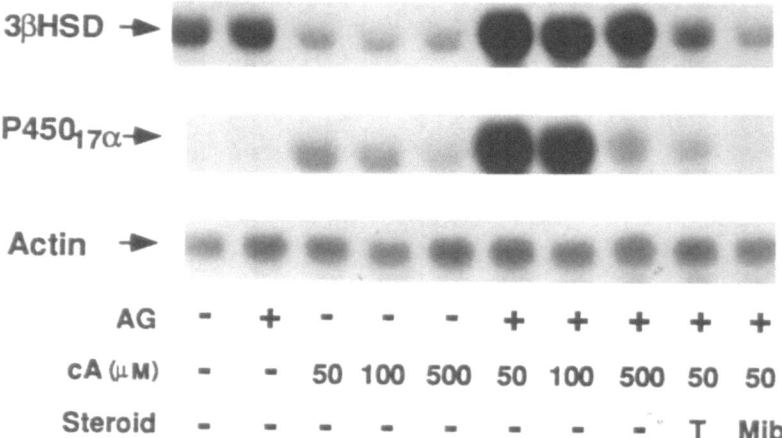

FIGURE 5.6. Effect of cAMP and androgens on 3βHSD and P450$_{17\alpha}$ mRNA levels. Mouse Leydig cells were incubated for 6 days before treatment for 24 h with increasing concentrations of cAMP (cA) in the presence or absence of 0.5-mM aminoglutethimide (AG) and, where indicated, 2-μM mibolerone (Mib) or 2-μM testosterone (T). Total cytoplasmic RNA was isolated, and 7 μg were subjected to Northern analysis and hybridized sequentially with 3βHSD, P450$_{17\alpha}$, and β-actin DNA probes. Reprinted with permission from Payne and Sha (13), © by The Endocrine Society, 1991.

after 24 h of treatment with cAMP (12, 13). Studies on the regulation of
$P450_{17\alpha}$ and 3βHSD mRNA levels demonstrate that in the absence of
cAMP, $P450_{17\alpha}$ mRNA levels are undetectable, while 3βHSD mRNA is
expressed constitutively (Fig. 5.6), similar to the expression of $P450_{scc}$
mRNA. Maximal induction by cAMP of $P450_{17\alpha}$ and 3βHSD mRNA
is observed at a concentration of 50 μM in the presence of aminoglutethimide to inhibit endogenous steroid production (Fig. 5.6 and section
below on effects of steroids).

The requirement for newly synthesized proteins in mediating cAMP
induction of $P450_{17\alpha}$ and 3βHSD mRNA differs from that of $P450_{scc}$
(Fig. 5.7). Inhibition of protein synthesis for 24 h by the addition of
cycloheximide to Leydig cell cultures completely suppressed both constitutive and cAMP-induced expression of 3βHSD mRNA. Cycloheximide
also markedly suppressed the cAMP induction of $P450_{17\alpha}$ mRNA. In
sharp contrast, cycloheximide did not suppress cAMP induction of
$P450_{scc}$ mRNA, while constitutive expression was reduced. To examine
in more detail the role of protein synthesis and the time required for

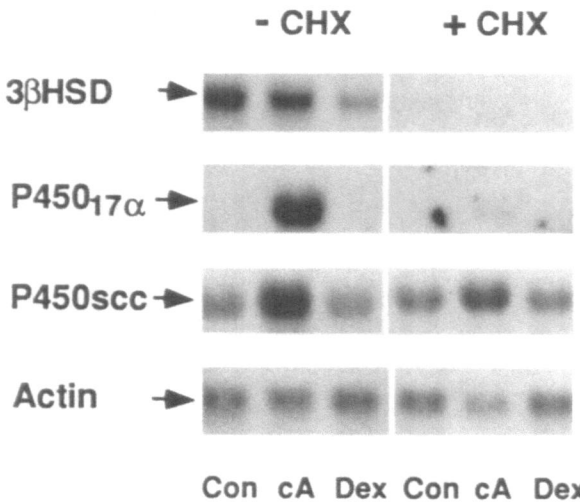

FIGURE 5.7. Effect of inhibition of protein synthesis on the expression of 3βHSD,
$P450_{17\alpha}$, and $P450_{scc}$ mRNA. Leydig cells were incubated and treated as described
in Fig. 5.6. Cycloheximide (CHX, 10 μg/mL) was added 30 min before addition of
50-μM cAMP (cA) or 100-nM dexamethasone (Dex). CHX was added to
untreated control cultures (Con) at the same time as it was added to the other
cultures. After 24 h, total RNA was isolated and subjected to Northern analysis
and hybridized sequentially with 3βHSD, $P450_{17\alpha}$, $P450_{scc}$, and β-actin cDNA
probes. Reprinted with permission from Payne and Sha (13), © by The
Endocrine Society, 1991.

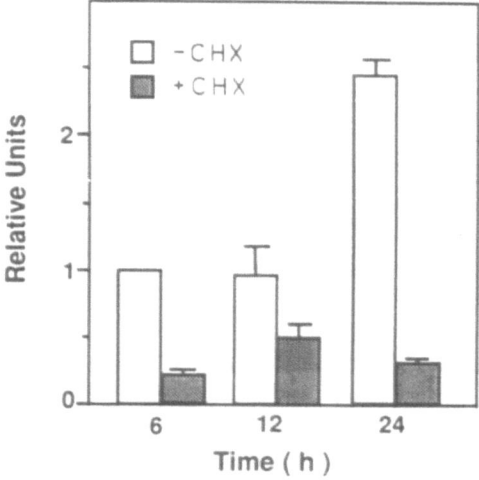

FIGURE 5.8. Time course for action of cAMP and cycloheximide on $P450_{17\alpha}$ mRNA levels. Leydig cell cultures were incubated for 6, 12, and 24 h in the absence or presence of 50-μM cAMP (cA) and in the absence or presence of cycloheximide (CHX, 10 μg/mL). CHX was added 30 min before the addition of cA. Total cytoplasmic RNA was isolated, subjected to Northern analysis, and hybridized sequentially with $P450_{17\alpha}$ and β-actin cDNA probes. $P450_{17\alpha}$ mRNA and β-actin mRNA were quantitated by laser densitometry, and the amount of $P450_{17\alpha}$ in each treatment group was corrected for the amount of β-actin in that treatment group. mRNA levels in cultures not treated with cAMP were undetectable and, therefore, are not illustrated. All values represent the mean ± range of 2 separate experiments and are expressed relative to the 6-h cA-treated-in-the-absence-of-cycloheximide (−CXH) sample, which was given the arbitrary value of 1. Reprinted with permission from Payne and Sha (13), © by The Endocrine Society, 1991.

maximal cAMP induction of $P450_{17\alpha}$ and $P450_{scc}$ mRNA, the effect of cycloheximide on basal and cAMP-stimulated mRNA levels was determined at 6, 12, and 24 h following treatment of cultures with cAMP. Inhibition of protein synthesis suppressed the cAMP induction of $P450_{17\alpha}$ mRNA at all time intervals examined (Fig. 5.8). Inhibition of protein synthesis for 24 h reduced cAMP-induced $P450_{17\alpha}$ mRNA levels to 12% of levels observed in the absence of cycloheximide. cAMP induction of $P450_{scc}$ mRNA levels does not appear to be dependent on newly synthesized proteins, while basal expression of $P450_{scc}$ mRNA requires newly synthesized proteins for optimal expression. Figure 5.9 illustrates that cAMP-induced increases in $P450_{scc}$ mRNA at 12 and 24 h are not prevented by the addition of cycloheximide, while basal expression is reduced by approximately 50%. During the first 6 h of exposure to cycloheximide, inhibition of protein synthesis had no effect on basal

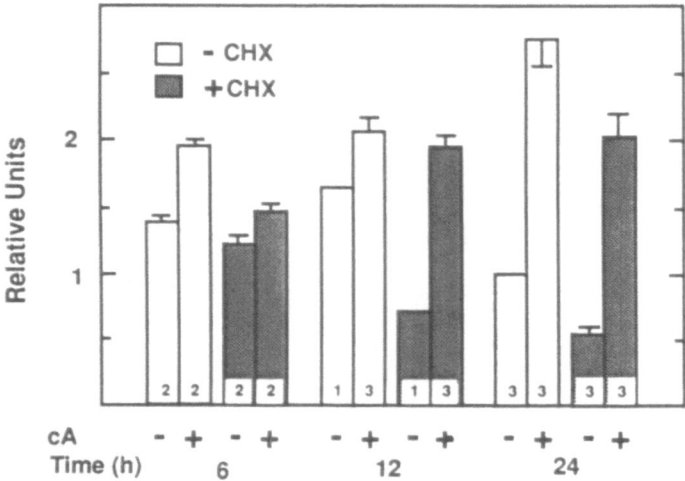

FIGURE 5.9. Time course for action of cAMP and cycloheximide on P450$_{scc}$ mRNA levels. All conditions were identical to those described in Fig. 5.8. Filters were hybridized with P450$_{scc}$ and β-actin cDNA probes and quantitated as described in Fig. 5.8. All values were expressed relative to the 24-h untreated control, which was given the arbitrary value of 1. The number in each bar is equal to the number of separate experiments. The value for 3 separate experiments represents the mean ± SE. The value for 2 separate experiments is the mean ± range. Reprinted with permission from Payne and Sha (13), © by The Endocrine Society, 1991.

expression, but prevented the 40% induced increase in P450$_{scc}$ mRNA levels (Fig. 5.7). Although the cAMP-induced increases in P450$_{scc}$ and P450$_{17α}$ mRNA levels were different, the pattern of the cAMP-induced increases with time were similar. cAMP-induced increases in both P450$_{17α}$ and P450$_{scc}$ mRNA levels were most marked between 12 and 24 h. This slow type of response to cAMP induction of mRNA levels is characteristic of genes whose induction by cAMP is mediated by newly synthesized proteins (14). The results of our studies, however, indicate that cAMP increases in P450$_{scc}$ mRNA do not require newly synthesized proteins, while cAMP-mediated induction of P450$_{17α}$ is highly dependent on newly synthesized proteins. The data suggest that cAMP-induced increases in these two P450 mRNAs in normal mouse Leydig cells occur by a different mechanism(s). The precise mechanism(s) involved requires further investigation.

To investigate further the cAMP induction of P450$_{17α}$ expression, we have isolated the mouse structural gene encoding P450$_{17α}$. Two genomic clones containing the entire coding region and approximately 10 kb of 5′ flanking sequences of the P450$_{17α}$ structural gene were isolated and

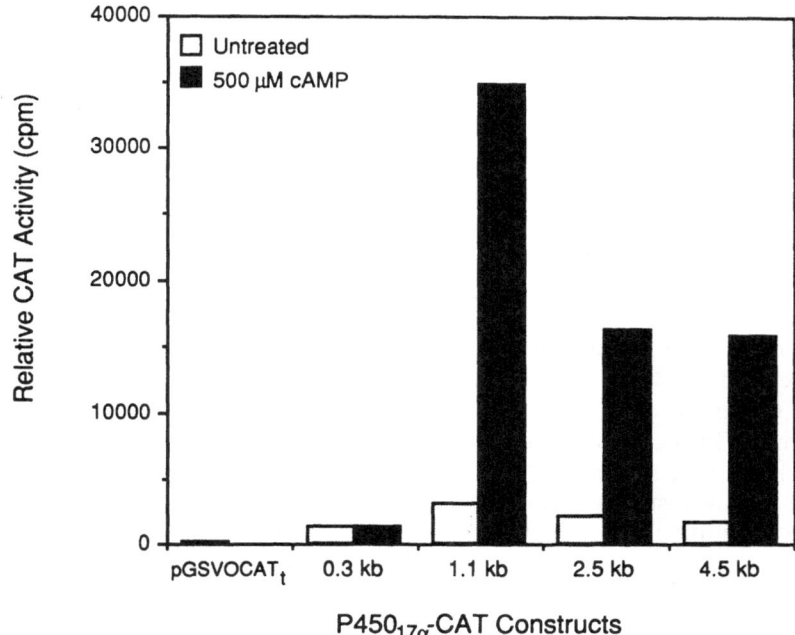

FIGURE 5.10. Localization of $P450_{17\alpha}$ 5' regulatory regions responsible for cAMP induction. Plasmids were constructed by subcloning different-size kilobase (kb) fragments 5' of the $P450_{17\alpha}$ coding region into the multiple cloning site of a promoterless plasmid, pGSVOCAT$_t$. pGSVOCAT$_t$ was constructed by inserting the chloramphenicol acetyltransferase (CAT) reporter gene into pGem7Zf$^+$. A terminator fragment (t) containing an SV40 poly(A) addition signal was inserted into the pGem-CAT upstream of the multiple cloning region. MA-10 Leydig tumor cells were transfected transiently with the indicated constructs (5-μg DNA) and treated for 12 h with 500-μM 8-Br-cAMP. CAT activity was measured in cell extracts by measuring the amount of [^{3}H] acetylated chloramphenicol produced during 2 h. All cultures were cotransfected with 5-μg SV2β-gal, and CAT activity is expressed relative to β-galactosidase activity (β-gal).

characterized (Youngblood, Payne, unpublished data). To identify sequences involved in the cAMP regulation of mouse $P450_{17\alpha}$ and to investigate further the mechanism by which cAMP induces transcription of $P450_{17\alpha}$, 5'-fragments of the coding region were subcloned into vectors containing the *chloramphenicol acetyl transferase* (CAT) reporter gene. The constructs, containing different lengths of 5' upstream sequences were transfected into MA-10 Leydig tumor cells and treated with cAMP. cAMP response of the constructs was determined by measuring CAT activity (amount of chloramphenicol acetylated with [^{3}H] acetate in the cell extracts). Maximal induction of $P450_{17\alpha}$ expression, as measured by CAT activity, occurred at 12 h following treatment with 500-μM 8-bromo-cyclic AMP. Data presented in Figure 5.10 demonstrate that the cAMP-responsive element is located between 0.3 and 1.1 kb upstream of the $P450_{17\alpha}$ coding region. Future studies should further delineate the sequences necessary for cAMP induction of $P450_{17\alpha}$ and resolve the mechanism by which cAMP induces transcription of the $P450_{17\alpha}$ gene. Identification of the sequences responsible for cAMP-induced increases in $P450_{scc}$ mRNA has not been successful to date.

Regulation by Steroid Hormones

Our laboratory has shown that steroid hormones negatively regulate the expression of $P450_{scc}$, $P450_{17\alpha}$, and 3βHSD in normal mouse Leydig cells. The repression is specific for a particular steroid and for a particular enzyme. Glucocorticoids repress protein synthesis and steady state levels of $P450_{scc}$ (12, 13), as well as 3βHSD mRNA (13), while endogenously produced testosterone represses cAMP induction of $P450_{17\alpha}$ (13, 16, 17) and constitutive and cAMP-induced 3βHSD mRNA (13). In contrast to the repressive effect of glucocorticoids on $P450_{scc}$ in normal mouse Leydig cell cultures, the glucocorticoid dexamethasone stimulates both de novo protein synthesis and steady state levels of $P450_{scc}$ mRNA in MA-10 tumor Leydig cells (15). This effect of dexamethasone in MA-10 cells is additive to cAMP stimulation of $P450_{scc}$ synthesis and mRNA levels. Both the negative and the positive effect of glucocorticoids on $P450_{scc}$ synthesis and mRNA levels are prevented by the glucocorticoid antagonist, RU486, and is not observed with either estradiol or testosterone (12, 15).

Testosterone produced during cAMP induction of $P450_{17\alpha}$ has a negative effect on $P450_{17\alpha}$ enzyme activities (16, 17) and de novo synthesis (17). The addition of aminoglutethimide, an inhibitor of cholesterol metabolism, markedly enhances the effect of cAMP on the induction of $P450_{17\alpha}$ enzyme activity and de novo synthesis (16, 17). Hales et al. demonstrated that the negative effect of testosterone on cAMP induction of $P450_{17\alpha}$ can be mimicked by the androgen agonist, mibolerone, and prevented by the addition of the androgen antagonist, hydroxyflutamide (17). These results indicate that testosterone produced during cAMP induction of $P450_{17\alpha}$ represses this induction by an androgen receptor-mediated mechanism. More recently, we demonstrated that testosterone produced during cAMP stimulation of Leydig cells represses cAMP induction of both $P450_{17\alpha}$ and 3βHSD mRNA levels (Fig. 5.6). cAMP induction of 3βHSD mRNA can be demonstrated only when increased testosterone production is prevented by the addition of aminoglutethimide to the Leydig cell cultures. The repression of $P450_{17\alpha}$ and 3βHSD mRNA by testosterone is not a general effect on all steroidogenic enzymes in Leydig cells. Testosterone does not repress de novo synthesis of $P450_{scc}$

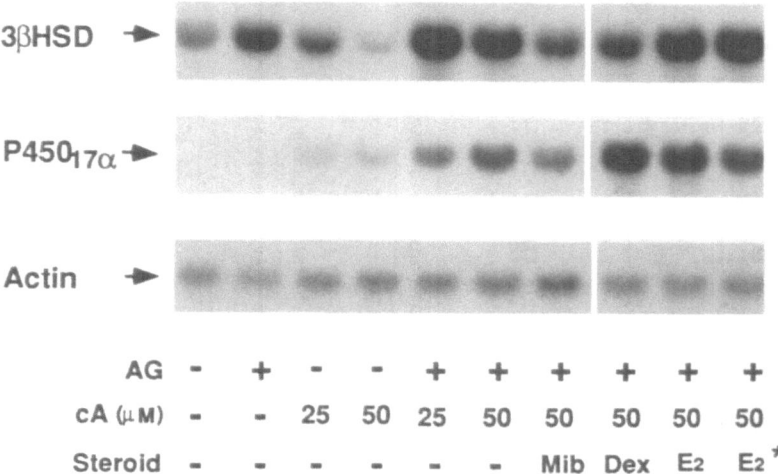

FIGURE 5.11. Effects of mibolerone, dexamethasone, and estradiol on 3βHSD and
$P450_{17\alpha}$ mRNA levels. Mouse Leydig cells were incubated and treated as
described in Fig. 5.6. Mibolerone (Mib, $2\,\mu M$), dexamethasone (Dex, $100\,nM$),
estradiol (E_2, $100\,nM$), and (E_2^*, $1\,\mu M$) were added to cAMP + aminoglutethimide (AG)-treated cultures where indicated. Total cytoplasmic RNA
was isolated, and $7\,\mu g$ was subjected to Northern analysis and hybridized, as
described in Fig. 5.6. Reprinted with permission from Payne and Sha (13), © by
The Endocrine Society, 1991.

in Leydig cells (12). Estrogen did not repress $P450_{17\alpha}$ or 3βHSD expression (Fig. 5.11) (17). Repression by dexamethasone of constitutive (Fig.
5.7) and cAMP-induced levels (Fig. 5.11) of 3βHSD mRNA, but not of
$P450_{17\alpha}$ mRNA, was observed (13). In summary, expression of $P450_{scc}$,
$P450_{17\alpha}$, and 3βHSD is differentially regulated by steroid hormones.
Endogenously produced testosterone negatively regulates expression of
cAMP-induced $P450_{17\alpha}$ and 3βHSD, but not $P450_{scc}$, while glucocorticoids negatively regulate 3βHSD and $P450_{scc}$, but not $P450_{17\alpha}$, in normal
mouse Leydig cells. Although glucocorticoids are not produced in Leydig
cells, glucocorticoid receptors have been demonstrated in interstitial cells
of the rat testis (18). Data from our laboratory demonstrating that the
effect of dexamethasone can be prevented by RU486, a glucocorticoid
antagonist, indicate the presence of glucocorticoid receptors in mouse
Leydig cells (12). Glucocorticoids are mediators of stress. Increased
production of glucocorticoids in such pathologic conditions of the adrenal
cortex as Cushing's syndrome can be associated with reproductive dys-
function, including decreased circulating testosterone levels (19). Our

```
                    10              20              30
I:    Met Ala Gly Trp Ser Cys Leu Val Thr Gly Ala Gly Gly Phe Val Gly Gln Arg Ile Ile Lys Met Leu Val Gln Glu Lys Glu Leu Gln
II:   --- *** *** *** *** *** *** *** *** *** *** *** *** *** *** *** *** *** *** *** *** *** *** *** *** *** *** *** *** ***
III:  --- Pro --- --- --- --- --- --- --- --- --- --- --- --- Leu --- --- --- --- --- Gln Leu --- --- --- --- Asp --- --- Glu

                    40              50              60
I:    Glu Val Arg Ala Leu Asp Lys Phe Val Arg Pro Glu Thr Lys Glu Phe Ser Lys Leu Gln Thr Val Thr Val Leu Val Thr Val Leu Glu
II:   --- --- --- --- --- --- --- --- --- *** *** *** *** *** *** *** *** *** *** *** *** *** *** *** *** *** *** *** *** ---
III:  Ile --- Ile --- Val --- Lys --- --- Arg --- --- Gly --- --- Gln --- --- Ser --- --- Phe Asn --- --- --- Gly --- Leu ---

                    70              80              90
I:    Gly Asp Ile Leu Asp Ala Gln Cys Leu Arg Arg Ala Cys Gln Gly Ile Ser Val Val Ile His Thr Ala Ala Val Ile Asp Val Thr Gly
II:   --- --- --- *** *** *** *** *** *** *** *** *** *** *** *** *** *** *** *** *** *** *** *** *** *** *** *** *** *** ***
III:  --- --- Thr --- --- Tyr --- --- --- --- --- --- --- --- --- --- --- --- --- --- --- Ile --- --- --- Ile --- --- --- ---

                    100             110             120
I:    Val Ile Pro Gln Thr Gln Thr Gly Thr Tyr Val Asp Leu Asn Leu Lys Tyr Lys Asn Leu Leu Glu Asn Ala Cys Val Gln Ala Ser Val Pro
II:   --- --- --- *** *** *** *** *** *** *** *** *** *** *** *** *** *** --- --- --- --- --- --- --- --- --- --- --- --- --- ---
III:  --- --- --- --- --- --- --- --- --- --- --- --- --- Glu --- Asp --- --- --- --- --- --- Ile Ile --- --- --- --- --- --- Phe

                    130             140             150
I:    Ile Phe Cys Ser Ser Val Asp Val Ala Gly Pro Asn Ser Tyr Lys Ile Leu Val Leu Asn Asn Gly His Glu Gln Gln Gln Asn His Glu Ser Thr
II:   --- --- --- --- --- --- --- --- --- --- --- --- --- --- --- --- --- --- --- --- --- --- --- --- --- --- --- --- --- --- --- ---
III:  --- Ser --- --- --- --- --- --- --- --- Asp --- --- --- Glu --- --- --- --- --- --- --- --- --- --- --- Asp Glu His Cys Arg

                    160             170             180
I:    Trp Ser Asp Pro Tyr Pro Tyr Ser Lys Lys Met Ala Glu Lys Ala Val Leu Ala Ala Asn Gly Ser Met Leu Lys Asn Gly Gly Thr Leu
II:   --- --- --- --- --- --- --- --- --- --- --- --- --- --- --- --- --- --- --- --- --- --- --- --- --- --- --- --- --- ---
III:  --- --- --- --- --- --- --- --- --- --- --- --- --- --- --- --- --- --- --- --- --- --- --- --- --- --- --- --- --- ---
```

```
                    190                 200                 210
I:   Asn Thr Cys Ala Leu Arg Pro Met Tyr Ile Tyr Gly Glu Arg Ser Pro Phe Ile Phe Asn Ala Ile Ile Ile Arg Ala Leu Lys Asn Lys Gly
II:  Gln  -   -   -   -   -   -   -   -   -   -   -   -   -   -   -  Leu Ser  -   -  Ile  -   -  Met  -   -   -  His Gly
III: Gln  -   -   -   -   -   -   -   -   -   -   -   -  Gln Phe Leu  -   -  Thr  -   -   -  Lys  -   -  Asn Phe

                    220                 230                 240
I:   Ile Leu Cys Val Thr Gly Lys Phe Ser Ile Ala Asn Pro Val Tyr Val Glu Asn Val Ala Trp Ala His Ile Leu Ala Ala Arg Gly Leu
II:   -   -  Arg Ser Phe  -   -   -  Asn Thr  -   -   -   -   -   -   -  Gly  -   -   -   -   -   -   -   -   -
III:  -   -  Arg Gly Gly  -   -   -  Ser Thr  -   -   -   -   -   -   -  Gly  -   -   -   -   -   -   -   -   -

                    250                 260                 270
I:   Arg Asp Pro Lys Lys Ser Thr Ser Ile Gln Gly Gln Phe Tyr Tyr Ile Ser Asp Asp Thr Pro His Gln Ser Tyr Asp Asp Leu Asn Tyr
II:   -  Asp  -   -   -   -  Pro Asn  -   -  Glu  -   -   -   -   -   -   -   -   -   -   -   -   -  Phe  -  Ile Ile
III:  -  Asn  -   -   -   -   -   -   -  Glu  -   -   -   -   -   -   -   -   -   -  Tyr  -  Leu Asn

                    280                 290                 300
I:   Thr Leu Ser Lys Glu Trp Gly Leu Arg Pro Asn Ala Ser Trp Ser Leu Pro Leu Tyr Leu Ala Phe Leu Leu Glu Thr
II:   -   -   -   -   -   -   -  Phe Cys Leu Asp Ser  -  Val  -   -   -   -   -   -   -   -   -
III:  -   -   -   -   -   -   -  Phe Cys Leu Asn Ser Arg Tyr Val  -   -   -   -   -   -   -   -

                    310                 320                 330
I:   Val Ser Phe Leu Arg Pro Val Tyr Arg Pro Leu Phe Asn Arg His Leu Ile Thr Leu Asn Ser Thr Phe Thr Phe Ser
II:   -   -   -   -  Ser  -  Ile  -  Ile  -   -   -   -   -   -  Val  -   -  Gly  -   -   -   -
III:  -   -   -   -  Ser  -  Ile  -  Ile  -   -   -   -   -   -  Val Thr Ala  -   -   -   -

                    340                 350                 360
I:   Tyr Lys Lys Ala Gln Arg Asp Leu Gly Tyr Glu Pro Leu Val Asn Trp Glu Glu Ala Lys Thr Ser Glu Trp Ile Gly Thr Ile
II:   -   -   -   -   -   -   -   -   -   -   -   -   -   -  Ser  -   -   -   -   -   -   -   -   -   -  Leu
III:  -   -   -   -   -   -   -   -   -   -   -   -   -   -  Ser  -   -   -   -   -   -   -   -   -   -  Leu

                    370
I:   Val Glu Gln His Arg Glu Ile Leu Asp Thr Lys Cys Gln End
II:   -   -   -   -   -   -   -  Thr  -   -   -  Ser  -
III:  -   -   -   -   -   -   -  Thr  -   -   -  Ser  -
```

FIGURE 5.12. Comparison of the predicted amino acid sequences of mouse 3βHSD types I, II and III. Asterisks indicate the region of 3βHSD-II that has not been isolated. Identical amino acids are indicated by (−).

studies suggest that increased circulating concentrations of glucocorticoids could decrease testicular testosterone production by decreasing the amount of two of the enzymes, P450$_{scc}$ and 3βHSD, necessary for the conversion of cholesterol to testosterone.

Tissue-Specific Expression

Isolation and Characterization of Different Forms of 3βHSD

Our laboratory has isolated 3 distinct 3βHSD cDNA clones, 1 from a mouse Leydig cell library, 3βHSD-I, and 2 from mouse liver libraries, 3βHSD-II and 3βHSD-III (20). A 3βHSD-I cDNA, isolated from a mouse Leydig cell library, is 1608 bp in length and contains a single 1122 nucleotide open reading frame encoding a protein of 373 amino acids with a molecular weight of 42,059. Two distinct clones were isolated from mouse liver libraries. One of these cDNA clones, 3βHSD-III, comprises 1553 bp and contains the complete coding region, a large 3' untranslated region and 51 bp of 5' untranslated region. The open reading frame is predicted to encode a protein with a molecular weight of 42,028. The longest 3βHSD-II clone isolated contains sequences corresponding to amino acids 109–373 of 3βHSD-I and 3βHSD-III (Fig. 5.12) and a complete 3' untranslated sequence including a poly(A) tail. Figure 5.12 compares the amino acid sequences encoded by the three 3βHSD cDNA clones. The coding region of 3βHSD-I and 3βHSD-III are 89.4% identical, while the coding region of 3βHSD-II is 90% and 94% identical to 3βHSD-I and 3βHSD-III, respectively. Within the 3' noncoding region, 3βHSD-I is 62% and 59% identical to 3βHSD-II and 3βHSD-III, respectively (20). Types II and III are 73% identical within this region. Within the 5' untranslated region of 3βHSD-III, there is an 8-nucleotide stretch of complete identity with 3βHSD-I immediately 5' of the coding region. Further 5' of this region, no identity between 3βHSD-I and 3βHSD-III is observed.

Expression of 3βHSD mRNAs in Steroidogenic and Nonsteroidogenic Tissues

Northern analysis using a 906-bp SacI-BglII fragment of the 3βHSD-I cDNA demonstrates the expression of a 1.7-kb mRNA in the adrenal gland, ovary, testis, liver, and kidney (Fig. 5.13). No 3βHSD mRNA was detected in spleen or brain tissue. In addition to the 1.7-kb mRNA, the kidney expresses a 1.9-kb mRNA that is not observed in the other tissues. The 1.7-kb mRNA is present at much higher levels in the adrenal gland

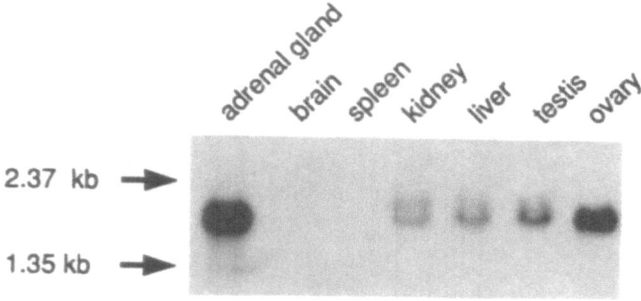

FIGURE 5.13. Northern blot analysis of 3βHSD transcripts in steroidogenic and nonsteroidogenic tissues. The probe used was a 906-bp SacI-BglII fragment of the clone 3βHSD-I. All tissues were from male mice except for the ovary. Each lane contains 20-μg total RNA except for the ovary and adrenal gland lanes, which contain 2 μg. Reprinted with permission from Bain, Yoo, Clarke, Hammond, and Payne (20).

and the ovary relative to the testis, liver, and kidney (Fig. 5.13). 3βHSD mRNA from the adrenal gland and the ovary was derived from 1/10 the amount of total RNA compared to the other tissues (2 μg vs 20 μg).

To further examine the expression of the different types of 3βHSD among murine tissues, a Northern blot of total RNA extracted from female and male liver, kidney, testis, and ovary was sequentially hybridized with, first, the SacI-BglII fragment of 3βHSD-I and, second, a BamHI-SacI fragment of 3βHSD-II. The two probes produced distinct patterns of hybridization in the different tissues (Fig. 5.14). The Leydig cell probe hybridized to a single 1.7-kb mRNA in all tissues from both sexes and also to a 1.9-kb mRNA in kidney (Fig. 5.14 top). Using the 3βHSD-I probe, the strongest signal was observed in total RNA from the ovary (2-μg ovarian RNA rather than 20 μg from the other tissues). The signal from testicular RNA is considerably less, but it must be taken into consideration that Leydig cells, which are the source of the 3βHSD cDNA in the testis, represent less than 5% of all the cells in the mouse testis (21). In the kidney lane, the 1.7-kb mRNA is more intense than the 1.9-kb mRNA in both male and female kidneys when hybridized with the Leydig cell cDNA probe. When 3βHSD-II is used as a probe, the pattern of hybridization is different (Fig. 5.14 bottom). The intensity of the signal from gonadal RNA is greatly reduced relative to the intensity of the 1.7-kb mRNA in the liver and kidney. Also, the intensity of the 1.9-kb mRNA in the kidney is increased relative to the 1.7-kb mRNA in all tissues. Furthermore, the 3βHSD-II probe recognizes a 1.9-kb mRNA in total liver RNA that was not detected with the 3βHSD-I probe.

To delineate tissue-specific expression of the 3 different types of 3βHSD, ribonuclease protection analysis was carried out with total RNA

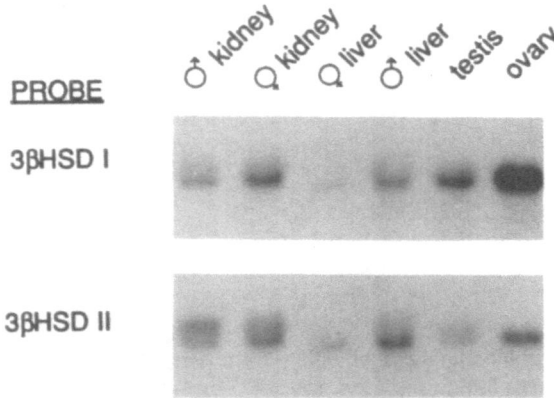

FIGURE 5.14. Expression of different forms and different amounts of 3βHSD mRNA in steroidogenic and nonsteroidogenic tissues. The Northern blot was first hybridized with probe 3βHSD-I and, after the removal of the first probe, hybridized with probe 3βHSD-II. Complete removal of probe 3βHSD-I was confirmed by autoradiography. Probe 3βHSD-I is a 906-bp *SacI-BglI* fragment of clone 3βHSD-I that is entirely included within the coding region of the cDNA. Probe 3βHSD-II is a 1069-bp *BamHI-SacI* fragment of the clone 3βHSD-II that includes a small portion of the pBluescript polylinker at the *BamHI* (5′) end. For each lane, 20-μg total RNA was applied to the gel except for the ovary and adrenal gland lanes, in which case, 2 μg was used.

isolated from gonads, liver, kidney, and adrenal glands from adult male and female mice. Antisense RNA probes from distinct regions of each 3βHSD form demonstrate that the 3βHSD-I specific probe protected a fragment of 104 nucleotides only in RNA from testes, ovaries, and adrenal glands (Fig. 5.15). No fragment of 104 nucleotides is protected by RNA from the liver or kidney of either sex (only 0.1-μg total RNA was used in the ovary and adrenal gland lanes compared to 10 μg from the other tissues). Figure 5.15 illustrates that the 3βHSD-II specific probe gives a 138-nucleotide fragment only with liver and kidney RNA from both sexes. The amount of 3βHSD-II RNA expressed in the kidney is always greater than in the liver. The 3βHSD-III specific probe generates a 188-nucleotide protected fragment also only with liver and kidney RNA, but no apparent difference between the amount of 3βHSD-III in liver and kidney was observed (Fig. 5.15). These data demonstrate that the different 3βHSD genes are expressed in a tissue-specific manner: 3βHSD-I is expressed only in steroidogenic tissues in both sexes, whereas 3βHSD-II and 3βHSD-III are found in liver and kidney of both sexes.

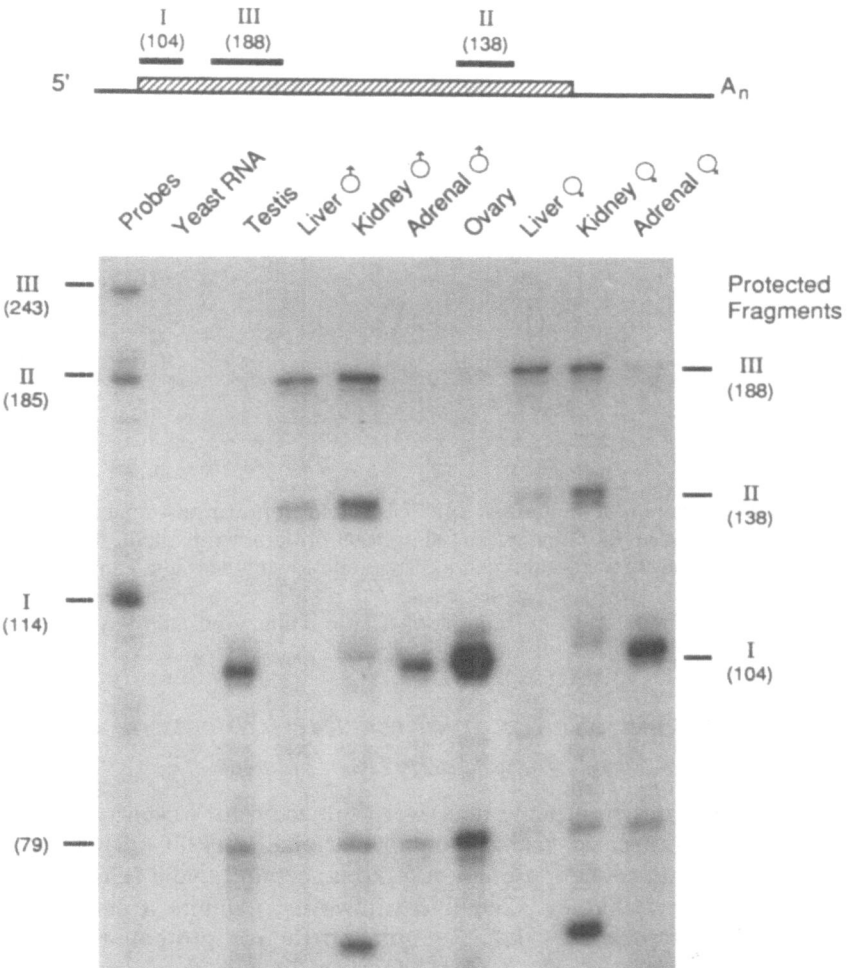

FIGURE 5.15. Ribonuclease protection analysis of RNA from mouse gonads, liver, kidney, and adrenal glands using probes specific for 3βHSD-I, -II, and -III. The diagram at the top of the figure is a generalized 3βHSD cDNA intended to show the positions and lengths of the 3 probes used in the analysis. The probes are indicated by black bars, and the number in parentheses is the length, in nucleotides, of the protected fragment produced by the probe. For the experiment depicted in this figure, all 3 probes were used simultaneously and were each added at a concentration of 1.0×10^5 cmp/30 μL hybridization reaction. Along the left of the figure, the positions of the full-length probes are indicated. The positions of the specific protected fragments are indicated along the right. The amount of RNA used in each lane was 10 μg except for the ovary and adrenal glands of both sexes, where 0.1 μg was used.

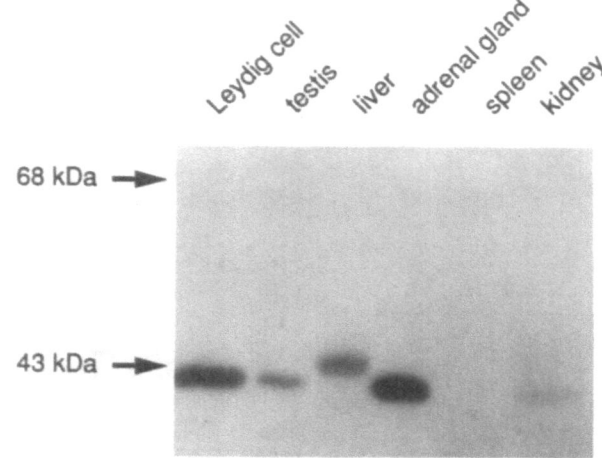

FIGURE 5.16. Immunoblot analysis of 3βHSD protein from various mouse tissues. For the testis, liver, and kidney lanes, 100-µg total protein was loaded on the gel; for spleen, 75 µg; and for adrenal gland, 30 µg. The Leydig cell lane contains the total protein from 5×10^4 Leydig cells. All tissues were from male mice. Reprinted with permission from Bain, Yoo, Clarke, Hammond, and Payne (20).

Expression of 3βHSD Immunoreactive Proteins in Steroidogenic and Nonsteroidogenic Tissues

Tissue homogenates from male mice were subjected to Western analysis using an antibody raised against human placental 3βHSD (22). Figure 5.16 illustrates that a single protein of 42 kd is seen in Leydig cells, testis, adrenal gland, and kidney. The liver, however, exhibits a single immunoreactive protein of 45 kd. No immunoreactive protein could be detected in spleen even after a 96-h exposure. In addition, exposure of the blot for 96 h did not detect an immunoreactive protein of 42 kd in liver, nor a 45-kd protein in any of the other tissues examined. These data demonstrate that the liver expresses a 3βHSD immunoreactive protein of a different size than the protein expressed in steroidogenic tissues. The data also suggest that either the 3βHSD immunoreactive protein in the kidney is distinct from the one expressed in the steroidogenic tissues and the liver or that there is decreased translation or increased degradation of the protein.

3βHSD Enzymatic Activity

3βHSD enzyme activity was determined in a variety of tissues from male mice. The activity was determined by measuring the conversion of two

Δ^5-3β-hydroxysteroids, pregnenolone and dehydroepiandrosterone, to the respective Δ^4-3-ketosteroids, progesterone and androstenedione, and by measuring the hydrogenation of the 5α-reduced steroid, dihydrotestosterone, to 5α-androstan-3β,17β-diol. Specific activity in the adrenal gland was more than 10-fold greater than that observed in the testis, which was 5-fold greater than the activity observed in the liver (Table 5.1). Specific activity in the kidney was considerably lower than the activity found in the liver. 3βHSD activity in the spleen was negligible.

When 3βHSD enzyme activity is expressed as total activity per organ, a very different pattern is observed. Liver exhibits by far the greatest amount of total 3βHSD enzyme activity for all 3 substrates (Table 5.2). Total 3βHSD activity was 5-fold greater in the liver compared to the testis and approximately 10-fold greater than what was observed in adrenal glands. Total enzyme activity in the kidneys was somewhat less than that observed in the adrenals. 3βHSD enzyme activity in the spleen was negligible. Comparison of 3βHSD enzyme activity among the 3 substrates indicates that activity with the 5α-reduced steroid as the substrate was 5-fold greater than with the Δ^5-3β-hydroxysteroids as the substrate. Measuring the dehydrogenation of 5α-androstan-3β,17β-diol gave similarly higher activity, as was observed with dihydrotestosterone as the

TABLE 5.1. Specific activity of 3βHSD in mouse tissues.

Tissue	Substrate pregnenolone	DHEA	DHT
Testis	0.502 ± 0.031	0.651 ± 0.026	3.940 ± 0.390*
Adrenal	8.390 ± 0.690	9.090 ± 0.510	44.900 ± 3.000
Liver	0.097 ± 0.015	0.121 ± 0.013	0.666 ± 0.070*
Kidney	0.016 ± 0.002*	0.030 ± 0.001	0.187 ± 0.009
Spleen	<0.002	0.007 ± 0.001	0.016 ± 0.001

Note: Enzyme activity is expressed as pmol min^{-1} µg protein^{-1}. Values are the mean of 3 mice ± SE. Asterisks (*) indicate the average of 2 mice ± the range.
Source: Data from Bain, Yoo, Clarke, Hammond, and Payne (20).

TABLE 5.2. Total activity of 3βHSD in mouse tissues.

Tissue	Substrate pregnenolone	DHEA	DHT
Testis	9.08 ± 0.29	11.90 ± 0.30	68.60 ± 3.20*
Adrenal	4.68 ± 0.68	5.31 ± 0.89	25.80 ± 4.10
Liver	44.80 ± 10.00	54.70 ± 9.10	301.00 ± 95.00*
Kidney	1.44 ± 0.30*	3.30 ± 0.13	18.10 ± 1.10
Spleen	<0.05	0.14 ± 0.02	0.36 ± 0.06

Note: Enzyme activity is expressed as nmol min^{-1} per total organ mass (paired testis and adrenal glands). Values are the mean of 3 mice ± SE. Asterisks (*) indicate the average of 2 mice ± the range.
Source: Data from Bain, Yoo, Clarke, Hammond, and Payne (20).

substrate (data not shown). From these studies in tissue homogenates, it cannot be determined whether the dehydrogenation and isomerization of the Δ^5-3β-hydroxysteroids and the reduction or dehydrogenation of the 5α-reduced steroids are catalyzed by the same enzyme. In preliminary studies, the coding region of 3βHSD-I or the coding region of 3βHSD-III was inserted into a pCMV5 expression vector and transfected into COS-1 cells. When enzyme activity was measured with each of the 3 substrates, proteins expressed from both the 3βHSD-I and 3βHSD-III vectors were able to catalyze the conversion of the Δ^5-3β-hydroxysteroids, as well as the 5α-reduced steroids (Payne, Bain, Clarke, Sha, unpublished data). These preliminary results demonstrate that the enzyme activities deter-mined in the tissue homogenates are catalyzed by the same protein. The high capacity of the liver for the conversion of Δ^5-3β-hydroxysteroids to Δ^4-3-ketosteroids suggests that the liver could play an important role in overall steroid hormone production.

Identification of Leydig Cell-Specific Sequences for the Expression of Mouse P450$_{scc}$

In a recent study on the analysis of the promoter region of the gene encoding mouse P450$_{scc}$, it was found that constructs containing 1500-bp 5' flanking region, when transfected into mouse Y-1 adrenocortical cells, exhibited high levels of expression of a growth hormone reporter gene (23). In contrast, the same construct showed very low expression in mouse MA-10 Leydig cells (23). To examine whether additional se-quences are necessary for the expression of the P450$_{scc}$ gene in Leydig cells, a ^{32}P-labeled probe containing +150 to −1300 of the mouse P450$_{scc}$ gene (a gift from Dr. Keith Parker) was used to screen an EMBL-3 mouse genomic library. A clone containing ~13,000-bp 5' flanking sequences plus 4000 bp of the structural gene was identified by restriction mapping and partial sequencing. Constructs containing different lengths of 5' flanking sequences were subcloned into vectors containing the CAT reporter gene. To determine if there are sequences further 5' than −1300 bp that are necessary for Leydig cell expression of the mouse P450$_{scc}$ gene, reporter constructs containing up to −8000-bp 5' flanking sequences were transfected into MA-10 Leydig tumor cells or Y-1 adrenocortical cells. Expression with the different constructs in each cell type was determined by measuring the CAT activity in the cell extracts. The results of this study are shown in Figure 5.17. Basal expression of P450$_{scc}$ in Y-1 adrenocortical cells was greatest in the construct containing −1200 bp. No increase in expression of P450$_{scc}$ was observed in Y-1 adrenocortical cells with constructs up to −5000-bp 5' flanking se-quences relative to a minimal promoter containing −198 bp. In Leydig cells, very little expression of the P450$_{scc}$ gene constructs was observed in

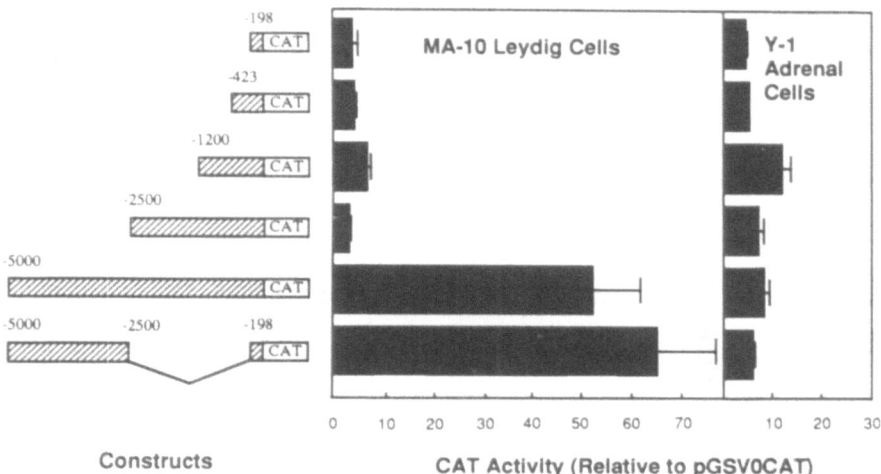

FIGURE 5.17. Identification of Leydig cell-specific sequences for the expression of mouse P450$_{scc}$. Different-size fragments 5' of the P450$_{scc}$ coding region were subcloned into the multiple cloning site of a promoterless plasmid (pGSVOCAT) as described in Fig. 5.10, except that pGSVOCAT does not contain the terminator fragment. The P450$_{scc}$ constructs are shown on the left-hand side of the figure. Ten μg of the indicated plasmid and 5 μg of SV2β-gal were transfected transiently into MA-10 Leydig tumor cells or Y-1 adrenocortical tumor cells. Cells were harvested 36 h following transfection, and CAT activity was determined in cell extracts by measuring the amount of [³H] acetylated chloramphenicol produced during 2 h for MA-10 cells or 1 h for Y-1 cells. Transfection efficiency was corrected for β-galactosidase activity. CAT activity is expressed relative to the promoterless pGSVOCAT plasmid.

plasmids containing up to −2500 bp 5' of the coding region. As shown in Figure 5.17, a marked increase in expression was observed in MA-10 Leydig cells with constructs that contained up to 5000 bp 5' of the coding region. To establish that sequences between −2500 and −5000 bp are essential for Leydig cell expression of P450$_{scc}$, a plasmid was constructed in which sequences between −198 bp and −2500 bp were deleted. The results demonstrate that sequences between 2.5 and 5.0 kb 5' of the coding region of the P450$_{scc}$ gene are essential for expression in Leydig cells, but not in adrenal cells (Fig. 5.17). Further deletion studies are under way to localize the Leydig cell-specific sequences more precisely.

Previous studies from this laboratory indicated that P450$_{scc}$ expression was regulated differently in Leydig cells and adrenal glands of the mouse (24). Using inbred strains of mice, we identified 2 strains, RF/J and SWR/J, whose Leydig cells exhibited a high amount of immunoreactive P450$_{scc}$ and 2 strains, DBA/2J and C3H/HeJ, whose Leydig cells exhibited a low amount of $P450_{scc}$. The pattern of expression of $P450_{scc}$ in adrenal glands from the same mice was different from that in Leydig cells. The amount of $P450_{scc}$ was highest in the adrenal glands of C3H/ HeJ and lowest in SWR/J (14). These findings suggested that tissue-specific factors in adrenal and/or Leydig cells influence the expression of this enzyme. In a subsequent study in which we investigated the relationship of the $P450_{scc}$ structural gene and the amount of $P450_{scc}$ protein, we demonstrated that quantitative differences in mouse Leydig cell $P450_{scc}$ protein are determined either by the structural gene for $P450_{scc}$ or by a closely linked locus (25). No evidence could be found that this locus determined quantitative differences in mouse adrenal $P450_{scc}$ (Nolan, Payne, unpublished data).

Summary

This chapter reviews the complexity of the regulation of expression of the enzymes involved in testosterone production in Leydig cells. The three major enzymes studied, $P450_{scc}$, $P450_{17\alpha}$, and 3βHSD, are differentially regulated. Maximal levels for mRNA of the three enzymes require cAMP. There is high constitutive expression of both $P450_{scc}$ and 3βHSD mRNA, while expression of $P450_{17\alpha}$ mRNA is absolutely dependent on chronic stimulation by cAMP. Endogenously produced testosterone acting via the androgen receptor negatively regulates expression of cAMP-induced $P450_{17\alpha}$, as well as 3βHSD mRNA; while glucocorticoids acting via the glucocorticoid receptor repress $P450_{scc}$ and 3βHSD mRNA levels. The mechanism by which cAMP increases expression of the three enzymes in normal mouse Leydig cells is different. Newly synthesized proteins are essential for cAMP induction of $P450_{17\alpha}$ and 3βHSD mRNA, but not for $P450_{scc}$ mRNA. In the absence of protein synthesis for a 24-h period, basal levels of 3βHSD mRNA are completely repressed, while basal levels of $P450_{scc}$ mRNA are reduced by 50%, indicating that rapidly turning over proteins are required for basal expression of 3βHSD and $P450_{scc}$ mRNA.

We present evidence for the presence of specific sequences in the 5' flanking region of the $P450_{scc}$ structural gene that are required for Leydig cell expression. These data indicate that there are tissue-specific factors that are necessary for expression of $P450_{scc}$ in Leydig cells and adrenal glands.

The presence of at least 3 distinct mouse cDNAs that encode 3 different isoforms of the enzyme 3βHSD is reviewed. The 3 forms share significant identity, but differ from each other by 5%–10%. 3βHSD-I was found to be specific for steroidogenic tissues and is only expressed in ovaries, testes, and adrenal glands of both sexes. 3βHSD-II and 3βHSD-III are only expressed in the liver and kidney of both sexes, but not in steroidogenic tissues. 3βHSD-II appears to be expressed preferentially in the kidney relative to the liver, whereas 3βHSD-III appears to be expressed to a similar extent in these two tissues. We report that the liver has a very high capacity for the conversion of the Δ^5-3β-hydroxysteroids, pregnenolone and dehydroepiandrosterone, to their respective Δ^4-3-ketosteroids, progesterone and androstenedione. This high capacity of the liver for the conversion of Δ^5-3β-hydroxysteroids to Δ^4-3-ketosteroids suggests that the liver could play an important role in overall steroid hormone production. Furthermore, expression of at least two 3βHSD gene products in the liver that are distinct from the gonadal and adrenal gene products may provide an explanation for the clinical manifestations of 3βHSD enzyme deficiency (26–29). A distinct liver protein that can convert Δ^5-3β-hydroxysteroids to Δ^4-3-ketosteroids, especially dehydroepiandrosterone that is produced in large amounts by the adrenal gland in patients exhibiting adrenal and gonadal 3βHSD deficiency, could account for the various degrees of virilization observed in these patients.

Acknowledgments. The authors gratefully acknowledge Dr. Maria Burgos-Trinidad for her input in preparing this manuscript. We also thank Rita Lemorie for typing the manuscript. These studies were supported by NIH Grants HD-08358 and HD-17916 to A.H. Payne. P.A. Bain and S.H. Hammond were supported by NIH Training Grant HD-07048.

References

1. Richards JS, Jahnsen T, Hedin L, et al. Ovarian follicular development: from physiology to molecular biology. Recent Prog Horm Res 1987;43:231–70.
2. Valladares LE, Payne AH. Induction of testicular aromatization by luteinizing hormone in mature rats. Endocrinology 1979;105:431–6.
3. Dorrington JH, Fritz IB, Armstrong DT. Control of testicular estrogen synthesis. Biol Reprod 1978;18:55–64.
4. Simpson ER, Merril JC, Hollub AJ, Graham-Lorence S, Mendelson CR. Regulation of estrogen biosynthesis by human adipose cells. Endocr Rev 1989;10:136–48.
5. Luu-The V, Labrie C, Simard J, et al. Structure of two in tandem human 17β-hydroxysteroid dehydrogenase genes. Mol Endocrinol 1990;4:268–75.
6. Youngblood GL, Nesbitt MN, Payne AH. The structural genes encoding P450$_{scc}$ and P450$_{arom}$ are closely linked on mouse chromosome 9. Endocrinology 1989;125:2784–6.
7. Youngblood GL, Sartorius C, Taylor BA, Payne AH. Isolation, characterization, and chromosomal mapping of mouse P450 17α-hydroxylase/C$_{17-20}$ lyase. Genomics 1991;10:270–5.
8. Berube V, Luu-The VL, Lachance Y, Gagne R, Labrie F. Assignment of the human 3β-hydroxysteroid gene (HSDB3) to the p13 band of chromosome 1. Cytogenet Cell Genet 1989;52:199–200.

9. Moseley WS, Seldin MF. Definition of mouse chromosome 1 and 3 gene linkage groups that are conserved on human chromosome 1: evidence that a conserved linkage group spans the centromere of human chromosome 1. Genomics 1989;5:899–905.

10. Kingsmore SF, Moseley WS, Watson ML, Sabina RL, Holmes EW, Seldin MF. Long-range restriction site mapping of a syntenic segment conserved between human chromosome 1 and mouse chromosome 3. Genomics 1990; 7:75–83.

11. Anakwe OO, Payne AH. Noncoordinate regulation of de novo synthesis of cytochrome P-450 cholesterol side-chain cleavage and cytochrome P-450 17α-hydroxylase/C_{17-20} lyase in mouse Leydig cell cultures: relation to steroid production. Mol Endocrinol 1987;1:595–603.

12. Hales DB, Payne AH. Glucocorticoid-mediated repression of $P450_{scc}$ mRNA and de novo synthesis in cultured Leydig cells. Endocrinology 1989;124: 2099–104.

13. Payne AH, Sha L. Multiple mechanisms for regulation of 3β-hydroxysteroid dehydrogenase/$\Delta^5 \rightarrow \Delta^4$-isomerase, 17α-hydroxylase/C_{17-20} lyase cytochrome P450, and cholesterol side-chain cleavage cytochrome P450 mRNA levels in primary cultures of mouse Leydig cells. Endocrinology 1991;129:1429–35.

14. Roesler WJ, Vandenbark GR, Hanson RW. Cyclic AMP and the induction of eukaryotic gene transcription. J Biol Chem 1988;263:9063–6.

15. Hales DB, Sha L, Paye AH. Glucocorticoid and cyclic adenosine 3′,5′-monophosphate-mediated induction of cholesterol side-chain cleavage cytochrome P450 ($P450_{scc}$) in MA-10 tumor Leydig cells. Increases in mRNA are cycloheximide sensitive. Endocrinology 1990;126:2800–8.

16. Rani CSS, Payne AH. Adenosine 3′,5′-monophosphate-mediated induction of 17α-hydroxylase and C_{17-20} lyase activities in cultured mouse Leydig cells is enhanced by inhibition of steroid biosynthesis. Endocrinology 1986;118: 1222–8.

17. Hales DB, Sha L, Payne AH. Testosterone inhibits cAMP-induced de novo synthesis of Leydig cell cytochrome P-$450_{17\alpha}$ by an androgen receptor-mediated mechanism. J Biol Chem 1987;262:11200–6.

18. Evain D, Morera AM, Saez JM. Glucocorticoid receptors in interstitial cells of the rat testis. J Steroid Biochem 1976;7:1135–9.

19. Griffin JE, Wilson JD. Disorders of the testes and male reproductive tract. In: Wilson JD, Foster DW, eds. Williams textbook of endocrinology, 7th ed. 1985:284.

20. Bain PA, Yoo M, Clarke T, Hammond SH, Payne AH. Multiple forms of mouse 3β-hydroxysteroid dehydrogenase/$\Delta^5 \rightarrow \Delta^4$-isomerase and differential expression in gonads, adrenal glands, liver and kidneys of both sexes. Proc Natl Acad Sci 1991:88.

21. Mori H, Shimizu D, Fukunishi R, Christensen AK. Morphometric analysis of testicular Leydig cells in normal adult mice. Anat Rec 1982;204:333–9.

22. Doody KM, Carr BR, Rainey WE, et al. 3β-hydroxysteroid dehydrogenase/isomerase in the fetal zone and neocortex of the human fetal adrenal gland. Endocrinology 1990;126:2487–92.

23. Rice DA, Kirkman MS, Aitken LD, Mouw AR, Schimmer BP, Parker KL. Analysis of the promoter region of the gene encoding mouse cholesterol side-chain cleavage enzyme. J Biol Chem 1990;265:11713–20.

24. Perkins LM, Payne AH. Quantification of $P450_{scc}$, $P450_{17\alpha}$ and iron sulfur protein reductase in Leydig cells and adrenals of inbred strains of mice. Endocrinology 1988;123:2675–82.
25. Nolan CJ, Payne AH. Genotype at the $P450_{scc}$ locus determines differences in the amount of $P450_{scc}$ protein and maximal testosterone production in mouse Leydig cells. Mol Endocrinol 1990;4:1459–64.
26. Bongiovanni AM, Kellenbenz G. The adrenogenital syndrome with deficiency of 3β-hydroxysteroid dehydrogenase. J Clin Invest 1962;41:2086–92.
27. Parks GA, Bermudez JA, Anast CS, Bongiovanni AM, New MI. Pubertal boy with the 3β-hydroxysteroid dehydrogenase defect. J Clin Endocrinol 1971;33:269–78.
28. Rosenfield RL, Rich BH, Wolfsdorf JI, et al. Pubertal presentation of congenital Δ^5-3β-hydroxysteroid dehydrogenase deficiency. J Clin Endocrinol Metab 1980;51:345–53.
29. Pang S, Levine LS, Stoner E, et al. Nonsalt-losing congenital adrenal hyperplasia due to 3β-hydroxysteroid dehydrogenase deficiency with normal glomerulosa function. J Clin Endocrinol Metab 1983;56:808–18.

6

Regulation of LH and FSH Receptor Gene Expression in the Ovary and Testis

KIMMO K. VIHKO, PHILIP S. LAPOLT, XIAO-CHI JIA, AND
AARON J.W. HSUEH

The action of gonadotropins is mediated through binding of these glyco-protein hormones to specific receptors on gonadal cells. In the testis, *follicle stimulating hormone receptors* (FSH-R) are located in Sertoli cells, while *luteinizing hormone receptors* (LH-R) are found exclusively in Leydig cells (reviewed in 1). In the ovary, granulosa cells contain the receptor for FSH, while granulosa, theca, and luteal cells contain LH-R (reviewed in 2). The binding of the respective ligand to the specific cell surface receptor initiates the transduction of the signal from the extra-cellular space into the cells, with resultant activation of specific genes (reviewed in 3).

Recent cloning of both LH-R and FSH-R (4–6) has made it possible to study the regulation of the mRNA coding for these proteins in both the ovary and the testis. Determination of the molecular structure of gonadotropin receptors revealed that they belong to the 7-transmem-brane, G-protein-coupled family of receptors. Based on the published cDNA sequences, we have amplified specific cDNA fragments of LH-R and FSH-R by *reverse transcriptase-polymerase chain reaction* (RT-PCR) and generated cRNA probes for studies of LH-R and FSH-R mRNA regulation. In addition, expression of these binding proteins in cells transfected with receptor cDNAs allows us to study the structure-function relationships of these proteins.

Analysis of gonadal mRNA has revealed multiple transcripts for these receptors both in the ovary and testis. In many instances, the regulation of LH and FSH mRNA transcripts correlates well with the respective ligand binding capacity. However, several studies also indicate differential regulation of individual mRNA species by gonadotropins and during postnatal development. In addition, some transcripts are smaller in size than the complete receptor cDNA, suggesting the existence of truncated forms of these receptors. In this chapter, recent progress in understanding the LH-R and FSH-R gene expression in the ovary and testis is reviewed.

Molecular Cloning of LH-R and FSH-R

The molecular cloning of rat (4) and porcine (5) LH-R was based on the partial amino acid sequence of the receptor protein and on the use of LH-R-specific antibodies, respectively. The rat Sertoli cell FSH-R was subsequently cloned based on the assumed homology of this receptor with LH-R (6). Hydropathic analyses revealed that all gonadotropin receptors belong to the G-protein-coupled, 7-transmembrane family of receptors, sometimes referred to as the "magnificent seven." As do other members of this family of receptors, the gonadotropin receptor molecules contain 7 hydrophobic transmembrane sequences, together with extracellular and intracellular domains (reviewed in 7, 8). In contrast to other members of this gene family, the gonadotropin receptors have a large extracellular ligand binding domain (4–6). This domain contains repetitive sequences showing significant homology to the leucine-rich glycoprotein family of proteins (9). The extracellular domain and the intracellular domain of LH-R and FSH-R contain several potential glycosylation sites, the physiological significance of which is presently unclear. In addition, the intracellular domain contains potential phosphorylation sites.

Based on the published rat and porcine LH-R cDNA sequences, it has been possible to clone the cDNA for the human LH-R using RT-PCR and cDNA library screening (10). Sequence analysis revealed that at the nucleotide level, the human receptor shows 89% and 82% homology with its porcine and rat counterparts, respectively. Comparison with the recently described human thyroid LH-R (11) revealed, however, significant differences. The cDNA cloning was followed with successful expression of the LH-R protein in a human fetal kidney cell line 293. The hormone specificity of this receptor was demonstrated by the inability of purified human FSH or recombinant human FSH to displace the $[^{125}I]$-hCG binding to human LH-R expressed by 293 cells, while recombinant human LH and hCG showed dose-dependent displacement (Fig. 6.1). Interestingly, the expressed receptor showed high species specificity since equine, rat, and ovine LH, as well as equine CG, did not compete with $[^{125}I]$-hCG for binding to the receptor (Fig. 6.1). The high species specificity of the human LH-R is distinct from the wide species specificity of the rat LH-R.

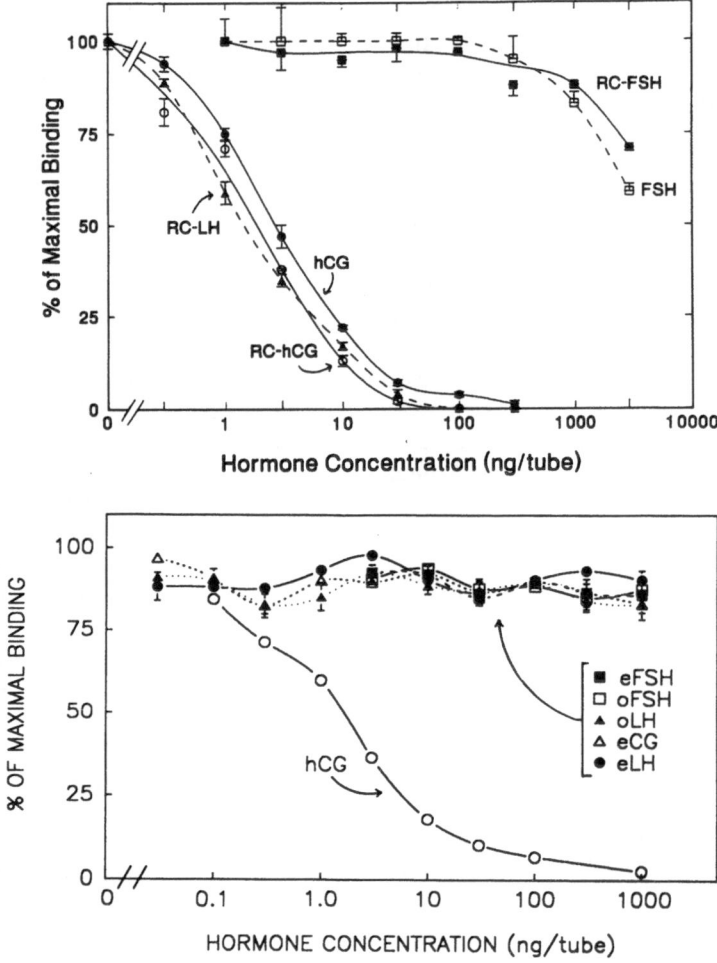

FIGURE 6.1. Interaction of gonadotropins with recombinant human LH-R expressed by transfected 293 cells. *Top:* Displacement of [125I]-hCG binding to LH-R by hCG, recombinant human LH (RC-LH), and human purified (FSH) and recombinant FSH (RC-FSH) is shown. *Bottom:* Displacement of [125I]-hCG binding to human LH-R expressed by transfected 293 cells by hCG, equine LH (eLH), rat LH (rLH), ovine LH (oLH), equine CG (eCG or PMSG), or FSH from different species is shown. Displacement is indicated as percentage of maximal binding at each dose of unlabeled hormone. Reprinted with permission from Jia, Oikawa, Bo, et al. (10), © by The Endocrine Society, 1991.

Application of RT-PCR for Amplifying Receptor cDNA Fragments

Based on the published rat gonadotropin receptor cDNA sequences, it has been possible to generate LH-R and FSH-R cDNA fragments using RT-PCR. Adult rat testis RNA was first reverse-transcribed to cDNA that in turn served as a template for oligonucleotide primers specific for the extracellular portion of the LH-R. Following PCR, the amplified cDNA product was sequenced and shown to be 100% homologous to the corresponding published LH-R cDNA. These analyses were followed by subcloning of the cDNA fragment to pGEM4Z plasmid. This construct has been used for cRNA probe production by using T7 RNA polymerase (12). A similar approach was used to generate a cDNA probe corresponding to the transmembrane region of the LH-R (13). Likewise, it was also possible to subclone a cDNA fragment corresponding to the extracellular portion of the FSH-R using RT-PCR (14). By using these molecular probes, it has been possible to study gonadotropin receptor mRNA regulation in the rat testis and ovary under different experimental conditions.

Regulation of Ovarian LH-R mRNA

By using the specific LH-R cRNA probe corresponding to the extracellular domain, it has been possible to demonstrate 4 mRNA transcripts with molecular size of 1.8, 2.5, 4.2, and 7.0 kb in the rat ovary (12) (Fig. 6.2). When immature female rats were injected with *pregnant mare serum gonadotropin* (PMSG) to induce follicular development, an ≈10-fold increase in LH-R mRNA level in the ovary was observed. When these animals received an ovulatory dose of *human chorionic gonadotropin* (hCG), a rapid decrease in the LH-R mRNA levels was found 24 h after the injection. These changes in the LH-R mRNA levels were closely correlated with changes in the respective receptor number. Furthermore, an increase in LH-R number 3 days after hCG treatment was associated with increases in LH-R transcripts.

In cultured granulosa cells, treatment with FSH increased the levels of LH-R mRNAs in a time- and dose-dependent manner (15). The increased level of LH-R mRNA was maintained by FSH, LH, and prolactin. In contrast, *basic fibroblast growth factor* (bFGF) or *epidermal growth factor* (EGF) suppressed the FSH induction of LH-R mRNA. Likewise, GnRH suppressed the FSH-induced LH-R mRNA production in a dose-dependent manner. The findings indicate that changes of granulosa cell LH-R content closely correlate with changes of the LH-R mRNA level.

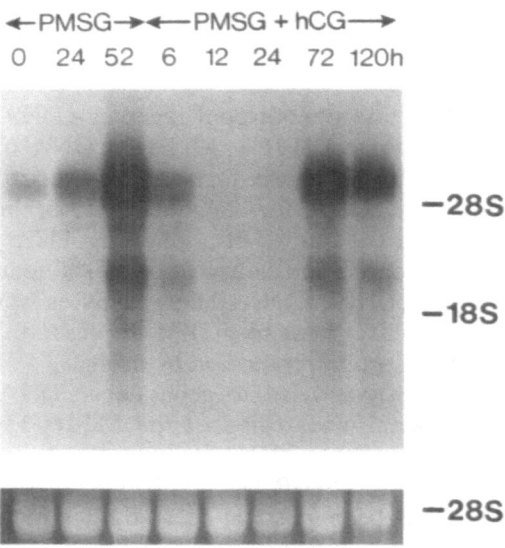

←PMSG→←—PMSG + hCG—→
0   24 52   6   12 24 72 120h

−28S

−18S

−28S

FIGURE 6.2. Northern blot analyses of ovarian LH-R mRNA at different time points following treatment with PMSG or PMSG + hCG. Samples were analyzed on a denaturing agarose gel, transferred to a nitrocellulose membrane, and hybridized with the extracellular LH-R cRNA probe. The positions of the 28S and 18S ribosomal RNAs are indicated. The lower panel shows the ethidium bromide staining pattern of the 28S rRNA. Reprinted with permission from LaPolt, Oikawa, Jia, Dargan, and Hsueh (12), © by The Endocrine Society, 1990.

Regulation of Testicular LH-R mRNA

When the extracellular LH-R cRNA probe was utilized, Northern blot analyses of rat testis RNA revealed 4 mRNA transcripts for LH-R with molecular size of 1.8, 2.5, 4.2, and 7.0 kb. When the transmembrane probe was used in Northern analyses, only the 7.0 and 2.5 kb mRNA transcripts were observed in the same samples (13) (Fig. 6.3).

In the testis, injection with hCG causes down-regulation of LH-R in both the adult and neonate rats. However, the LH-R are recovered more rapidly in the neonate than in the adult rat testis (16). In the adult rat testis, the decrease in receptor number correlates well with decreased levels of LH-R mRNA, while, in the neonate, increased LH-R mRNA levels are observed following injection with hCG (17). In Northern blot analyses of testicular RNA extracted from adult rat testis at various time points after injection with hCG, differential regulation of individual LH-R transcripts was observed (13). The level of the 7.0-, 4.2-, and 2.5-kb mRNA transcripts decreased following hCG injection, while the 1.8-kb transcript was maintained (Fig. 6.4). This transcript is of special interest

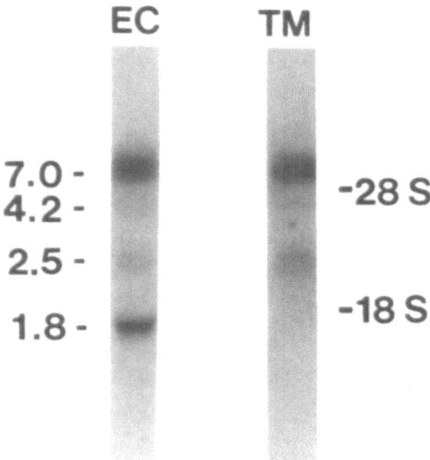

FIGURE 6.3. Northern blot analyses of testis LH-R mRNA using the cRNA probes corresponding to the extracellular (EC) or transmembrane (TM) domains. Extracted testicular RNA was fractionated on denaturing agarose gels, transferred to nitrocellulose, and hybridized with the cRNA probes. The migration position of the 28S and 18S ribosomal RNA is indicated on the right. Reprinted with permission from Lapolt, Jia, Sincich, and Hsueh (13), © by The Endocrine Society, 1991.

because it possibly represents a truncated form of the LH-R based on its smaller molecular size when compared to the full-length cDNA (4). In addition, it is not detectable in Northern analyses when the transmembrane cRNA probe is used (13), suggesting that it does not contain part of the transmembrane region and may code for a soluble form of the LH-R. Experiments are presently performed to characterize the molecular nature of this mRNA transcript.

During postnatal development, the testicular LH-R mRNA transcripts show an interesting age-dependent pattern (18). In the 5-day-old rat, only the 1.8- and 4.2-kb LH-R mRNA transcripts were observed. Additional 7.0- and 1.2-kb transcripts appeared at 10 and 15 days, respectively. From the age of 25 days through adulthood, 2 additional LH-R transcripts (molecular size 2.5 and 7.8 kb) were observed.

Regulation of FSH-R mRNA Levels

Treatment of immature rats with PMSG to induce follicular development increased the FSH-R mRNA 24 h after treatment, with a further increase at 52 h, concomitant with increases in FSH-R content (14). Subsequent

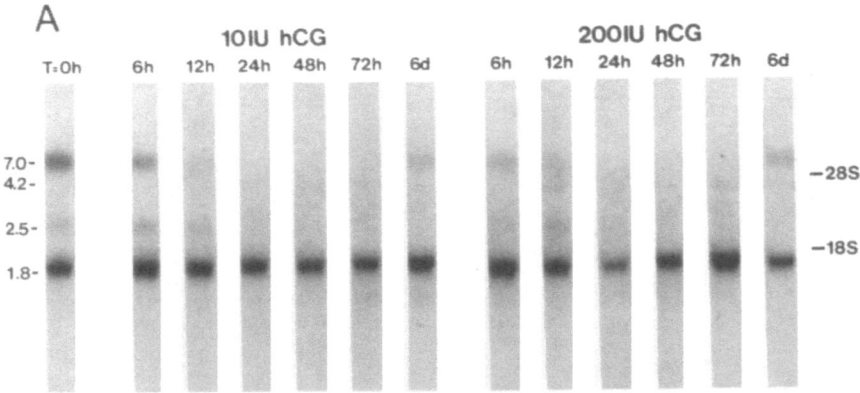

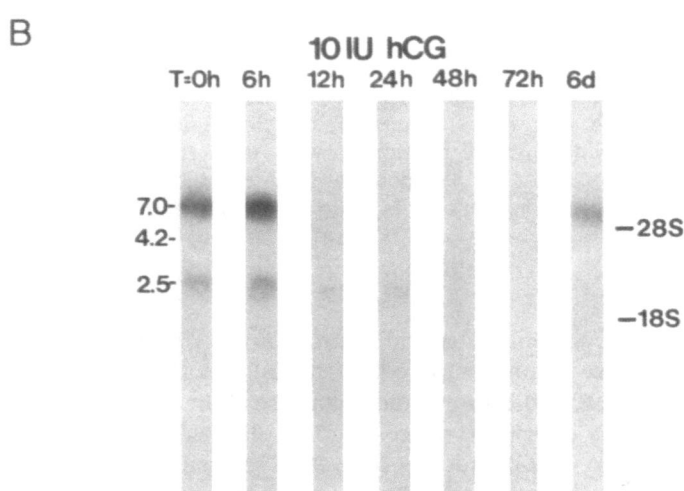

FIGURE 6.4. Northern blot analyses of testicular LH-R mRNA transcripts at various time points following injection with 10-IU or 200-IU hCG. Total RNA of individual testes was extracted and fractionated on denaturing agarose gels, transferred to nitrocellulose membranes, and hybridized with the cRNA probe corresponding to the extracellular (*A*) or transmembrane (*B*) domain of the LH-R. Reprinted with permission from Lapolt, Jia, Sincich, and Hsueh (13), © by The Endocrine Society, 1991.

treatment with an ovulatory dose of hCG decreased [^{125}I]-FSH binding and FSH-R mRNAs by 6h. In luteinized ovaries obtained 3 and 5 days after ovulation induction, FSH-R mRNA increased again, but no concomitant increase in FSH-R content was observed (14).

In the testis, Northern blot analyses of fractionated RNA revealed 4 mRNA transcripts with molecular weight of 1.7, 2.3, 4.0, and 7.5 kb, respectively (18). Analyses of RNA from testes of rats at different ages revealed that all these transcripts were present in testes obtained from rats from the age of 10 days onwards. High levels of mRNAs were observed in the adult (70 days) and prepuberal (40 days) rats. In a recent study (19), high levels of FSH-R mRNA were observed in testes obtained from animals at ages of 10, 20, and 30 days, while adult rat testis showed low levels of FSH-R mRNA. Neither of the studies on FSH-R mRNA levels during postnatal development (18, 19) shows close correlation to FSH-R number during testis development, suggesting the existence of posttranslational regulation of FSH-R gene expression during onto-genesis.

Summary

Based on the published rat LH-R and FSH-R cDNA sequences, it has been possible to generate cRNA probes for studies of gonadal LH-R and FSH-R mRNA levels. In addition, based on the published sequence of rat and porcine receptors, it has been possible to obtain the human LH-R cDNA sequence, together with successful expression of the human LH-R protein in vitro for analyses of its structure-function relationships. Northern blot analyses of LH-R and FSH-R mRNAs in the ovary and the testis revealed multiple mRNA transcripts with diverse molecular sizes. Northern blot analyses of LH-R mRNAs obtained from testes of rats at different ages revealed interesting developmental regulation of specific transcripts. In adult rats, treatment with hCG results in a significant decrease in testicular LH-R mRNA levels. However, a 1.8-kb mRNA transcript, the molecular nature of which is under investigation, was not down-regulated with hCG. In immature female rats, where follicular development was induced by PMSG treatment followed by treatment with a dose of hCG to induce ovulation and luteinization, both the increases and decreases in LH-R and FSH-R mRNA levels correlated well with the observed changes in respective receptor number. In cultured granulosa cells, FSH was able to increase LH-R mRNA levels. This stimulatory effect could be inhibited by bFGF, EGF, and GnRH, suggesting the existence of paracrine-type regulation. Further molecular characterization of the LH-R and FSH-R mRNA transcripts will allow elucidation of the structure and function of individual receptor transcripts. Analysis of the receptor genes will provide understanding about the mechanisms of regulation of LH-R and FSH-R gene expression by hormones and during gonadal development.

References

1. Parvinen M, Vihko KK, Toppari J. Cell interactions during the seminiferous epithelial cycle. Int Rev Cytol 1986;104:115–51.
2. Hsueh AJW, Adashi EY, Jones PBC, Welsh TH. Hormonal regulation of the differentiation of cultured ovarian granulosa cells. Endocr Rev 1984;5:76–127.
3. Richards JS, Hedin L. Molecular aspects of hormone action in ovarian follicular development, ovulation, and luteinization. Annu Rev Physiol 1988;50:441–63.
4. McFarland KC, Sprengel R, Phillips HS, et al. Lutropin-choriogonadotropin receptor: an unusual member of the G protein-coupled receptor family. Science 1989;245:494–9.
5. Loosfelt H, Misrahi M, Atger M, et al. Cloning and sequencing of porcine LH-hCG receptor cDNA: variants lacking transmembrane domain. Science 1989;245:525-8.
6. Sprengel R, Braun T, Nikolics K, Segaloff DL, Seeburg PH. The testicular receptor for follicle stimulating hormone: structure and functional expression of cloned cDNA. Mol Endocrinol 1990;4:525–30.
7. Metsikko MK, Petaja-Repo UE, Lakkakorpi JT, Rajaniemi HJ. Structural features of the LH/CG receptor. Acta Endocrinol (Copenh) 1990;122:545–52.
8. Hsueh AJW, LaPolt PS. Molecular basis of gonadotropin receptor regulation. Trends Endocr Metab 1991 (submitted).
9. Takahashi N, Takahashi Y, Putnam FW. Periodicity of leucine and tandem repetition of a 24-amino acid segment in the primary structure of leucine-rich α_2-glycoprotein in human serum. Proc Natl Acad Sci USA 1985;82:1906–10.
10. Jia X-CJ, Oikawa M, Bo M, et al. Expression of human luteinizing hormone (LH) receptor: interaction with LH and chorionic gonadotropin from human but not equine, rat, and ovine species. Mol Endocrinol 1991;5:759–68.
11. Frazier AL, Robbins LS, Stork PJ, Sprengel R, Segaloff DL, Cone RD. Isolation of TSH and LH/CG receptor cDNAs from human thyroid: regulation by tissue-specific splicing. Mol Endocrinol 1990;4:1264–76.
12. LaPolt PS, Oikawa M, Jia X-C, Dargan C, Hsueh AJW. Gonadotropin-induced up- and down-regulation of rat ovarian LH receptor message during follicular growth, ovulation and luteinization. Endocrinology 1990;126:3277–9.
13. LaPolt PS, Jia X-C, Sincich C, Hsueh AJW. Ligand-induced down-regulation of testicular and ovarian luteinizing hormone (LH) receptors is preceded by tissue-specific inhibition of alternatively processed LH receptor trarnscripts. Mol Endocrinol 1991;5:397–403.
14. LaPolt PS, Sincich C, Aihara T, Nishimori K, Hsueh AJW. Up- and down-regulation of ovarian FSH receptor gene expression by PMSG, hCG, and recombinant FSH [Abstract 430]. Proc 38th annu meet Soc Gynecologic Investigation, March 20–23, 1991, San Antonio, TX.
15. Piquette GN, LaPolt PS, Oikawa M, Hsueh AJW. Regulation of luteinizing hormone receptor messenger RNA levels by gonadotropins, growth factors and gonadotropin-releasing hormone in cultured rat granulosa cells. Endocrinology 1991:2449–56.

16. Huhtaniemi IT, Nozu K, Warren DW, Dufau ML, Catt KJ. Acquisition of regulatory mechanisms for gonadotropin receptors and steroidogenesis in the maturing rat testis. Endocrinology 1982;111:1711–20.
17. Pakarinen P, Vihko KK, Voutilainen R, Huhtaniemi I. Differential response of luteinizing hormone receptor and steroidogenic enzyme gene expression to human chorionic gonadotropin stimulation in the neonatal and adult rat testis. Endocrinology 1990;127:2469–74.
18. Vihko KK, Nishimori K, LaPolt PS. LH and FSH receptor mRNA in the rat testis: developmental regulation of multiple transcripts during postnatal life [Abstract 476]. Proc annu meet Soc Study of Reproduction, July 25–31, 1991, Vancouver, BC, Canada.
19. Heckert LL, Griswold MD. Expression of follicle-stimulating hormone receptor mRNA in rat testes and Sertoli cells. Mol Endocrinol 1991;5:670–7.

7

Hormonal Control of Gene Expression During Ovarian Cell Differentiation

JoAnne S. Richards, Jeffrey W. Clemens, Jean Sirois, Susan L. Fitzpatrick, Winona L. Wong, and Richard C. Kurten

Growth of ovarian follicles, ovulation, and luteinization involve sequential changes in the response of granulosa cells and theca cells to the gonadotropins, FSH and LH. Some of the genes that are differentially regulated by FSH/LH/cAMP as follicles develop are RIIβ, the regulatory subunit of A-kinase type II, *aromatase cytochrome P450* (P450$_{arom}$), *cholesterol side-chain cleavage cytochrome P450* (P450$_{scc}$), *LH receptor* (LH-R), and *prostaglandin endoperoxide synthase* (PGS). Whereas RIIβ (1), P450$_{arom}$ (2) and LH-R (3) are induced in growing follicles by low concentrations of hormone, P450$_{scc}$ is maximally induced by high concentrations of cAMP and under conditions in which the granulosa cells are stimulated to luteinize (4). In contrast, one isoform of PGS is rapidly but transiently induced by elevated gonadotropins/cAMP preceding ovulation (Fig. 7.1). To analyze the different molecular mechanisms by which FSH/LH/cAMP may be regulating some of these genes, we have isolated genomic 5' flanking DNA of the RIIβ (5), P450$_{scc}$ (4), and P450$_{arom}$ (6) genes, determined the nucleotide sequences and the transcriptional start sites of each putative promoter-enhancer region, and are analyzing the ability of these promoters to confer cAMP regulation to reporter genes in transfected granulosa cells (4–8). Sequences similar to *cAMP regulatory elements* (CREs) (Fig. 7.1 solid circles) but not consensus CREs have been identified in *cis*-acting DNA of the RIIβ, P450$_{scc}$, and P450$_{arom}$ genes (4–6). However, promoter regions of each gene do confer cAMP regulation on the reporter CAT gene and exhibit specific binding activities in electrophoretic mobility shift assays using granulosa cell nuclear extract proteins (4–8). The *trans*-acting factors binding to these promoter elements remain to be determined. Based on the lack of sequence homology among these promoters, the *trans*-acting factors are likely to be different (Fig. 7.1). Lastly, ion exchange chromatography,

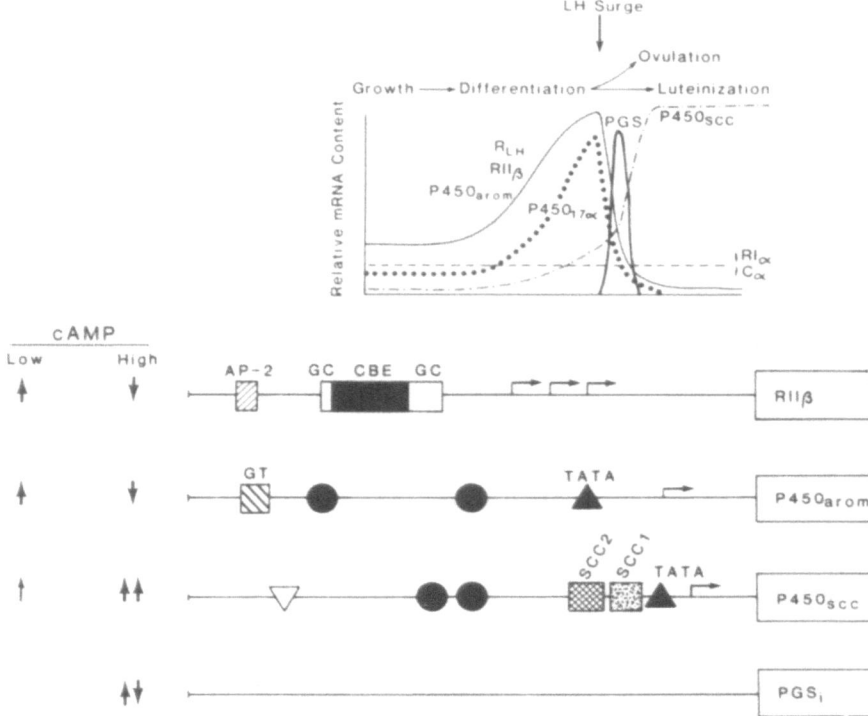

FIGURE 7.1

immunoblot analyses, and amino acid sequence data indicate that the rat ovary contains 2 distinct forms of PGS, each of which is selectively regulated by hormones, localized to specific cell types, and differentially sensitive to inhibitors of transcription/translation (9, 10). Thus, the form of PGS induced by FSH/LH in granulosa cells of preovulatory follicles appears to be a novel product of a distinct gene, different in structure and expression from that originally cloned (12).

RIIβ, Type IIβ Regulatory Subunit of A-Kinase

RIIβ, the regulatory subunit of type IIβ cAMP-dependent *protein kinase A* (PKA) is one of a family of mammalian R-subunits, encoded by a single gene and selectively expressed in neuroendocrine, adipose, erythroleukemic, and gonadal cells (1, 5). At defined stages of ovarian cell differentiation, RIIβ is a major cellular protein that exists as 3 isoelectric (phospho) variants, each of which can be further phosphorylated by the *catalytic* (C) subunit of PKA (1, 5). At stages of granulosa cell function when RIIβ is expressed in abundance, it is not only the major R-subunit found in the endogenous type II holoenzyme ($RII\beta_2C\alpha_2$), but is also present as a free R-subunit (11, 12). $RII\beta_2C\alpha_2$ tetramers are also the predominant form of the A-kinase holoenzyme in another ovarian tissue, the corpus luteum (12), which expresses low levels of RIIβ mRNA and protein (1), but contains both RIα mRNA (1) and high levels of free RIα protein (11, 12). Thus, in the ovary RIIβ appears to be the dominant holoenzyme phenotype, even in the presence of an abundance of RI. This situation is similar to that produced experimentally in studies conducted by McKnight and colleagues (13). The fact that the ovary is highly dependent on gonadotropins for maintaining its myriad functions, including growth, differentiation, and steroidogenesis, and that the gonadotropins mediate their effects primarily by elevating cAMP levels suggests that there may be specific functional relevance to the hormonally related expression and regulation of the RIIβ gene in this tissue (1). The preferential expression of RIIβ in rat gonadal cells becomes even more intriguing because this R-subunit is one of several genes regulated by FSH in granulosa cells of rat ovarian follicles (1–10). In vitro and in vivo studies have clearly shown that the effects of FSH on RIIβ expression in these gonadal cells can be mimicked by cAMP and are mediated, at least in part, by increased transcription of the RIIβ gene (1, 5).

Comparative observations of FSH/LH regulated genes in the rat ovary (Fig. 7.1) indicate that some genes, of which RIIβ is an example, are induced in granulosa cells by low concentrations of hormone/cAMP and are subsequently repressed by elevated amounts of hormone (LH) and high intracellular levels of cAMP. Another group of genes, of which $P450_{scc}$ is an example, is induced by the LH surge (1, 4), high levels of cAMP, or longer exposure to increased cAMP levels (1, 4). The ability of cAMP to dictate these differential responses in granulosa cells has led us to propose that specific transcription factors and *cis*-acting DNA elements are likely to be involved.

We have previously demonstrated cAMP-inducible expression of a reporter gene fused to a CRE-containing promoter in granulosa cells (14), suggesting that CREB or a CREB-like protein is present in granulosa cells. Pei et al. (15) have recently shown that the α-inhibin gene contains a CRE element to which CREB can bind and that confers cAMP inducibility in granulosa cells. However, other ovarian genes analyzed to date do not appear to contain consensus CREs, despite the fact that they are hormone- and cAMP-regulated (1–7). Similarly, although FSH has been shown to increase c-*fos* mRNA in granulosa cells (16), functional AP-1 sites have not yet been identified in the FSH-induced genes mentioned above. In other cells, AP-2 has also been shown to mediate cAMP-induced transcription (17), either alone or in concert with CRE/AP-1 elements. Because PKA plays a pivotal role in mediating the gonadotropin-induced responses of ovarian cells to cAMP and because RIIβ is itself regulated in a highly specific manner by gonadotropins/ cAMP, we have sought to determine which, if any, of the known transcriptional factors might be implicated in regulating the RIIβ gene. Our results (5) indicate that the RIIβ gene belongs to a class of genes with promoters that are rich in *granulosa cells* (GC), lack TATA boxes, and initiate transcription at multiple sites (Fig. 7.1). The RIIβ gene also appears to belong to a subclass of these TATA-less genes that are expressed selectively in specific tissues and that can be regulated by hormones/cAMP. Band-shifting and DNase 1 footprinting experiments using 400 bp of the RIIβ promoter detected several related, specific, DNA protein complexes formed using crude and fractionated nuclear extracts from rat ovary, brain, adrenal gland, and liver. All binding in these experiments mapped to a domain (*complex binding element* [CBE]) (Fig. 7.1) within the same region found to confer cAMP inducibility to a CAT reporter gene when transfected into primary cultures of rat granulosa cells. Although putative GC boxes (SP1 binding sites) and AP-2 elements were present in this functional region, and although expression vectors containing AP-2 sites conferred cAMP regulation of the CAT gene in cultured ovarian cells, neither the GC boxes nor the AP-2 sites were protected by footprint analyses or required for band-shift activity of nuclear extract protein. However, purified AP-2 does bind the RIIβ promoter sequence containing the AP-2 consensus site. These known regulatory elements, therefore, may be involved in functional activity of the RIIβ promoter, but additional *cis*-acting DNA and *trans*-acting factors (yet to be characterized) also appear to interact with the functional promoter of the RIIβ gene and regulate the hormone-specific expression of the A-kinase subunit in ovarian and neuronal cells.

Cholesterol Side-Chain Cleavage Cytochrome P450 (P450$_{scc}$)

The entire rat P450$_{scc}$ gene has been cloned, positions/sequences of the exon-intron boundaries (I–IX) have been described, and 940 bp of 5' flanking DNA have been sequenced, compared to mouse, bovine, and human genes, and analyzed by functional assays (4). Primer extension analysis mapped the transcription start site 42 bp upstream of the initiator methionine codon (4). To determine the *cis*-acting DNA sequences involved in cAMP induction of the P450$_{scc}$ gene, −898 bp of rat P450$_{scc}$, 5' flanking DNA were fused to the bacterial CAT reporter gene and 5' deletions (−379 CAT, −101 CAT, and −73 CAT) were prepared. These constructs, when transfected into primary rat granulosa cell cultures, exhibited *forskolin* (FSK) (7.5 μM) inducibility of CAT activity (7). The highest basal and FSK-stimulated activities were observed in the −379

CAT and −101 CAT constructs, suggesting that negative regulatory elements may reside in the −898 to −379 bp region and that deletion to −73 bp may remove key regulatory elements of the promoter. These activities reflect activation of the endogenous gene since FSK also induced secretion of progesterone into the media. Promoter regions of human, bovine, murine, and rat P450$_{scc}$ genes contain 2 highly conserved regions (4) that in the rat are designated SCC1(−58/−38) and SCC2(−83/−64). Either 1 or 2 copies of an SCC1 oligonucleotide enhanced FSK stimulation of CAT activity when placed upstream of a thymidine kinase CAT reporter gene (7). Nuclear extracts prepared from hormonally primed rat granulosa cells demonstrated sequence-specific binding to −101/+1-, SCC1-, and SCC2-labeled DNA probes using electrophoretic mobility shift assays. Specific DNA/protein complexes were not competed by the presence of unlabeled oligonucleotide consensus sequences for CRE, COUP, or the RIIβ CBE. Furthermore, CREB does not bind SCC1 or SCC2. These observations suggest that some of the *cis*-acting DNA elements involved in transcriptional activation of the P450$_{scc}$ gene by cAMP reside in the −101/+1 promoter region and do not involve consensus CRE regulatory domains or CREB transactivation. The nature and role of the proteins binding to specific sequences within these regions of the P450$_{scc}$ gene remain to be determined.

Aromatase Cytochrome P450 (P450$_{arom}$)

Induction of aromatase by FSH was influenced by the age of the rats from which the cells were obtained, the addition of specific steroids, and the presence of serum in the culture media. Under a variety of culture conditions tested, changes in aromatase mRNA were consistently associated with parallel changes in aromatase activity, suggesting that activity reflects enzyme levels (2). These results contrast with those observed previously in the pregnant rat corpus luteum in which elevated levels of mRNA and protein were associated with low enzyme activity (18).

Results from in vivo and in vitro studies also document that *estrogen* (E) enhances the ability of FSH to increase aromatase mRNA and activity in rat granulosa cells, but has no effect alone. The time required for FSH to induce aromatase mRNA is relatively long (24–48 h) compared to the initial increase in cAMP (1) and induction of other proteins, such as c-*fos* (16). Although the rat aromatase gene contains CRE-like elements in its 5′ flanking region (Fig. 7.1 solid circles), there is no consensus CRE sequence in the 500-bp promoter analyzed thus far (6), and none of the CRE-like sequences exhibit intense binding of CREB (8). Thus, both the time course of aromatase mRNA induction by FSH and the known structure of the gene suggest that factors in addition to

CREB and CREB-like proteins may be involved in the transcriptional activation of the aromatase gene.

The factors mediating cAMP action in granulosa cells become even more complex because it is now clear that low amounts of FSH increase expression of aromatase, whereas high levels of hormone/cAMP acting on differentiated granulosa cells can cause a rapid loss of steady state $P450_{arom}$ mRNA in vivo (18) or in vitro (2) (Fig. 7.1). Because cycloheximide prevented the rapid LH/cAMP-mediated loss of $P450_{arom}$ mRNA between 42–48 h of culture (2), the steady state level of aromatase mRNA appears to be regulated by rapid changes in the synthesis of a negative regulatory protein, the nature of which remains to be determined. It is important to note that the decline in aromatase mRNA in the differentiated granulosa cells was not associated with a corresponding loss of aromatase enzyme activity (2) or enzyme protein (19). Thus, the half-life of the enzyme is much longer than that of its mRNA.

Notably, E is not the only steroid capable of enhancing the induction of aromatase mRNA by FSH in cultured granulosa cells. *Testosterone* (T), DHT (a nonaromatizable androgen), and Dex were all as effective as E in their ability to enhance FSH action in the absence or presence of serum. The greater ability of T compared to E to enhance the actions of FSH and the ability of DHT to synergize with FSH as effectively as E indicate that aromatase expression is regulated by androgen- as well as estrogen-specific pathways (2). During follicular development, theca cell-derived T would initially be the most physiologically relevant steroid because it could enhance the actions of FSH and serve as a substrate for the aromatase enzyme (induced at low levels by FSH alone). Once E is synthesized by granulosa cells, T and E can further synergize with FSH to amplify specific responses.

As mentioned above, $P450_{arom}$ mRNA was drastically reduced in differentiated cells by high concentrations of LH in vitro. Because cycloheximide not only prevented the loss of aromatase mRNA, but also blocked the increase in progesterone biosynthesis without altering cAMP, the data suggested that there might be an inverse relationship between progesterone and aromatase mRNA. This, along with observations of others that pharmacological doses of progesterone or progestin analogs decreased E biosynthesis in cultured granulosa cells, led us to hypothesize that progesterone itself might act to inhibit FSH induction of aromatase mRNA in a time- and dose-dependent manner. However, neither a low dose of progesterone (20 nM) nor a higher dose (100 nM), when present for 48 h of culture, altered the effects of FSH. At the highest dose (1 μM; comparable to that stimulated by the ovulatory doses of LH), progesterone enhanced FSH stimulation of cAMP, aromatase mRNA, and enzyme activity (2). Because granulosa cells of small follicles would never be exposed to this amount of progesterone, the physiological significance is unclear. In differentiated cells cultured with FSH plus T, progesterone

(20, 100, and 1000 nM) had no effect on cAMP, aromatase mRNA, or enzyme activity, perhaps because these cells were capable of synthesizing significant amounts (20–40 ng/mL) of their own progesterone between 24–48 h of culture. The differential responses of cultured rat granulosa cells to progesterone could depend on many factors: the metabolism of added progesterone, the presence or absence of receptors, or the stage of granulosa cell differentiation. At the pharmacological doses (3 μM) used in some previous studies (20), progesterone or a metabolite might bind the glucocorticoid as well as progesterone receptor or may act via other pathways. Overall, our data and that of others suggest that progesterone is unlikely to play a major physiological role in regulating follicular development and expression of aromatase in rat granulosa cells of small antral and preovulatory follicles. Lastly, addition of the high dose (1 μM) of progesterone to approximate that produced during the last 6 h of culture in response to ovulating levels of LH failed to mimic the inhibitory effects of LH or FSK on aromatase mRNA. Thus, increased synthesis of cAMP, rather than elevated amounts of progesterone, appears to be the primary mediator of the negative regulatory effects on aromatase mRNA.

Prostaglandin Endoperoxide Synthase (PGS)

Prostaglandins comprise a class of potent regulatory molecules that are produced by many tissues, are involved in myriad biological processes (inflammation, vasodilation/constriction, renal function, and implantation and ovulation), and are implicated in various clinical disorders, such as endometriosis, arthritis, emphysema, and cancer. The rate-limiting enzyme for prostaglandin biosynthesis is PGS, which has both cyclooxygenase and hydroperoxidase activities, converting arachidonic acid to *prostaglandin H2* (PGH2) (21). Based on the isolation of highly homologous cDNAs encoding sheep (22, 23), mouse (24), and human (25) PGS, investigators have strongly believed that only one gene encodes this enzyme. However, we have recently obtained immunological, biochemical, and functional data that provide the first evidence that the rat ovary contains 2 immunologically distinct forms and molecular weight variants of PGS (9, 10). Specifically, 2 affinity purified polyclonal antibodies have been generated (using purified ovine PGS as the antigen) that differentially recognize 2 molecular weight variants of PGS in the rat ovary: AntiPGS-2 recognizes PGS mol wt = 72,000 (PGS72), and antiPGS-3 recognizes PGS mol wt = 69,000 (PGS69). Immunoblot analyses show that PGS72 (also designated PGS$_i$ as seen in Fig. 7.1) is rapidly induced by LH/hCG in granulosa cells of preovulatory follicles and is associated with the increased production of prostaglandins (PGE and PGS$_{2\alpha}$) obligatory for ovulation (9). PGS72 is low (negligible) in other ovarian tissues, including preovulatory follicles, corpora lutea, and interstitium. In contrast, PGS69 is constitutively present in small antral and preovulatory follicles (primarily in theca cells), is unaffected by LH, and is found at higher levels in corpora lutea throughout pregnancy and in the ovarian interstitium (9). PGS69 (but not PGS72) is also detected by immunoblots in rat uterus and kidney, as well as in human dermal fibroblasts (9). Immunofluorescent localization of PGS72 and PGS69 to ovarian tissue sections confirmed the cell-specific distribution of PGS observed by immunoblot analyses of cell extracts (9).

Purification of these 2 ovarian forms of PGS using ion exchange chromatography and a pH step gradient indicates that each form elutes at a distinct pH: Induced PGS72 eluted at pH ~7.0, and PGS69, at pH ~4.5 (10). Other biochemical properties of rat PGS72 distinguish it from oPGS and its human/mouse homologues. Specifically, tryptic digests of PGS72 and oPGS are distinctly different based on 1D SDS-PAGE and immuno-blotting. Furthermore, the N-terminal amino acid sequences of PGS72 (10) differ (60% similarity) from those of the mouse, sheep, and human enzymes for which cDNAs have been cloned (22–25), but are similar (90%) to a PGS-like cDNA cloned from RSV-transformed chicken embryo fibroblasts (26). Peroxidase activity of PGS72 also exhibits a lower K_m for its substrate (H_2O_2) than does purified ovine PGS (10). Collectively, these studies provide the first evidence that the rat ovary contains 2 immunologically distinct forms and molecular weight variants of PGS, each of which is selectively regulated by hormones, localized to specific cell types, differentially sensitive to inhibitors of transcription/translation, and differentially solubilized for immunocytochemical localization (9, 10).

The rapid and transient induction of PGS72 (PGS_i in Fig. 7.1) indicates that transcriptional controls of this PGS gene and translational control of mRNA stability may combine to enforce stringent control of the synthesis and degradation of this key ovarian regulatory molecule.

References

1. Richards JS, Jahnsen T, Hedin L, Lifka J, Ratoosh SL, Goldring NB. Ovarian follicular development: from physiology to molecular biology. Recent Prog Horm Res 1987;43:231–76.
2. Fitzpatrick SL, Richards JS. Regulation of cytochrome P450 aromatase mRNA and activity by steroids and gonadotropins in rat granulosa cells. Endocrinology 1991.
3. Segaloff DL, Wang H, Richards JS. Hormonal regulation of luteinizing hormone/chorionic gonadotropin receptor mRNA in rat ovarian cells during follicular development and luteinization. Mol Endocrinol 1990;4:1856–65.
4. Oonk RB, Parker KL, Gibson JL, Richards JS. Rat cholesterol side-chain cleavage cytochrome P-450 (P450scc) gene. Structure and regulation by cAMP in vitro. J Biol Chem 1990;265:22392–401.

5. Kurten RC, Levy LO, Shey JL, Durica JM, Richards JS. Identification and characterization of the GC-rich and cAMP-inducible promoter of the type IIβ cAMP-dependent kinase regulatory subunit gene. J Biol Chem 1991 (submitted).

6. Hickey GJ, Krasnow JS, Beattie WG, Richards JS. Aromatase cytochrome P450 in rat ovarian granulosa cells before and after luteinization: adenosine 3′,5′-monophosphate-dependent and independent regulation. Cloning and sequencing of rat aromatase cDNA and 5′ genomic DNA. Mol Endocrinol 1990;4:3–12.

7. Clemens JW, Levy LO, Richards JS. Characterization of cAMP regulatory domains within the rat cholesterol side-chain cleavage P450 gene. Serono Symposium on Molecular Basis of Reproductive Endocrinology. Serono Symposia, USA, Vancouver, 1991.

8. Fitzpatrick SL, Richards JS. Characterization of the rat aromatase promoter (unpublished).

9. Wong WYL, Richards JS. Evidence for two immunodistinct forms and isoelectric variants of prostaglandin endoperoxide synthase. Mol Endocrinol 1991.

10. Sirois J, Richards JS. Purification and characterization of a novel, distinct isoform of prostaglandin endoperoxide synthase induced by hormones in the rat ovary. J Biol Chem 1991 (submitted).

11. Hunzicker-Dunn M, Culter RE Jr, Maizels ET, et al. Isozymes of cAMP-dependent protein kinase present in the rat corpus luteum. J Biol Chem 1991;266:7166–75.

12. Hunzicker-Dunn M, Lorenzini NA, Lynch LL, West DE. Coelution of the type II holoenzyme form of cAMP-dependent protein kinase with regulatory subunits of the type I form of cAMP-dependent protein kinase. J Biol Chem 1985;260:13360–9.

13. McKnight GS, Cadd GG, Clegg CH, Otten AD, Correll LA. Expression of wild-type and mutant subunits of the cAMP-dependent protein kinase. Cold Spring Harbor Symposia on Quantitative Biology 1988;53:111–9.

14. Kurten RC, Richards JS. An adenosine 3′,5′-monophosphate-responsive deoxyribonucleic acid element confers forskolin sensitivity on gene expression by primary rat granulosa cells. Endocrinology 1989;125:1345–57.

15. Pei L, Dodson R, Schoderbek WE, Maurer RA, Mayo KE. Regulation of the α inhibin gene by cyclic adenosine 3′,5′-monophosphate after transfection into rat granulosa cells. Mol Endocrinol 1991;5:521–34.

16. Delidow BC, White BA, Peluso JJ. Gonadotropin induction of c-fos and c-myc expression and DNA synthesis in rat granulosa cells. Endocrinology 1990;126:2302–6.

17. Williams T, Admon A, Lusdrer B, Tjian R. Cloning and expression of AP-2, a cell-type specific transcription factor that activates inducible enhancer elements. Genes Dev 1988;2:1557–69.

18. Hickey GJ, Chen S, Besman MJ, et al. Hormonal regulation, tissue distribution, and content of aromatase cytochrome P450 messenger ribonucleic acid and enzyme in rat ovarian follicles and corpora lutea: relationship to estradiol biosynthesis. Endocrinology 1988;122:1426–36.

19. Wong WYL, DeWitt DL, Smith WL, Richards JS. Rapid induction of prostaglandin endoperoxide synthase in rat preovulatory follicles by lutein-

izing hormone and cAMP is blocked by inhibitors of transcription and translation. Mol Endocrinol 1989;3:1714–23.

20. Fortune JE, Vincent SE. Progesterone inhibits the induction of aromatase activity in rat granulosa cells in vitro. Biol Reprod 1983;28:1078–89.

21. Smith WL. The eicosanoids and their biochemical mechanisms of action. Biochem J 1989;259:315–24.

22. DeWitt DL, Smith WL. Primary structure of prostaglandin G/H synthase from the sheep vesicular gland determined from the complementary DNA sequence. Proc Natl Acad Sci USA 1988;85:1412–6.

23. Merlie JP, Fagan D, Mudd J, Needleman P. Isolation and characterization of the complementary DNA for sheep seminal vesicle prostaglandin endoperoxide synthase (cyclooxygenase). J Biol Chem 1988;263:3550–3.

24. Dewitt DL, Kraemer SA, Meade EA. Serum induction and superinduction of PGG/H synthase mRNA levels in 3T3 fibroblasts. Adv Prostaglandin Thromboxane Leukotriene Res 1990;21:65–8.

25. Yokoyama C, Tanabe T. Cloning of human gene encoding prostaglandin endoperoxide synthase and primary structure of the enzyme. Biochem Biophys Res Commun 1989;165:888–94.

26. Xie W, Chipman JG, Robertson DL, Erickson RL, Simmons DL. Expression of a mitogen-responsive gene encoding prostaglandin synthase is regulated by mRNA splicing. Proc Natl Acad Sci USA;88:2692–6.

8

Structure and Control of Expression of the 3βHSD and 17βHSD Genes in Classical Steroidogenic and Peripheral Intracrine Tissues

F. Labrie, J. Simard, V. Luu-The, G. Pelletier, C. Labrie,
E. Dupont, C. Martel, J. Couët, C. Trudel, E. Rhéaume,
N. Breton, Y. de Launoit, M. Dumont, H.-F. Zhao,
and Y. Lachance

It is remarkable that humans, in addition to possessing a highly sophisticated endocrine system, have largely vested sex steroid formation in peripheral tissues (1). In fact, while the ovaries and testes are the exclusive sources of androgens and estrogens in the lower mammals, the situation is very different in higher primates, where active sex steroids are in a large part or whole synthesized locally, thus providing autonomous control to target tissues that are thus able to adjust formation and metabolism of sex steroids to local requirements. The situation of a high secretion rate of adrenal precursor sex steroids in men and women is thus completely different from current animal models used in the laboratory; namely rats, mice, guinea pigs, and all others except monkeys, where the secretion of sex steroids takes place exclusively in the gonads (1–3). Primates are thus unique in having adrenals that secrete large amounts of the precursor steroids *dehydroepiandrosterone* (DHEA) and especially *DHEA-sulfate* (DHEA-S) that are converted into *androstenedione* (Δ^4-dione) and then into potent androgens and estrogens in peripheral tissues (2, 4).

While approximately 40% of androgens in adult men are synthesized in peripheral tissues (2), our best estimate of the intracrine formation of estrogens in peripheral tissues in women is on the order of 75% before menopause and close to 100% after menopause. The testes and ovaries are responsible for gamete formation, but the peripheral tissues themselves play a key, and frequently the predominant, role in the formation of the androgens and estrogens required for specific sex steroid target tissue function. Such data stress the importance of studying steroidogenesis, not only in the classical steroidogenic tissues—namely, the testis, ovary, adrenal, and placenta—but also in the peripheral intracrine tissues that are important sites of biosynthesis of active steroids, especially androgens and estrogens.

A key enzyme in steroidogenesis is *3β-hydroxysteroid dehydrogenase/Δ⁵-Δ⁴-isomerase* (3βHSD), the enzyme required for the biosynthesis of all classes of hormonal steroids; namely, progesterone, glucocorticoids, mineralocorticoids, androgens, and estrogens. The 3βHSD enzymatic system is present in the adrenals, testes, ovaries, and placenta, as well as in many peripheral tissues, including the prostate, breast, liver, and skin (5–7). Congenital deficiency of 3βHSD activity causes severe depletion of steroid formation by the adrenals and gonads and can be lethal in early life (8).

Since the structure of 3βHSD was not known, we have cloned cDNAs encoding human, rat, macaque, and bovine 3βHSD, and we have deduced the amino acid sequences of the corresponding proteins (9–15). We have also elucidated the structure of 2 human 3βHSD genes (16; Lachance et al., unpublished data) localized to the p13 band of chromosome 1 (17). Moreover, we have studied the localization and the ontogeny of 3βHSD in human adrenals, testis, ovary, and placenta, as well as in the same rodent tissues (18–22; Dupont et al., unpublished data). We have also performed studies on the control of 3βHSD mRNA levels, expression, and activity in the rat ovary, testis, and adrenal (23–26), as well as during the estrous cycle in the bovine ovary (27).

Since it is well recognized that sex steroids are synthesized in peripheral tissues from circulating precursors, it is of major importance to gain precise information on the enzymatic systems responsible for the formation of these active sex steroids in peripheral target tissues. The key enzyme that synthesizes *androst-5-ene-3β,17β-diol* (Δ^5-diol) from DHEA, *testosterone* (T) from Δ^4-dione, *17β-estradiol* (E₂) from *estrone* (E₁), and *dihydrotestosterone* (DHT) from Δ^4-dione is *17β-hydroxysteroid dehydrogenase* (17βHSD). This enzyme is thus required for the synthesis of all androgens and all estrogens. We recently reported the molecular cloning and sequencing of cDNAs encoding human 17βHSD (28). Using human 17βHSD cDNA (clone hpE₂DH216) as a probe, we have isolated, sequenced, and characterized 2 in-tandem 17βHSD genes that reside within a 13-kpb genomic DNA fragment (29). The availability of this cDNA offers the opportunity of studying in detail the factors controlling the expression of this crucial enzyme, not only in gonadal tissue, but also in several peripheral estrogen target tissues.

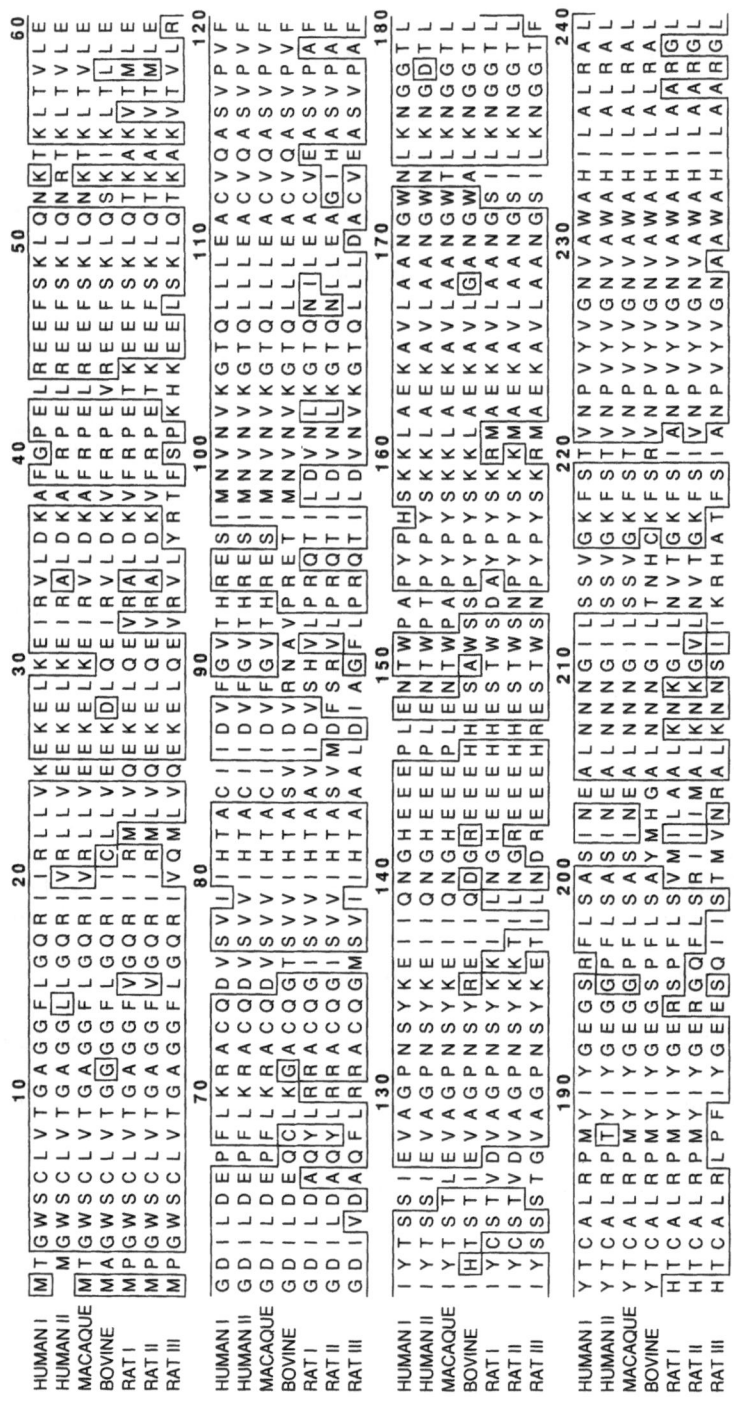

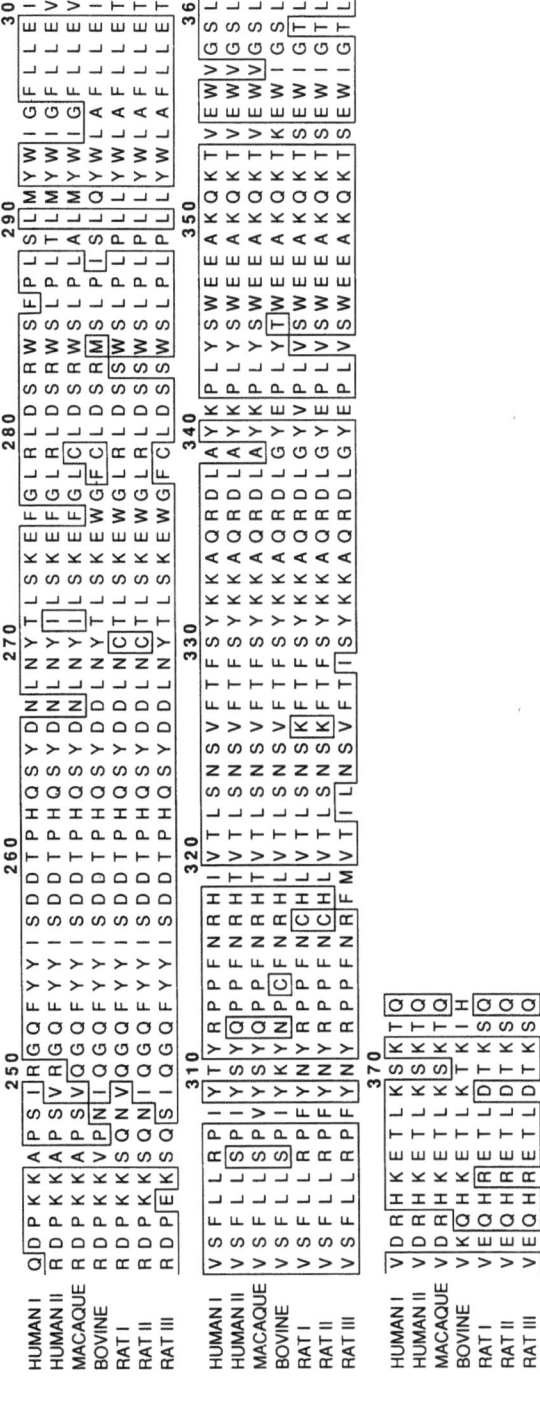

FIGURE 8.1. Comparison of the deduced amino acid sequences of human type I, human type II, macaque, bovine, and rat types I, II, and III 3βHSD proteins. Amino acid sequences are designated by the universal single-letter code. Amino acid residues are numbered relative to the first NH2-terminal methionine. Residues common to at least 4 of the 7 predicted sequences are boxed. Reprinted with permission from Rhéaume, Lachance, Zhao, et al. (15), © by The Endocrine Society, 1991.

Human 3βHSD Isoenzymes

Structure of Human Types I and II 3βHSD cDNAs

We first reported characterization of a human 3βHSD cDNA isolated from a placental library (9) and the corresponding gene (16) located in the p11–p13 region of chromosome 1 (17), which encodes a deduced protein of 372 amino acids. This cDNA has been named *human type I 3βHSD*. However, the detection of multiple, unexpected DNA fragments by Southern blot analysis of human genomic DNA (16) and the heterogeneous clinical picture in 3βHSD-deficient patients suggest the presence of multiple 3βHSDs in the human. We have thus screened a human adrenal λgt22A cDNA library with human placental cDNA clone hp3βHSD63 (9). This new type of 3βHSD cDNA has been chronologically designated *human type II 3βHSD*.

The cDNA sequence of type II 3βHSD includes an open reading frame of 1116 nucleotides, compared to 1119 nucleotides for type I 3βHSD cDNA. The nucleotide sequence of human type II 3βHSD cDNA displays 93.6% similarity with that of human type I 3βHSD. The nucleotide sequence of the expected coding region of human type II 3βHSD shares 95.4%, 81.0%, 77.8%, 77.4%, and 75.4% similarity with that of macaque (14), bovine (11), and rat types I, II, and III, respectively (12, 13). The human type II 3βHSD cDNA thus predicts a 41,921-d protein having 371 amino acid residues (excluding the first methionine), while human type I, macaque, bovine, and rat types I, II, and III 3βHSD cDNAs all encode a deduced protein containing 372 amino acids. The deduced amino acid sequence of type II 3βHSD shares 93.5% similarity with that of human type I 3βHSD, which differs by only 23 residues. The similarity of the human type II 3βHSD amino acid sequence to that of macaque, bovine, and rat types I, II, and III is 96.2%, 78%, 72%, 71%, and 66.7%, respectively (Fig. 8.1).

Enzymatic Characteristics of Expressed Human Types I and II 3βHSD Isoenzymes

In order to verify that human types I and II 3βHSD cDNAs encode proteins that effectively catalyze 3β-hydroxysteroid dehydrogenation as well as Δ^5-Δ^4 isomerization and to characterize potential functional differences between the two types, plasmids derived from pCMV containing either type I (pCMV type I h3βHSD) or type II (pCMV type II h3βHSD) 3βHSD cDNA inserts driven by the CMV promoter were transiently expressed in HeLa cells. In vitro incubation with homogenate from cells transfected with pCMV type I h3βHSD or pCMV type II h3βHSD in the presence of 1 mM NAD$^+$ and ^{3}H-labeled substrates shows that the type I enzyme possesses a 3βHSD/Δ^5-Δ^4 isomerase activity higher than type II with respective Km values of 0.24 and 1.2 μM for PREG (pregnenolone) and 0.18 and 1.6 μM for DHEA, while the specific activity (V_{max}) of both types is equivalent when standardized for the estimated amount of corresponding translated proteins. We have also observed that following incubation of cell homogenate in the presence of NADH and ^{3}H-labeled DHT, the 3β-hydroxysteroid oxidoreductase activity, measured by the formation of 5α-androstane-3β,17β-diol, is higher for type I compared to the type II 3βHSD protein, with K_m values of 0.26 μM and 2.7 μM, respectively. The present data also show that the affinity of the human type I as well as type II 3βHSD proteins is similar for the 3 substrates tested. In fact, analysis of the kinetic properties of both expressed 3βHSD proteins reveals that the relative enzymatic activity (V_{max}/K_m) of type I is 5.9-, 4.5-, and 2.8-fold higher than that of the type II 3βHSD protein using PREG, DHEA, and DHT as substrate, respectively (15).

Structure of Human Types I and II 3βHSD Genes

Human placental α-^{32}P-labeled 3βHSD cDNA (hp3βHSD63) was used to screen a human leucocyte genomic DNA library constructed in the λ-EMBL3 phage vector. As illustrated in Figure 8.2, the human 3βHSD genes corresponding to the human cDNAs types I and II contain 4 exons and 3 introns included within a total length of 7,8 kbp. The first exon of both genes contains 53 nucleotides in the 5' noncoding region. Exon 2 contains 85 or 89 nucleotides in the 5' noncoding region of the type I and type II genes, respectively, the nucleotide sequence of the first 48 amino acids of type I 3βHSD or of the first 47 residues of type II 3βHSD, and the first nucleotide of the following Lys codon. Exon 3 contains the last 2 nucleotides of the Lys codon and encodes the 53 following residues and the first nucleotide of the following Gly codon. In fact, exon 4 contains the last 2 nucleotides of the following Gly codon, the nucleotides encoding

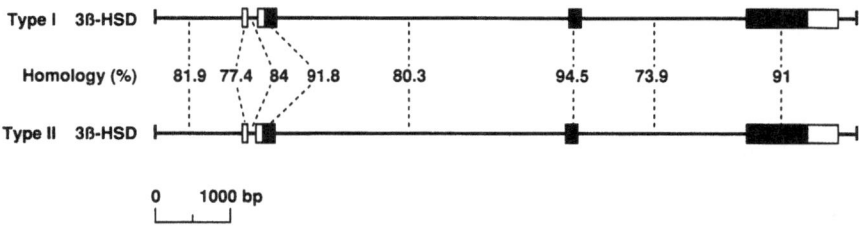

FIGURE 8.2. Comparison of the exon-intron organization and percentage of homology of human type I and type II 3βHSD genes. Exons are represented by boxes and numbered from I to IV. Black box areas correspond to coding regions, while open box areas indicate noncoding regions. Intron and flanking regions are represented by solid lines. The scale bar represents 1000 bp.

the following 269 amino acids, and the stop codon, as well as 414 and 413 nucleotides in the 3' untranslated region of the type I and type II genes, respectively. Comparison of the nucleotide sequences of the two genes indicates that they share 77.4%, 91.8%, 94.5%, 91.0%, 84.0%, 80.3%, and 73.9% homology in exons 1, 2, 3, and 4, as well as in introns 1, 2, and 3, respectively. Moreover, it is also of interest to mention that the 1250 nucleotides in the 5' flanking region share 81.9% homology.

Tissue-Specific Expression of Human Types I and II 3βHSD mRNA Species

In order to determine the tissue-specific expression of human type I and type II 3βHSD genes and the relative abundance of the two types of 3βHSD mRNA populations, we performed a ribonuclease protection assay that offers the opportunity to discriminate accurately a few base pair mismatches occurring after annealing of the type I or type II cRNA probe to the type I or type II mRNA species. Somewhat surprisingly, using specific cRNA probes, it can be seen in Figure 8.3A that the type II 3βHSD mRNA population is the almost exclusive species detectable in the human adrenal gland, testis, and ovary, as revealed by the presence of the expected full-length (220 nucleotides) protected fragment using the specific type II cRNA probe, as well as by the occurrence of the expected small fragments (about 98 and 64 nucleotides) when the type I cRNA probe was used. It was possible, however, to detect the presence of type I 3βHSD protected mRNA fragments after a longer time exposure of testis and ovary mRNA. With overexposed autoradiographs, it was not possible to detect human type I 3βHSD mRNA in total RNA from either human adrenal or human type II 3βHSD mRNA in human placenta. It can be seen in Figures 8.3B and 8.3C that the human type I 3βHSD mRNA population corresponds to the only detectable species in human placenta

-->

FIGURE 8.3. RNase protection analysis of the distribution of human types I and II 3βHSD mRNAs in classical steroidogenic as well as peripheral intracrine tissues in the human. Samples of total RNA from the human adrenal (3 μg), ovary (10 μg), placenta (20 μg), skin (from the breast region) (20 μg) or poly(A)$^+$ for the testis (10 μg) or breast (normal mammary gland) (10 μg) were hybridized to the type I or type II cRNA probe for 14 h at 37°C and then digested with ribonuclease A and T1. The protected fragments were resolved on 6% denaturing polyacrylamide-7M urea sequencing gels. With either probe (315 nucleotides), the longest protected fragment (220 nucleotides), which included nucleotides +688 to +909 for human type I 3βHSD and +685 to +904 for human type II 3βHSD, corresponds to the homologous RNA species protected by the cRNA probe.

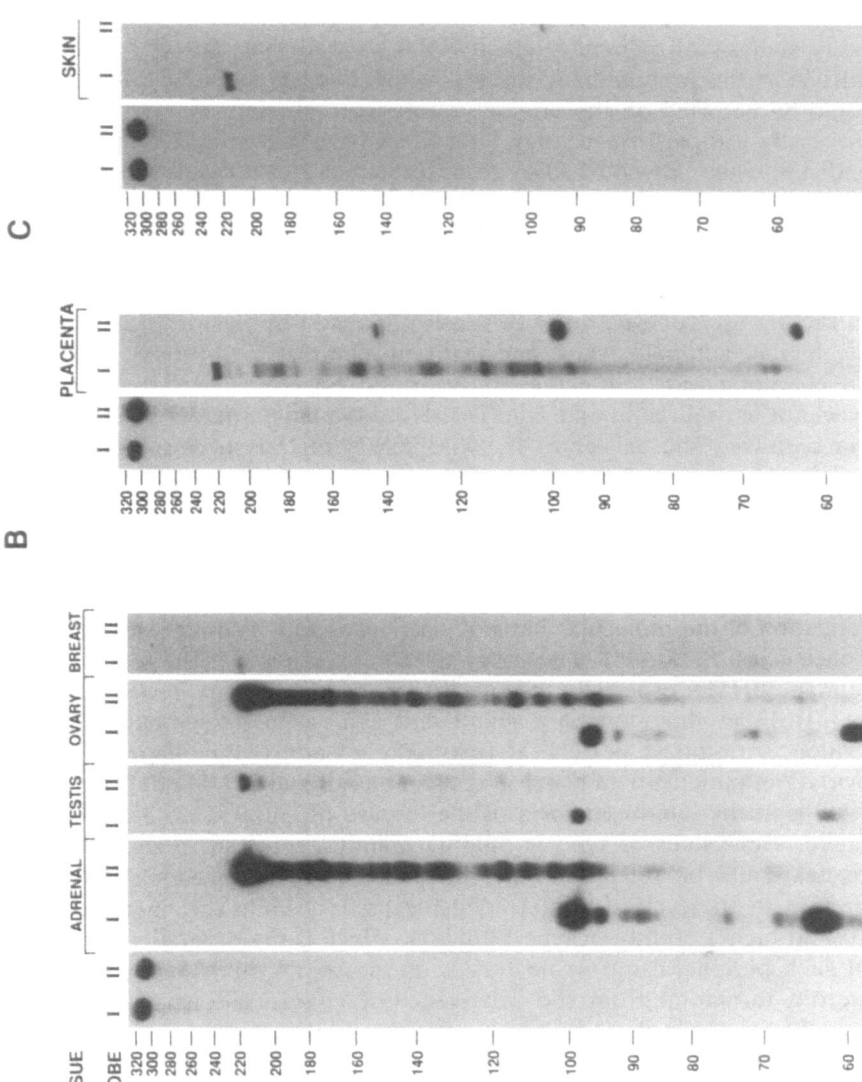

Protected fragments of 98 and 64 nucleotides correspond to predicted well-recognized mismatches for RNase A. Lanes corresponding to the adrenal and placenta were exposed to X-ray film for 14 h, while those from testis, breast, and ovary were exposed for 7 days, and the skin sample was exposed for 9 days. Lanes corresponding to adrenal, ovary, and placenta are overexposed in order to further demonstrate the type I- or II-specific mRNA expression in these tissues. Reprinted with permission from Rhéaume, Lachance, Zhao, et al. (15), © by The Endocrine Society, 1991.

and skin under the experimental conditions used. In addition, ribonuclease assay analysis of mammary gland RNA showed that the type I 3βHSD mRNA is the predominant species, while some type II 3βHSD mRNA could be detected on the original autoradiograph (Fig. 8.3A). Note that this study was performed using total RNA from placenta, adrenal, ovary, and skin while poly(A)$^+$ RNA from testis and breast tissue was required to detect the hybridization signal. The present data indicate that the human type I 3βHSD mRNA levels in skin are markedly higher than those of the human type II 3βHSD mRNA species in the testis.

The present data show that the type II 3βHSD transcripts correspond to the almost exclusive mRNA species detected by RNase protection in the human adrenal, ovary, and testis, in contrast to the type I 3βHSD mRNA population that is the almost exclusive mRNA species revealed in placenta as well as in skin and is the predominant species in mammary gland tissue.. The existence of two 3βHSD mRNA species offers many possibilities for the tissue-specific regulation of the activity of these isoenzymes.

The present elucidation of the structure of human type II 3βHSD, which represents the almost exclusive isoenzyme expressed in the adrenals and gonads, should provide the necessary tools for studies on the characterization of the molecular basis of classical as well as nonclasical 3βHSD deficiencies (8, 30–38). The tissue-specific expression of the two types of human 3βHSD, especially the specific or predominant expression of type I 3βHSD in the mammary gland and skin is in agreement with the evidence for intact peripheral (extraadrenal and extragonadal) 3βHSD activity in patients with classical congenital adrenal 3βHSD deficiency, as well as in the late-onset form of the disease (8, 30, 31, 33–35, 38). The higher Km values of type II 3βHSD mainly expressed in steroidogenic tissues could be related to the higher levels of endogenous substrates present in these classical steroidogenic tissues; while the approximately 10-fold higher affinity of type I 3βHSD, which is preferentially expressed in such peripheral intracrine tissues as the skin, could greatly facilitate steroid formation from the relatively low concentrations of substrates usually present in these tissues.

The physiological importance of 3βHSD gene expression in the skin is supported by the observation that DHEA can stimulate sebaceous gland secretion in humans (39) following its conversion into the potent androgens T and DHT (40), thus indicating, in addition, the presence of 17βHSD and 5α-reductase activity in human skin.

The novel 3βHSD isoenzyme has been arbitrarily designated as type II by reference to our previously characterized human type I 3βHSD obtained from a placental library (9, 16) without any relationships with rat type I and type II, which are both expressed in the adrenals and gonads (13). The finding that the human type II 3βHSD differs from the 3βHSD protein deduced from macaque ovarian cDNA by only 13 residues while 23 different residues distinguish the human type II and type I 3βHSD proteins could suggest that human type II 3βHSD evolved from the same duplicated ancestor gene as that of the macaque 3βHSD protein, while the human type I gene may have evolved from another member of this gene family that diverged after a duplication took place earlier in evolution.

Immunocytochemical Localization of 3βHSD in Human Ovary

Immunocytochemical localization of 3βHSD was achieved using specific polyclonal antibodies developed against purified human placental 3βHSD (9, 10). In the fetal ovary of 28–34 weeks, specific immunostaining for 3βHSD was exclusively detected in thecal cells surrounding primary follicles and in interstitial cells. Immunostaining was first observed at 28 weeks of gestation and persisted throughout the third trimester of gestation (20). Such results support the suggestion that human fetal ovaries may be involved in sex steroid synthesis during the last trimester of gestation knowing that other steroid-synthesizing enzymes, such as aromatase, are also expressed in fetal ovaries. This finding has to be interpreted in the light that previous reports have indicated that only minimal amounts of estrogens can be synthesized by fetal ovaries in several mammalian species, including humans (41, 42). The high levels of estrogens found in the fetal circulation are in fact thought to originate from placenta (41). The development of primordial follicles to the stage of primary follicles, as well as the expression of 3βHSD observed from the 28th week of gestation, could be related to the increase in fetal plasma FSH and LH levels that occurs at the end of the second trimester of gestation (43). HCG, which has been shown to increase 3βHSD mRNA levels in rat ovaries (24) as well as in luteinized porcine granulosa cells (44), might also be responsible for induction of the expression of the enzyme during fetal ovarian development.

From birth until puberty, no significant immunostaining for 3βHSD could be observed in human ovaries, while from puberty to menopause, staining was detected in theca interna cells as well as in granulosa cells of growing follicles. The intensity of staining in the theca interna cells was always much higher than in granulosa cells. Immunostaining was also found in the cytoplasm of luteinized granulosa and theca interna cells of the corpus luteum (Fig. 8.4). Interestingly, there was an absence of immunoreactivity for 3βHSD in one to several layers of theca interna cells lying just beneath the basement membrane. These negative cells may correspond to fibroblast-like cells that are devoid of 3βHSD activity. In menopausal and postmenopausal women, 3βHSD immunoreactivity was found in dispersed interstitial cells, thus indicating that active steroidogenesis could occur after menopause (20). These results are in

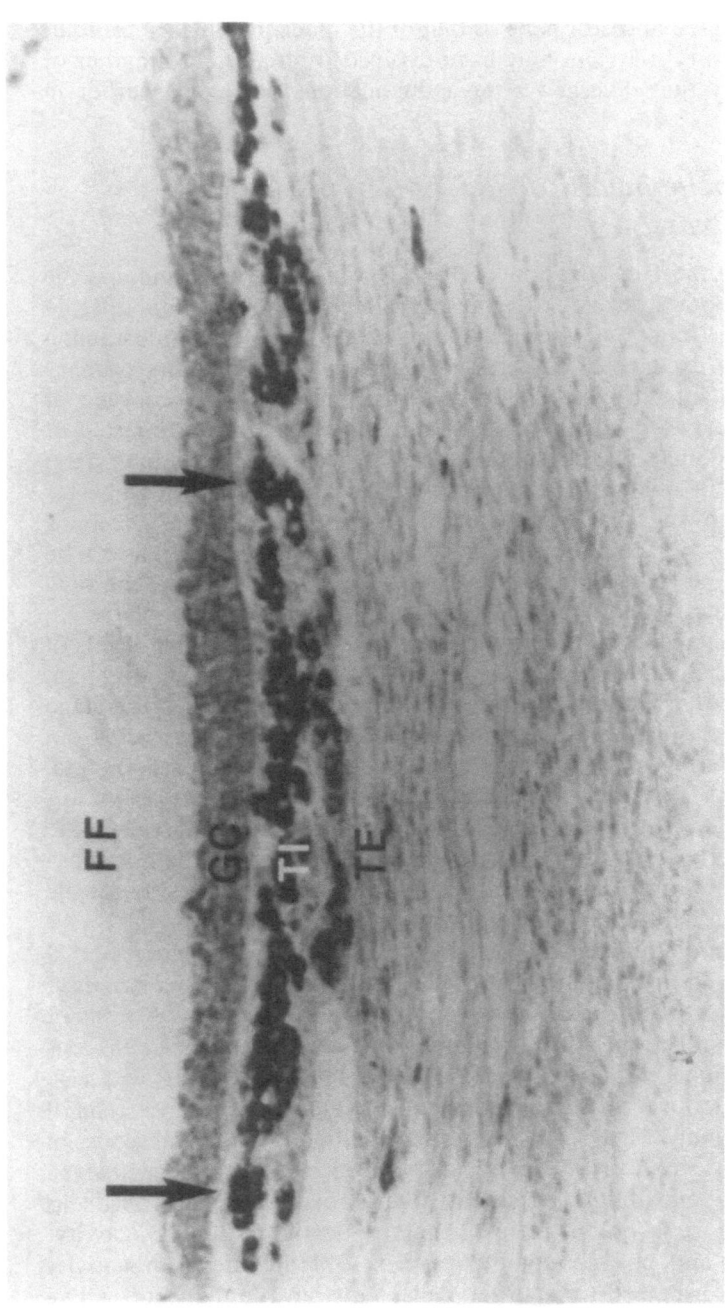

FIGURE 8.4. Immunostaining for 3βHSD located in the theca interna (TI) and in the granulosa cells (GC) in a well-developed follicle. The labeling is less intense in granulosa cells than in thecal cells. No immunoreactivity can be observed in the follicular fluid (FF) or theca externa (TE). Interestingly, one to several layers of the theca interna cells that were lying directly beneath the basement membrane did not show any immunoreactivity (200×). Reprinted with permission from Dupont, Labrie, Luu-The, and Pelletier (20), © by The Endocrine Society, 1991.

agreement with the findings of Aiman et al. (45) who reported the presence of 17β-estradiol in ovarian venous blood in postmenopausal women. However, the main steroid secreted by the human ovary after menopause is androstenedione. In fact, in postmenopausal women, about 30% of circulating androstenedione is of ovarian origin, the remaining being of adrenal sources (46, 47).

Ontogeny of 3βHSD in Human Testis

To correlate possible changes in 3βHSD expression and activity with the well-known variations in testosterone secretion during development, we localized this enzyme by immunocytochemistry during fetal and postnatal periods of development in the human testis (19). In the fetal testis, 3βHSD was detected in Leydig cells during the second and third trimester of gestation (Fig. 8.5). In 8-month-old and 11-year-old boys, however, no immunoreactivity could be detected in the testis. In pubertal boys, Leydig cells appeared well developed and immunopositive. Since the fluctuations in 3βHSD immunoreactivity are similar to those already observed for androgen secretion, activation of 3βHSD by trophic hormones may play an important role in androgen production during fetal and postnatal development. Our findings, which indicate that 3βHSD is present in the fetal testis during the second and third trimesters of gestation (22, 28, and 31 weeks), agree with previous data demonstrating that plasma testosterone levels are high during the second and third trimesters of gestation and decrease during the last weeks of gestation in the human fetus (48).

By 2 to 3 months after birth, circulating levels of gonadotropins and sex steroids reach a peak and then begin to drop to very low levels that persist until the onset of puberty (49). The current results indicating the absence of immunoreactive 3βHSD in the testis of an 8-month-old infant and an 11-year-old boy agree with these observations (19). Moreover, it has been shown that Leydig cells involuted a few weeks after birth and almost completely disappeared during the following months (50).

At the onset of puberty, there is an increase in LH and FSH secretion and a corresponding increase in plasma T and DHT (51). We have observed that the interstitial cells in the pubertal testis are well developed and contain 3βHSD, while the seminiferous tubules are considerably enlarged (19). These morphologic changes are likely to be a consequence of the increase in gonadotropin secretion.

Immunocytochemical Localization of 3βHSD and 17βHSD in Human Placenta

We have studied the localization of the 3βHSD enzyme in the human placenta and its development during pregnancy. At 7 weeks of pregnancy,

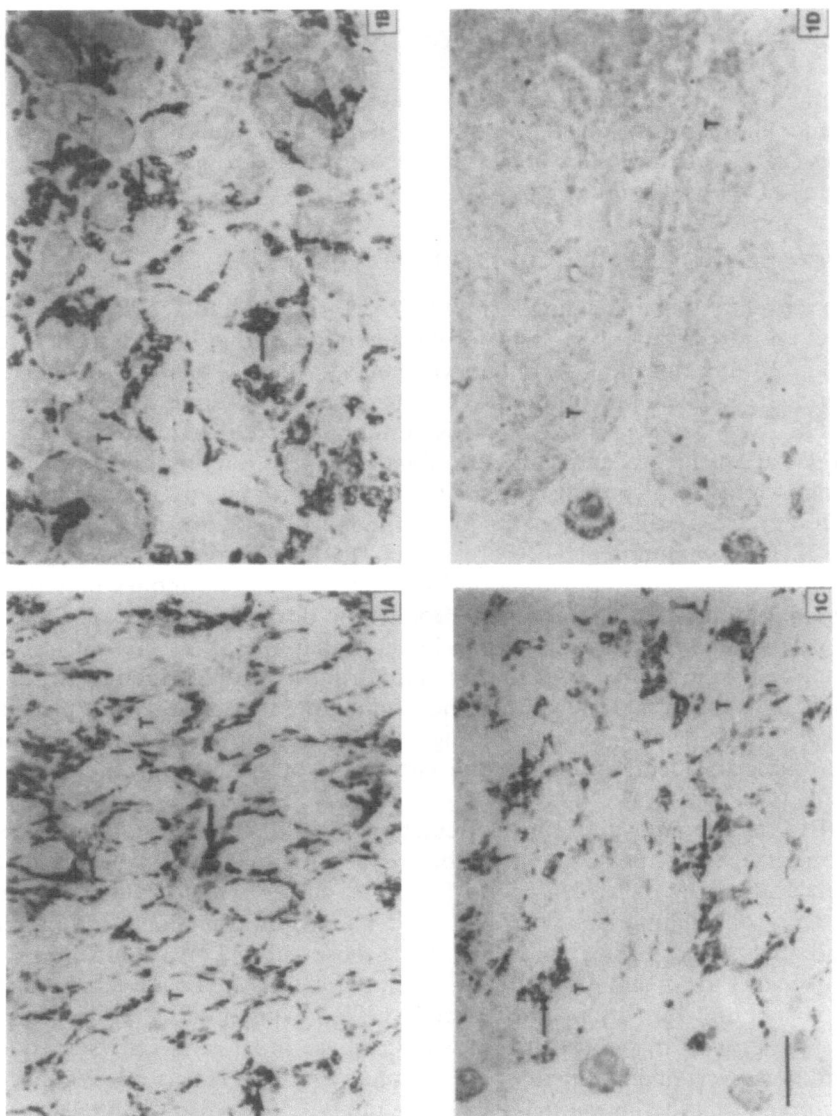

FIGURE 8.5. *1A:* Section of human fetal testis at 22 weeks of gestation. Immunostaining can be observed in the cytoplasm of interstitial cells (arrows). Tubules (T) are unstained. (Scale bar = 100 μm.) *1B:* Section of human fetal testis at 28 weeks of gestation (*1A*). (T = tubules; scale bar = 100 μm.) *1C:* Section of human fetal testis at 31 weeks of gestation. Heterogeneous intensity of immunostaining of interstitial cells (arrows) can be observed. (T = tubules; scale bar = 100 μm.) *1D:* Control section of human fetal testis adjacent to that shown in *1C*. Immunoabsorption with excess of human 3βHSD (10^{-6} mol/L) has completely prevented labeling. (T = tubules; scale bar = 100 μm.) Reprinted with permission from Dupont, Luu-The, Labrie, and Pelletier (19), © by J.B. Lippincott Company.

immunostaining was observed in the cytoplasm of syncytiotrophoblastic cells in about half of the villi. The amount of labeled villi increases with maturation of the placenta such that at 14 weeks of gestation, all the syncytiotrophoblastic cells contain immunoreactive material. In term placenta, immunostaining is similarly observed in the cytoplasm of all the syncytiotrophoblastic cells (data not shown). In conclusion, the present data show that 3βHSD is present in syncytiotrophoblastic cells, this enzyme being responsible for a key step in the biosynthesis of all classes of biologically active steroids. Moreover, the present results suggest that from 14 weeks of gestation, all the syncytiotrophoblastic cells are involved in steroid biosynthesis.

In order to obtain information about the age-specific expression of 17βHSD in the human placenta, we have localized this enzyme by immunocytochemistry at the light microscopic level at different periods of gestation. In the 7- and 9-week-old placenta, immunostaining was detected exclusively in the cytoplasm of the syncytiotrophoblast. Between the 10th and 14th week of gestation, immunolabeling was also observed in the cytoplasm of the cytotrophoblastic cells, thus suggesting that these cells could transiently be involved in the biosynthesis of sex steroids. Interestingly, between the 14th and 25th week of gestation, 17βHSD immunoreactivity was observed in both the cytoplasm and nucleus of the syncytiotrophoblast. The reaction product was much more intense in nuclei than in cytoplasm. During the last trimester of gestation, a strong immunocytochemical staining was observed in all the nuclei of the syncytiotrophoblast, the cytoplasm being unstained (data not shown). The meaning of this nuclear staining for 17βHSD is still unclear.

Ontogeny of 3βHSD in Human Adrenal Gland

In the fetal adrenal from midgestation (22 weeks), immunostaining was observed exclusively in the cytoplasm of cells of the outer cortex. Six weeks later, at the beginning of the third trimester (28 weeks), immuno-labeling was more widely distributed and extended to all cells of the neocortex. In term fetal adrenals, on the other hand, strong staining was observed in the presumptive zona glomerulosa (18).

Contrary to the observation of Murakoshi et al. (52), who localized 3βHSD in the inner zone of the human 20-week fetal adrenal by histochemistry, our data clearly identify 3βHSD exclusively in the outer zone at this period of fetal life. During early to midgestation, it is generally thought that 3βHSD activity in fetal adrenal tissue is minimal and that placental progesterone must be used as a precursor for the synthesis of cortisol. At midgestation and approaching term, our data indicate an increase in 3βHSD levels, a finding that could well explain the increase in circulating cortisol at midgestation (53). Thus, the increase in 3βHSD immunoreactivity in the outer zone of the adrenal cortex could provide the basis for increased cortisol secretion by the fetal adrenal independently from placental transfer of cortisol (54, 55). Such data also suggest that the fetal adrenal is actively involved in progesterone synthesis.

In 2-month-old adrenals, immunostaining was seen in both the zona glomerulosa and the zona fasciculata, while no staining could be detected in the zona reticularis, which is easily identified (18). Similar 3βHSD localization was seen in the 8-month-old adrenals. In 2-year-olds and adults, on the other hand, immunostaining was observed in the three layers of the cortex, the medulla being devoid of any reaction. Moreover, the intensity of staining was similar in the zona reticularis, fasciculata, and glomerulosa. At the subcellular level, the labeling was exclusively observed in the cytoplasm (18).

Rat 3βHSD Isoenzymes

Since rat is the best-known model for studies of endocrine regulation, the availability of rat 3βHSD probes would make possible detailed investigations of the tissue-specific expression and regulation of 3βHSD in steroidogenic as well as in peripheral intracrine tissues. For this reason, we have used the human type I 3βHSD cDNA to first screen a rat ovary λgt11 cDNA library.

Structure of Rat Types I and II 3βHSDs

The sequence of type I and type II cDNAs have an open reading frame of 1119 nucleotides, with only 33 mismatches between the two sequences (13). The overall nucleotide similarity between the two types of rat 3βHSD cDNAs is 96.5%. The deduced amino acid sequences of rat types I and II 3βHSD (Fig. 8.1) share 93.8% similarity, with only 23 nonidentical residues.

Enzymatic Characteristics of Expressed Rat Types I and II 3βHSDs

In view of the existence of two types of rat 3βHSD cDNA clones, plasmids derived from pCMV that contained either the type I or the type II 3βHSD cDNA inserts driven by the CMV promoter were transiently expressed in HeLa human cervical carcinoma cells in order to verify that both proteins were functional and that they effectively mediated 3β-hydroxysteroid dehydrogenation as well as Δ^5-Δ^4 isomerization (13, 56). In vitro incubation with homogenates from cells transfected with pCMV type I 3βHSD or pCMV type II 3βHSD in the presence of 1 mM NAD$^+$ and ^{3}H-labeled PREG clearly showed that type I had a 3βHSD/Δ^5-Δ^4

isomerase relative specificity, as reflected by relative V_{max}/K_m ratio, 64-fold higher than type II with a K_m value of 0.74 mM compared with 14.3 mM for type II (56). The higher relative specificity of rat type I 3βHSD compared to type II 3βHSD was confirmed using DHEA as labeled substrate with respective K_m values of 0.68 and 12.9 mM, while their relative specificities were 147 and 3.2, respectively. These original data indicate that the lower activity of type II 3βHSD results primarily from a 95% decreased affinity for both PREG and DHEA (56).

The hydropathy profiles of rat type I and rat type II 3βHSD proteins are also quite superimposable (13). Since 3βHSD is a membrane-bound protein, it is of interest to note that computer analysis of the amino acid sequences of rat types I and II 3βHSD proteins predicts one additional membrane-spanning segment in the type I protein compared with the type II isoform (13). We have recently demonstrated the crucial role of the predicted membrane-spanning domain between residues 75 and 91 for the enzymatic specificity of type I 3βHSD, while the absence of this putative MSD in type II 3βHSD can explain its much lower activity (56).

Tissue-Specific Expression of Rat Types I and II 3βHSD mRNA Species

In view of our finding of two cDNAs encoding rat 3βHSD and the impossibility of distinguishing the corresponding mRNAs by conventional RNA blot analysis, we proceeded to study the distribution of 3βHSD mRNAs in rat tissues using the sensitive and specific RNase protection assay. Using the cRNA probes specific for type I and type II 3βHSD mRNAs, we have demonstrated that both types I and II mRNAs are present in classical steroidogenic tissues, namely, the rat ovary, testis, and adrenal. We have also examined the distribution of these two mRNA species in various peripheral tissues of the rat. Both mRNA species were detected in female rat adipose tissue. On the other hand, only type I mRNA could be detected in both male and female kidney poly(A)$^+$ RNA (13).

Liver-Specific Member of Rat 3βHSD Family

Following screening of a male rat liver λgt11 cDNA library, a third type of 3βHSD cDNA could be isolated (12). Rat type III 3βHSD cDNA has an open reading frame of 1119 nucleotides that displays 85.7% similarity with that of rat type I or type II 3βHSD cDNA (13). The predicted rat type III 3βHSD protein having a calculated molecular mass of 42,080 d also contains 372 amino acids and shares 80% similarity with type I and II 3βHSD proteins (Fig. 8.1). Type III 3βHSD cDNA contains a 736-bp 3′ untranslated region before the poly(A) tail, while the corresponding region in types I and II cDNAs contains 355 and 369 bp, respectively (13).

In agreement with a longer 3' untranslated region in type III mRNA, RNA blot analysis using poly(A)$^+$ RNA from male rat adrenal, testis, and liver clearly illustrates liver-specific expression of a 2.1-kb mRNA species in male liver, while the crosshybridization signal obtained in adrenal and testis reveals a 1.7-kb mRNA species that corresponds to type I and II 3βHSD mRNA (13). It is noteworthy that no signal was detected in poly(A)$^+$ RNA from intact female liver, while a strong hybridization signal was detected in total RNA from female liver obtained from animals hypophysectomized 24 days earlier (12).

We have recently observed after transient expression of type II cDNA in HeLa cells that this isoform possesses an almost exclusive 3β-keto steroid reductase activity using DHT and androstanedione as substrates while PREG, DHEA, $Δ^5$-diol, and 3β-diol do not act as substrates (de Launoit, Zhao, Labrie, Simard, unpublished data).

Using the cRNA probes specific for types I, II, or III 3βHSD mRNA, we have demonstrated that both types I and II mRNAs are expressed in the rat adrenal and ovary but not in liver, while the type III mRNA species is detected specifically in the liver (12). The present finding of an apparently pituitary hormone-induced repression of gene expression of a liver-specific 3βHSD is in agreement with the previously reported masculinization of rat liver enzyme activity following hypophysectomy in female rats (57).

The presence of 3βHSD mRNAs in a wide variety of peripheral rat tissues—such as adipose tissue and kidney—as well as the demonstration of 3βHSD activity in hormone-sensitive organs—such as the skin, breast, uterus, ventral prostate, and seminal vesicle (13)—indicate that these tissues can form active sex steroids from circulating $Δ^5$-3β-hydroxysteroid precursors. In fact, we have previously demonstrated that the admin-

————————————————————————————————▶

FIGURE 8.6. Modulation of rat ovary 3βHSD mRNA levels (top) and total ovarian 3βHSD content (bottom) by gonadotropins and PRL. Adult female rats hypophysectomized 15 days previously received twice-daily injections of hCG (10 IU), oFSH (0.5 μg), or oPRL (1.0 mg), singly or in combination, for 10 days. RNA was extracted from individual whole ovaries and blotted onto nylon membranes. Ovarian 3βHSD mRNA levels were measured by dot blot hybridization using the 32P-random primer-labeled full-length rat ovary 3βHSD cDNA (ro 3b-HSD 56), as described in reference 13. The amounts of ovarian 3βHSD mRNA were calculated relative to those levels observed in hypophysectomized rat ovaries, using RNA from intact ovaries as standard and expressed in units (top) or corrected for ovarian weight and expressed in units/ovary (bottom). Data are expressed as means ± SEM ($n = 6–7$), and statistically significant variations were determined by Duncan-Kramer. (* = $P < 0.05$; ** = $P < 0.01$ vs hypophysectomized control.) Reprinted with permission from Martel, Labrie, Dupont, et al. (23), © by The Endocrine Society, 1990.

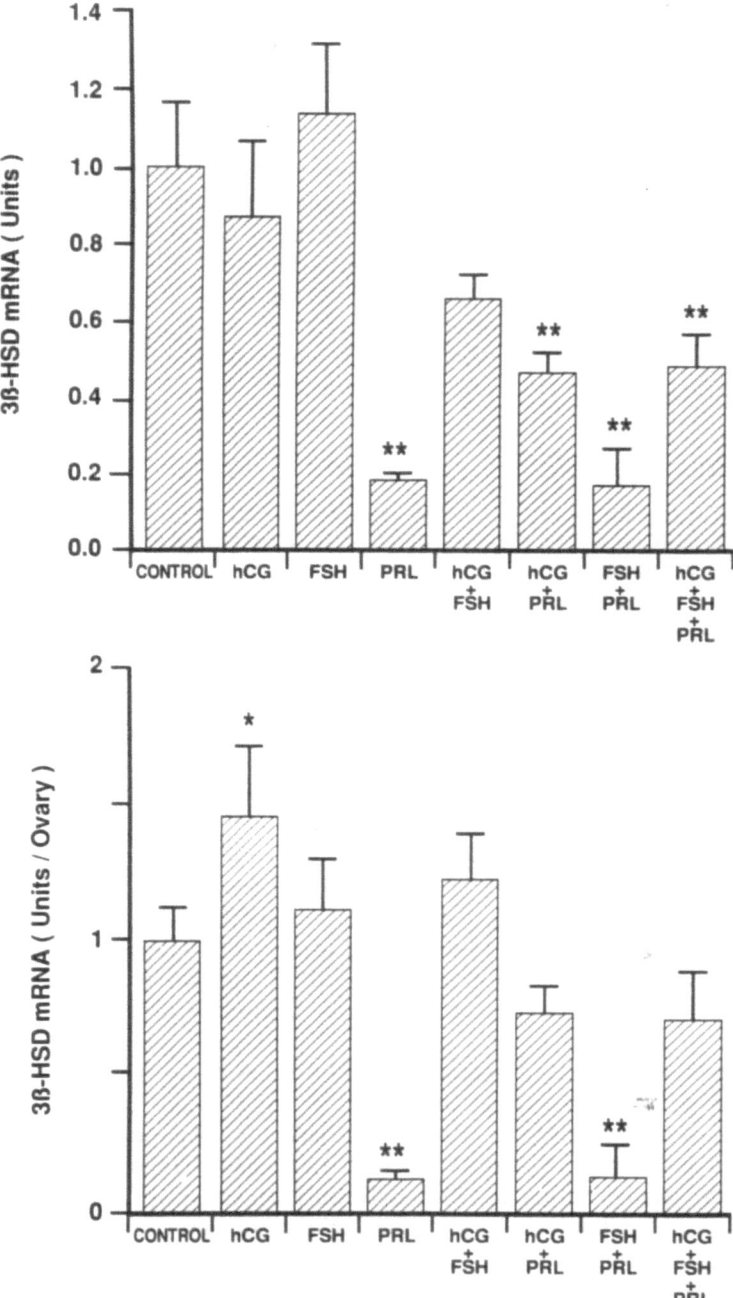

istration of DHEA to castrated rats caused a marked increase in ventral prostate weight accompanied by increased prostatic DHT and androgen-dependent mRNA levels (58, 59). Moreover, considering their relatively large size, the skin and adipose tissue are likely to be important sites of extragonadal sex steroid formation.

In summary, the characterization of rat 3βHSD cDNAs should greatly further our understanding of steroidogenesis in both classical steroidogenic tissues and peripheral intracrine tissues. In fact, the availability of cDNAs, as well as specific cRNA probes for rat 3βHSD isoenzymes, should make possible detailed in vivo as well as in vitro studies on the regulation of the expression of this crucial enzyme. For example, characterization of the gene(s) encoding rat 3βHSD(s) should shed light on the possible existence of tissue-specific promoters.

Regulation of 3βHSD Expression in Rat Ovary

As illustrated in Figure 8.6 (top), whereas hCG alone exerted no effect on the steady state levels of whole ovarian 3βHSD mRNA, oPRL exerted a potent inhibitory effect ($P < 0.01$) on the same parameter as compared to the values found in control hypophysectomized animals. Simultaneous treatment with hCG partially ($P < 0.01$) reversed the inhibitory effect of oPRL, increasing 3βHSD mRNA levels from 19% (oPRL alone) to 55% of the value found in control hypophysectomized animals. Ovine FSH, on the other hand, had no significant effect when administered alone or in combination with hCG, oPRL, or hCG + oPRL. Since hCG and oPRL exerted significant effects on ovarian weight, it is of interest to express 3βHSD mRNA content following correction for relative changes in total ovarian weight. It can be seen in Figure 8.6 (bottom) that hCG caused a 45% ($P < 0.05$) increase in total 3βHSD mRNA content, whereas oPRL exerted an 88% inhibition of total ovarian 3βHSD mRNA content ($P < 0.01$) as compared to the values found in control hypophysectomized rat ovaries. The potent inhibitory effect of oPRL on ovarian 3βHSD mRNA content was also observed on the levels of 3βHSD protein and 3βHSD activity (23).

Since measurement of ovarian 3βHSD mRNA by dot blot hybridization reflects overall changes in ovarian 3βHSD mRNA and does not account for variations occurring in distinct ovarian cell subpopulations, sections of fixed ovaries were hybridized with the same cDNA probe (labeled with ^{35}S) used for the dot blot hybridization assay in order to permit examination of the cellular distribution and to determine possible cell-specific changes in ovarian 3βHSD mRNA levels. The X-ray autoradiographs pictured in Figure 8.7 are representative of the effects of hCG, oFSH, and oPRL in hypophysectomized rat ovaries. As shown in the upper left-hand panel, the corpora lutea found in the ovaries of rats hypophysectomized 25 days previously still express large quantities of

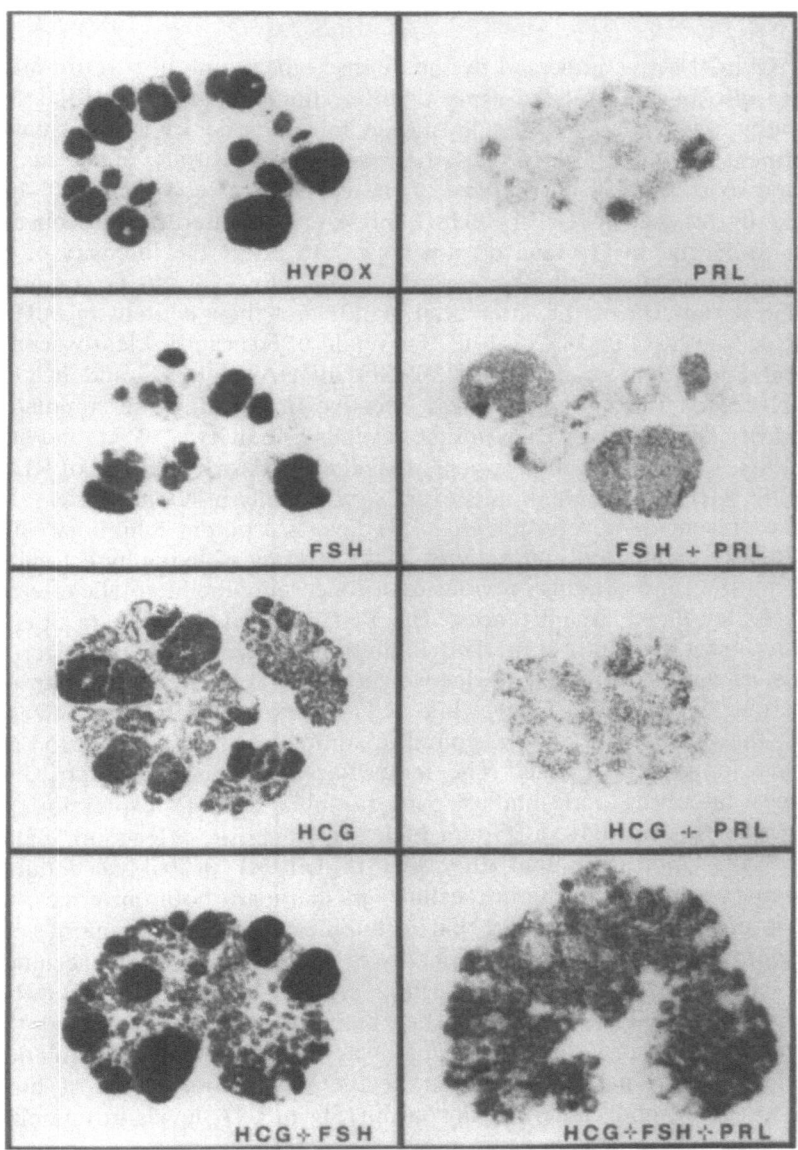

FIGURE 8.7. X-ray autoradiograph illustrating the effects of gonadotropins and PRL on ovarian 3βHSD mRNA levels evaluated by in situ hybridization. Fixed sections of ovaries from left to right and top to bottom: Control hypophysectomized rats (HYPOX) or rats treated with oPRL; oFSH; oFSH + oPRL; hCG; hCG + oPRL; hCG + oFSH; and hCG + oFSH + oPRL, were hybridized to the ^{32}S-labeled rat ovary 3βHSD cDNA (ro3b-HSD 56) as described in reference 23. Treatment with hCG also increases the labeling associated with the interstitial elements. (Magnification: 15×; exposure time: 3 days.) Reprinted with permission from Martel, Labrie, Dupont, et al. (23), © by The Endocrine Society, 1990.

3βHSD mRNA, as indicated by the intense labeling of all corpora lutea. It can also be seen on the same tissue section that the interstitial cells are very weakly labeled. As illustrated in the upper right-hand panel, treatment with oPRL alone caused a reduction in ovarian volume mainly related to a decrease in the size of the corpora lutea, which also show markedly reduced levels of 3βHSD mRNA. It is interesting to observe that oFSH and hCG alone do not appear to affect the intensity of the labeling associated with the corpora lutea. However, hCG treatment increases the size of the interstitial glands and their content in 3βHSD mRNA (see hCG and hCG + FSH). While oPRL can be clearly seen to inhibit corpora lutea-associated 3βHSD mRNA in hCG- and hCG + oFSH-treated rats, it appears less effective in blocking the stimulatory effect of hCG on the ovarian interstitial cells (see hCG + oPRL and hCG + oFSH + oPRL). It thus appears that the inhibitory effect of oPRL on 3βHSD mRNA accumulation is exerted principally in luteal cells.

The present demonstration that PRL exerts a potent inhibitory effect on 3βHSD expression and activity in the ovaries of hypophysectomized rats confirms and extends previous data concerning the luteolytic effect of PRL under these circumstances. The PRL-induced decrease in 3βHSD activity translates into a marked inhibition of circulating progesterone levels, as well as decreased uterine weight, which reflects reduced estrogen secretion. On the other hand, while hCG reverses the inhibitory effect of PRL, the same treatment was found to stimulate 3βHSD expression and activity in interstitial cells. The molecular mechanisms whereby PRL induces luteolysis and inhibits corpora lutea 3βHSD expression are unknown. Interestingly, hCG and PRL exert opposite effects on 3βHSD expression, indicating that the 3βHSD gene(s) probably contain(s) elements responsive to the intracellular mediators of both hormones.

It is of interest to mention that in luteinized porcine granulosa cells, gonadotropins and agents that increase intracellular cAMP accumulation—cholera toxin, forskolin, and (Bu)2 cAMP—increased 3βHSD mRNA levels (44, 60). Furthermore, it has recently been demonstrated that activation of the protein kinase C pathway induces cAMP accumulation, but leads to a marked inhibition of the stimulatory effect of hCG, LH, forskolin, and cholera toxin on 3βHSD mRNA levels in luteinized porcine granulosa cells in culture (60).

Since it is well known that PRL exerts inhibitory effects on LH secretion (61), the present data indicate that PRL, in addition to its inhibitory effect at the hypothalamo-pituitary level, can also act directly on ovarian cells to suppress steroidogenic enzyme gene expression, thus providing an additional explanation for the antifertility effects of hyperprolactinemia in women (62, 63).

While much progress has been made towards a better understanding of the regulation of expression of other steroidogenic enzyme genes in the rat ovary, the recent availability of 3βHSD cDNA probes (9, 11–15) permits detailed studies on the control of this crucial enzyme, thus offering the possibility of obtaining a more complete overall picture of the regulation of ovarian steroidogenesis. The close correlation observed between 3βHSD mRNA, protein content, and activity levels under a wide range of hormonal conditions—namely, inhibition by PRL and stimulation by hCG—indicates that the modulation of 3βHSD activity by both hCG and PRL is largely, if not exclusively, related to parallel changes in 3βHSD gene expression and/or 3βHSD mRNA stability.

Changes in 3βHSD Gene Expression and Activity During the Estrous Cycle in the Bovine Ovary

Understanding of the pattern of ovarian estrogen and progesterone secretion during the estrous cycle requires precise knowledge of the mechanisms controlling the expression and activity of each steroidogenic enzyme involved. Previous studies have described the enzymatic activity and mRNA levels of two important steroidogenic enzymes, $P450_{ssc}$ and $P450_{17\alpha}$ (64, 65). Whereas $P450_{ssc}$ catalyzes the conversion of cholesterol into pregnenolone and $P450_{17\alpha}$ converts pregnenolone and progesterone into the precursors of androgens and estrogens, no information was available about 3βHSD, an essential step in the formation of progesterone as well as of all androgens and estrogens.

We then observed that there is a parallel increase in 3βHSD mRNA, protein content, and enzymatic activity levels from days 1–3 after estrus to maximal values at 50%–100% above control on days 8–11 after estrus. Thereafter, all values decreased progressively until days 16–17 before a dramatic fall to 5% or less than maximal values on days 18–20 after estrus (Fig. 8.8) (27). Almost superimposable results of enzymatic activity were obtained with the three substrates—PREG, DHEA, and Δ^5-diol—thus suggesting a predominant 3βHSD or parallel changes in the activity of multiple 3βHSDs. The above-described changes observed during the luteal phase are almost exclusively due to variations taking place in corpora lutea. In fact, 3βHSD activity in ovarian follicles was approximately 10,000 lower than that measured in corpora lutea.

Such changes in 3βHSD mRNA levels, enzyme content, and enzymatic activity offer an explanation for the marked changes in progesterone secretion observed during the estrous cycle (66). A large increase in the secretion of progesterone occurs after ovulation, and maximal levels are reached 8–10 days after estrus in the bovine species. In parallel, there is a marked increase in $P450_{ssc}$ activity after ovulation (64) from undetectable levels in bovine ovarian follicles (64). Moreover, the activity of HMGCoA reductase, a rate-limiting enzyme in cholesterol biosynthesis, is also greatly elevated in corpora lutea (67), thus providing high levels of precursor cholesterol; whereas the activity of 17α-hydroxylase originally present in theca interna cells decreases to undetectable levels in the

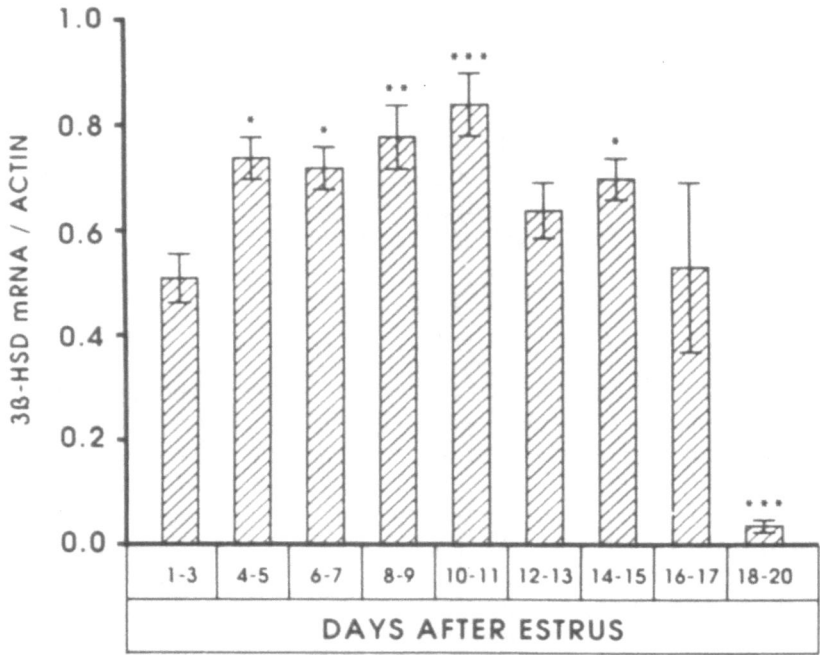

FIGURE 8.8. Relative levels of 3βHSD mRNA in bovine corpora lutea during the luteal phase of the estrous cycle. Total RNA from bovine corpora lutea obtained from the indicated time periods after estrus was extracted by the guanidinium isothiocyanate-CsCl method. RNA was blotted onto nylon membranes after serial 2-fold dilutions (from $10-0.078\,\mu g$ RNA). The ^{32}P-labeled bovine 3βHSD cDNA or chick β-actin probes were added at a concentration of $3 \times 10^6\,cpm/mL$ in hybridization buffer, and the incubation was performed for 16h at 42°C. Spot intensities were measured with an image analyzer, and data are expressed as meants ± SEM of ratios of 3βHSD and actin hybridization intensities. (* = P < 0.05; ** = P < 0.01; *** = P < 0.001, experimental vs days 1–3 after estrus). Reprinted with permission from Couet, Martel, Dupont, et al. (27), © by The Endocrine Society, 1990.

corpus luteum (64), thus blocking androgen and estrogen formation and providing the precursors required for maximal progesterone secretion (64).

Structure and Expression of Human 17βHSD Gene

The nucleotide sequence of the cDNAs encoding *human estradiol 17β-dehydrogenase* (E$_2$DH) predicts a protein of 327 amino acids (28). Placental E$_2$DH contains the 17- and 20-amino acid sequences of the suggested coenzyme binding and catalytic sites, respectively. It also contains the sequence of the 23-amino acid N-terminus determined directly by Edman degradation of the purified enzyme (28). The h17βHSDII gene (29) consists of 6 exons and 5 introns, and its exonic sequence is identical to the hpE$_2$DH216 cDNA sequence previously reported (28). We thus determined the complete nucleotide sequence of the human h17βHSDII and deduced the amino acid sequence of its coding region (29). Comparison of the nucleotide sequence of h17βHSDI with that of the exons and introns of the h17βHSDII gene shows 89% homology. In analogy with gene h17βHSDII, gene h17βHSDI could potentially be transcribed, spliced at the same exon-intron junctions and translated from the first corresponding in-frame ATG codon position. However, due to a change from a G to a T, thus creating a TAA stop codon rather than encoding for the amino acid Gln at position 218, gene h17βHSDI potentially encodes a protein of 214 amino acids (including the first Met) (29).

Two major mRNA species have been identified in poly(A)$^+$ RNA from human placenta: a major species of 1.3 kb and a minor one of 2.2 kb (28, 29). In placenta, S$_1$ analysis indicates that the major mRNA starts 9 nucleotides upstream from the starting codon, while the minor mRNA species contains approximately 971 nucleotides upstream from the same in-frame ATG-initiating codon (29). The 2.2-kb mRNA species is predominant in myometrium and abdominal fat tissues and also has been detected in all other tissues tested, including placenta, testis, ovary, prostate, and breast cancer cells (29). The 1.3-kb h17βHSD mRNA, on the other hand, is very abundant in the placenta and ovary, but present at a lower level in testicular tissue.

Measurement of estrogenic 17βHSD activity in 15 human tissues is illustrated in Figure 8.9A. Such data are complementary to those describing the presence of 17βHSD activity in a series of human tissues (7, 29, 68–76). It can be seen that estrogenic 17βHSD activity was detected in all tissues examined. The highest rates of estrogenic 17βHSD activity were found in the placenta, liver, ovary, endometrium, prostate, testis, and adipose tissue. It can be seen that the differences observed between the reductive and the oxidative pathways are less important in human tissues than in rat tissues (77, data not shown). However, the formation of E$_1$ and E$_2$ is biased toward, or equal to, the formation of E$_2$ into E$_1$ in all tissues except in adipose tissue (78), liver, skin, and testis (68), where the formation of E$_2$ is favored.

Androgenic 17βHSD activity has also been demonstrated in the same 15 human tissues. The rates of testosterone formation (white bars) or Δ^4-dione formation (dark bars) are illustrated in Figure 8.9B. The highest level of androgenic 17βHSD activity was found in the placenta and liver, followed by the testis, endometrium, prostate, adipose tissue, adrenal, prostate, and skin. In human tissues, contrary to the rat, the oxidative

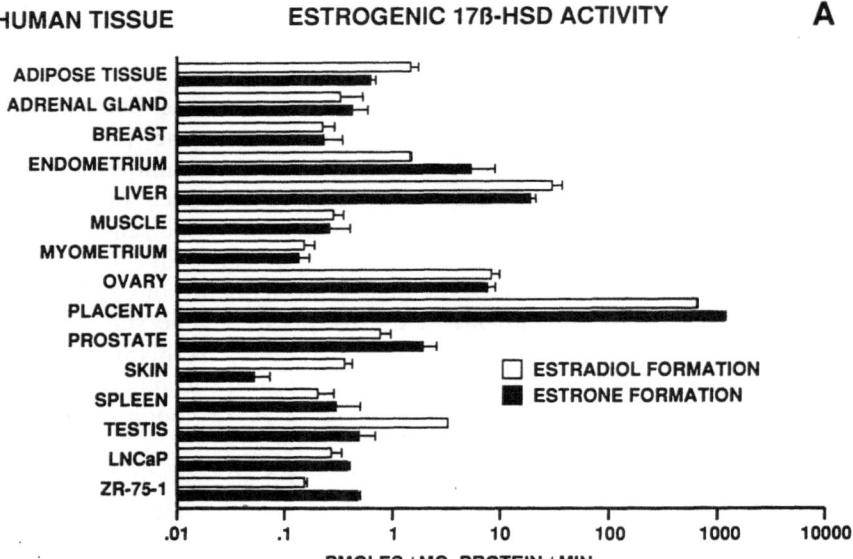

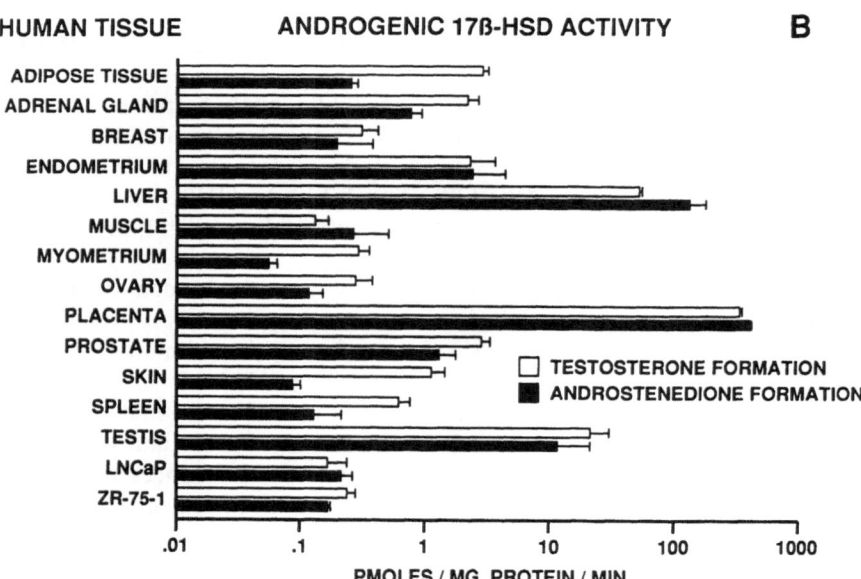

FIGURE 8.9. Tissue distribution of 17βHSD enzymatic activity in a series of human steroidogenic and peripheral intracrine tissues. Estrogenic 17βHSD activity was measured by the formation of E_1 from $[^{14}C]$-E_2 or E_2 from $[^{14}C]$-E_1 *(A)*, while androgenic 17βHSD activity was measured by the formation of Δ^4-dione from $[^{14}C]T$ or T from $[^{14}C]\Delta^4$-dione *(B)*. Incubations were performed at 37° for 1 h with 0.5-μM-labeled substrate and 1 mM of the appropriate cofactors. Data are presented as means ± SEM in pmol of product formed/mg protein/min (log scale).

pathway of Δ^4-dione formation is slightly favored only in the liver and placenta, while all other tissues favor testosterone formation (adipose tissue, adrenal, myometrium, ovary, prostate, skin, spleen, and ZR75-1) or show no significant difference (77).

We have also demonstrated the presence of 17βHSD mRNA by the RNase protection assay performed with poly(A)$^+$ RNA obtained from steroidogenic as well as peripheral human tissues using a cRNA probe specific for 17βHSD mRNA encoded by human 17βHSDII (29). The presence of 17βHSD mRNA has been shown in the RNA from placenta, breast, ovary, endometrium, ZR75-1 cells, LNCaP prostate cancer cells, adipose tissue, skin and prostate (77).

Estrogenic as well as androgenic 17βHSD activity was observed in all tissues examined. However, some tissues, especially the rat adrenal, kidney, liver, ovary and skin, have higher levels of enzymatic activity with estrogenic compared to androgenic substrates, while other tissues, such as the testis, are most specific for androgenic substrates. Such characteristics of specificity for estrogens or androgens suggest the existence of more than one enzyme having 17βHSD activity. Multiple forms of 17βHSD have also been suggested by other groups (79, 81). However, it is also possible that the differences observed could be attributed to conformational changes induced by interaction of the enzyme with different membranes or other specific cell components.

Conclusion

In addition to the classical steroidogenic tissues—namely, the ovaries, testis, adrenals, and placenta—a large series of human peripheral tissues possess all the enzymatic systems required for the formation of active androgens and estrogens from a relatively constant supply of precursor steroids provided by the adrenals. While, so far, most therapeutical approaches have been aimed and limited at controlling steroid formation by the classical steroidogenic tissues, it is clear that most efforts should now be turned towards intracrinology in order to better understand the physiological mechanisms controlling local steroid formation. We would thus be in a position to develop novel therapeutical approaches that take into account the high proportion of steroids that are made locally and that are responsible for the growth and function of normal as well as cancerous tissues. It is important to mention that one-third of all cancers—breast, prostate, ovarian, and uterine cancers—are hormone sensitive and are thus prime candidates for approaches based upon control of intracrine activity. In addition to their therapeutical efficiency, an advantage of endocrine therapies is their excellent tolerance and the usual absence of serious side effects. The field of intracrinology should generate major interest in the pharmaceutical industry in order to develop specific inhibitors of enzymatic activity that could increase the efficacy of more potent and specific antiestrogens and antiandrogens.

References

1. Labrie F. Intracrinology. Mol Cell Endocrinol 1991;78:C113–8.
2. Labrie F, Dupont A, Bélanger A. Complete androgen blockade for the treatment of prostate cancer. In: De Vita V, Hellman S, Rosenberg SA, eds. Important advances in oncology. Philadelphia: J.B. Lippincott, 1985:193–217.
3. Bélanger B, Bélanger A, Labrie F, Dupont A, Cusan L, Monfette G. Comparison of residual C-19 steroids in plasma and prostatic tissue of human, rat and guinea pig after castration: unique importance of extratesticular androgens in men. J Steroid Biochem 32;1989:695–8.
4. Adams JB. Control of secretion and function of C19-delta-5-steroids of the human adrenal gland. Mol Cell Endocrinol 41;1985:1–17.
5. Ferre F, Breuiller M, Cedard L, Duchesnes MJ, Saintot M, Descomps B, Crastes de Paulet A. Human placental Δ5-3β-hydroxysteroid dehydrogenase activity Δ5-3β-HSD; intracellular distribution, kinetic properties, retroinhibition and influence of membrane delipidation. Steroids 26:1975:551–70.
6. Tamaoki B. Steroidogenesis and cell structures. Biochemical pursuit of sites of steroid biosynthesis. J Steroid Biochem 4;1973:89–118.
7. Lacoste D, Bélanger A, Labrie F. Biosynthesis and degradation of androgens in human prostatic cancer cell lines. In: Bradlow H, Castagnetta L, d'Aquino S, Labrie F, eds. Steroid formation, degradation and action in peripheral, normal and neoplastic tissues. Ann New York Acad Sci 1990;595:389–92.
8. Bongiovanni AM, Kellenbenz G. The adrenogenital syndrome with deficiency of 3β-hydroxysteroid dehydrogenase. J Clin Invest 1962;41:2086–92.
9. Luu-The V, Lachance Y, Labrie C, et al. Full length cDNA structure and deduced amino acid sequence of human 3β-hydroxy-5-ene steroid dehydrogenase. Mol Endocrinol 1989;3:1310–2.
10. Luu-The V, Takahashi M, Labrie F. Purification of mitochondrial 3β-hydroxysteroid dehydrogenase/Δ5-Δ4 isomerase from human placenta. Ann NY Acad Sci 1990;595:386–8.
11. Zhao HF, Simard J, Labrie C, Breton N, Rhéaume E, Luu-The V, Labrie F. Molecular cloning, cDNA structure and predicted amino acid sequence of bovine 3β-hydroxy-5-ene steroid dehydrogenase/Δ5-Δ4 isomerase. FEBS Lett 1989;259:153–7.
12. Zhao HF, Rhéaume E, Trudel C, Couet J, Labrie F, Simard J. Structure and sexual dimorphic expression of a liver-specific rat 3β-hydroxysteroid dehydrogenase/isomerase. Endocrinology 1990;127:3237–9.
13. Zhao HF, Labrie C, Simard J, et al. Characterization of 3β-hydroxysteroid dehydrogenase Δ5-Δ4 isomerase cDNA and differential tissue-specific expression of the corresponding mRNAs in steroidogenic and peripheral tissues. J Bio Chem 1991;266:583–93.
14. Simard J, Melner MH, Breton N, et al. Characterization of macaque 3β-hydroxy-5-ene steroid dehydrogenase/Δ5-Δ4 isomerase: structure and

expression in steroidogenic and peripheral tissues in primates. Mol Cell Endocrinol 1991;75:101–10.

15. Rhéaume E, Lachance Y, Zhao HF, et al. Structure and expression of a new cDNA encoding the almost exclusive 3β-hydroxysteroid dehydrogenase/Δ5-Δ4 isomerase present in human adrenals and gonads. Mol Endocrinol 1991;5:1147.

16. Lachance Y, Luu-The V, Labrie C, et al. Characterization of human 3β-hydroxysteroid dehydrogenase/Δ5-Δ4 isomerase gene and its expression in mammalian cells. J Biol Chem 1990;265:20469–75.

17. Bérubé D, Luu-The V, Lachance Y, Gagné R, Labrie F. Assignment of the human 3β-hydroxysteroid dehydrogenase gene to the p13 band of chromosome 1. Cytogenet Cell Genet 1989;52:199–200.

18. Dupont E, Luu-The V, Labrie F, Pelletier G. Ontogeny of 3β-hydroxysteroid dehydrogenase/Δ5-Δ4 isomerase (3β-HSD) in human adrenal gland performed by immunocytochemistry. Mol Cell Endocrinol 1990;74:R7–10.

19. Dupont E, Luu-The V, Labrie F, Pelletier G. Ontogeny of 3β-hydroxysteroid dehydrogenase/Δ5-Δ4 isomerase (3β-HSD) in human testis as studied by immunocytochemistry. J Androl 1991;12:161–4.

20. Dupont E, Labrie F, Luu-The V, Pelletier G. Immunochemical localization of 3β-hydroxysteroid dehydrogenase/Δ5-Δ4 isomerase (3β-HSD) in human ovary. J Clin Endocrinol Metab 1991.

21. Dupont E, Zhao HF, Rhéaume E, et al. Localization of 3β-hydroxysteroid dehydrogenase/Δ5-Δ4 isomerase in the rat gonads and adrenal glands by immunocytochemistry and in situ hybridization. Endocrinology 1990; 127:1394–1403.

22. Dupont E, Luu-The V, Labrie F, Pelletier G. Immunocytochemical localization of 3β-hydroxysteroid dehydrogenase/Δ5-Δ4 isomerase in the gonads and adrenal glands of the guinea pig. Endocrinology 1990;126:2906–9.

23. Martel C, Labrie C, Dupont E, et al. Regulation of 3β-hydroxysteroid dehydrogenase/Δ5-Δ4isomerase expression and activity in the hypophysectomized rat ovary: interactions between the stimulatory effect of human chorionic gonadotropin and the luteolytic effect of prolactin. Endocrinology 1990;127:2726–37.

24. Martel C, Labrie C, Couet J, et al. Effect of human chorionic gonadotropin (hCG) and prolactin (PRL) on 3β-hydroxysteroid dehydrogenase/Δ5-Δ4 isomerase (3β-HSD) expression and activity in the rat ovary. Mol Cell Endocrinol 1990;72:R7–13.

25. Trudel C, Couet J, Martel C, Labrie C, Labrie F. Regulation of adrenal 3β-hydroxysteroid dehydrogenase/Δ5-Δ4 isomerase expression and activity by adrenocorticotropin and corticosterone in the rat. Endocrinology 1991;129.

26. Labrie C, Martel C, Trudel C, et al. Hormonal regulation of testicular 3β-hydroxysteroid dehydrogenase/Δ5-Δ4 isomerase expression and activity in intact rats (submitted).

27. Couet J, Martel C, Dupont E, et al. Changes in 3β-hydroxysteroid dehydrogenase/Δ5-Δ4 isomerase messenger ribonucleic acid, activity and protein levels during the estrous cycle in the bovine ovary. Endocrinology 1990;127:2141–8.

28. Luu-The V, Labrie C, Zhao HF, et al. Characterization of cDNAs for human estradiol 17β-dehydrogenase and assignment of the gene to chromosome 17: evidence of two mRNA species with distinct 5′-termini in human placenta. Mol Endocrinol 1989;3:1301–9.

29. Luu-The V, Labrie C, Simard J, et al. Structure of two in tandem human 17β-hydroxysteroid dehydrogenase genes. Mol Endocrinol 1990;4:268–75.

30. Bongiovanni AM. Acquired adrenal hyperlasia: with special reference to 3β-hydroxysteroid dehydrogenase. Fertil Steril 1981;35:599–608.

31. Cara JF, Moshang T, Bongiovanni AM, Marx BS. Elevated 17-hydroxyprogesterone and testosterone in a newborn with 3β-hydroxysteroid dehydrogenase deficiency. New Engl J Med 1985;313:618–21.

32. Craviato M, Ulloa-Aguirre A, Bermudez JA, et al. A new inherited variant of the 3β-hydroxysteroid dehydrogenase-isomerase deficiency syndrome: evidence for the existence of two isoenzymes. J Clin Endocrinol Metab 1986;63:360–7.

33. de Peretti E, Forest MG, Feit JP, David M. Endocrine studies in two children with male pseudohermaphroditism due to 3β-hydroxysteroid (3β-HSD) dehydrogenase defect. In: Genazzani AR, Thijssen JHH, Siiteri PK, eds. Adrenal androgens. New York: Raven Press, 1980:141–5.

34. Pang S, Levine LS, Stoner E, et al. Nonsalt-losing congenital adrenal hyperplasia due to 3β-hydroxysteroid dehydrogenase deficiency with normal glomerulosa function. J Clin Endocrinol Metab 1983;56:808–18.

35. Zerah M, Schram P, New MI. The diagnosis and treatment of nonclassical 3β-HSD deficiency. Endocrinologist 1991;1:75–81.

36. Eldar-Geva T, Hurwitz A, Becsei P, Palti Z, Milwidsky A, Rosler I. Secondary biosynthetic defects in women with late-onset congenital adrenal hyperplasia. New Engl J Med 1990;323:855–63.

37. Lobo RA, Goebelsmann U. Evidence for reduced 3β-ol-hydroxysteroid dehydrogenase activity in some hirsute women thought to have polycystic ovary syndrome. J Clin Endocrinol Metab 1981;53:394–400.

38. Pang S, Lerner AJ, Stoner E, et al. Late-onset adrenal steroid 3β-hydroxysteroid dehydrogenase deficiency, I. A cause of hirsutism in pubertal and postpubertal women. J Clin Endocrinol Metab 1985;60:428–39.

39. Pochi P, Strauss JS. Sebaceous gland response in man to the administration of testosterone, Δ4-androstenedione, and dehydro-isoandrosterone. J Invest Dermatol 1969;52:32–6.

40. Hay JB, Hodgins MB. Distribution of androgen metabolizing enzymes in isolated tissues of human forehead and axillary skin. J Endocrinol 1978;79:29–39.

41. George FW, Wilson JD. Sex determination. In: Knobil E, Neil JD, eds. The physiology of reproduction. New York: Raven Press, 1988:3.

42. Vigier B, Forest MG, Eychenne B, et al. Antimüllerian hormone produced endocrine sex reversal of fetal ovaries. Proc Natl Acad Sci USA 1989; 86:3684–9.

43. Kaplan SL, Grumach MM. Pituitary and placental gonadotrophin and sex steroids in the human and sub-human primate fetus. J Clin Endocrinol Metab 1978;7:487–511.

44. Chedrese PJ, Luu-The V, Labrie F, Juorio AV, Murphy BD. Evidence for the regulation of 3β-hydroxysteroid dehydrogenase messenger RNA by

human chorionic gonadotrophin in luteinized porcine granulosa cells. Endocrinology 1990;26:2228–30.

45. Aiman J, Forney JP, Parker CR. Secretion of androgens and estrogens by normal and neoplastic ovaries in postmenopausal women. Obstet Gynecol 1986;68:1–5.

46. Vermeulen A. The hormonal activity of the postmenopausal ovary. J Clin Endocrinol Metab 1976;42:247–53.

47. Labrie F, Simard J, de Launoit Y, et al. Androgens and breast cancer. In: Proc Int Symposium on Management of Benign Breast Diseases and Chemoprevention of Breast Cancer, 1991.

48. Tapanainen J, Kellokumpar-Lehtinen P, Pellinemi L, Huhtaniemi I. Age-related changes in endogenous steroids of human fetal testis during early and midpregnancy. J Clin Endocrinol Metab 1981;52:98–102.

49. Winter JSD, Hughes IA, Reyes FI. Pituitary-gonadal relations in infancy, II. Patterns of serum gonadal steroid concentrations in man from birth to two years of age. J Clin Endocrinol Metab 1976;42:679–86.

50. Mancini RE, Vilar O, Lavieri JC, Androda JA, Heinrich JJ. Development of Leydig cells in the normal human testis. Am J Anat 1963;112:203–10.

51. August GP, Grumbach MM, Kaplan SL. Hormonal changes in puberty, III. Correlation of plasma testosterone, LH, FSH, testicular size, and bone age with male pubertal development. J Clin Endocrinol Metab 1972;34:319–26.

52. Murakoshi M, Osamura Y, Watanabe KK, Kuroshima V. Enzyme histochemical studies in human fetal adrenal glands. Tokai J Exp Clin Med 1983;8:89–96.

53. Migeon CJ, Bertrand J, Wale PE. Physiologic disposition of [^{14}C] cortisol during late pregnancy. J Clin Invest 1957;36:1350–62.

54. Beitins IZ, Bayard F, Ances IG, Kowanski A, Migeon CJ. The metabolic clearance rate, blood production, interconversion and trans-placental passage of cortisol and cortisone in pregnancy near term. Pediatr Res 1973;7:509.

55. Murphy BEP. Human fetal cortisol levels related to gestational age: evidence of a mid-gestational fall and a step late gestational rise, independent of sex or mode of delivery. Am J Obstet Gynecol 1982;144:276–82.

56. Simard J, de Launoit Y, Labrie F. Characterization of the structure-activity relationships of rat type I and type II 3β-hydroxysteroid dehydrogenase/Δ5-Δ4 isomerase by site directed mutagenesis and expression in HeLa cells. J Biol Chem 1992 (in press).

57. Gustafsson JA, Stenberg A. Masculinization of rat liver enzyme activities following hypophysectomy. Endocrinology 1974;95:891–6.

58. Labrie C, Bélanger A, Labrie F. Androgenic activity of dehydroepiandro-sterone and androstenedione in the rat ventral prostate. Endocrinology 1988; 123:1412–7.

59. Labrie C, Simard J, Zhao H-F, Bélanger A, Pelletier G, Labrie F. Stimulation of androgen-dependent gene expression by the adrenal precursors dehydroepiandrosterone and androstenedione in the rat ventral prostate. Endocrinology 1989;124:2745–54.

60. Chedrese PJ, Zhang, Luu-The V, Labrie F, Juorio AV, Murphy BD. Regulation of the mRNA expression of 3β-hydroxy-5-ene steroid dehydrogenase in porcine granulosa cells in culture: a role for the protein kinase C pathway. Mol Endocrinol 1991;4:1532–8.

61. Marchetti B, Labrie F. Prolactin inhibits pituitary luteinizing hormone-releasing hormone receptors in the rat. Endocrinology 1982;111:1209–16.
62. Schultz KD, Geiger W, Del Pozo E, Lose KH, Kunzig HJ, Lancranjan I. The influence of the prolactin inhibitor bromocriptine (CB154) on human luteal function in vivo. Arch Gynecol 1976;221:93–9.
63. Micic S, Svenstoup B, Neilsen J. Treatment of hyperprolactinemic luteal insuffciency with bromocriptine. Acta Obstet Gynecol Scand 1979;58:379–83.
64. Rodgers RJ, Waterman MR, Simpson ER. Cytochromes P450ssc, P45017α adrenodoxin and reduced nicotinamide adenine dinucleotide phosphate-cytochrome P450 reductase in bovine follicles and corpora lutea; changes in specific contents during the ovarian cycle. Endocrinology 1986;118:1366.
65. Rodgers RJ, Waterman MR, Simpson ER. Levels of messenger ribonucleic acid encoding cholesterol side-chain cleavage cytochrome P-450, 17α-hydroxylase cytochrome P-450, adrenodoxin and low density lipoprotein receptor in bovine follicles and corpora lutea throughout the ovarian cycle. Mol Endocrinol 1987;1:274.
66. Hansel W, Convey EM. Physiology of the estrous cycle. J Anim Sci 1983;57:404.
67. Rodgers RJ, Mason JI, Waterman MR, Simpson ER. Regulation of the synthesis of 3-hydroxy-3-methylglutamyl coenzyme A reductase in the bovine ovary in vivo and in vitro. Mol Endocrinol 1987;1:172.
68. Ryan KJ, Engel LL. The interconversion of estrone and estradiol by human tissue slices. Endocrinology 1953;52:287–91.
69. Inano H, Tamaoki B. Testicular 17β-hydroxysteroid dehydrogenase: molecular properties and reaction mechanism. Steroids 1986;48:3–26.
70. Tseng L, Stolee A, Gurpide E. Quantitative studies on the uptake and metabolism of estrogens and progesterone by human endometrium. Endocrinology 1972;90:390–404.
71. Weinstein GD, Frost P, Hsia SL. In vitro interconversion of estrone and 17β-estradiol in human skin and vaginal mucosa. J Invest Dermatol 1968;5:4–10.
72. Milewich L, Hendricks TS, Romero LH. Interconversion of estrone and estradiol-17β in lung slices of the adult human. J Steroid Biochem 1982;17:669–74.
73. Breuer H, Knuppen R, Haupt M. Metabolism of oestrone and oestradiol-17β in human liver in vitro. Nature 1966;212:76.
74. Bleau G, Roberts KD, Chapdelaine A. The in vitro and in vivo uptake and metabolism of steroid in human adipose tissue. J Clin Endocrinol Metab 1974;39:236–46.
75. Jacobson GM, Hochberg RB. 17β-hydroxysteroid dehydrogenase from human red blood cells. J Biol Chem 1968;243:2985–94.
76. Poulin R, Poirier D, Mérand Y, Thériault C, Bélanger A, Labrie F. Extensive esterification of adrenal C_{19}-Δ^5 sex steroids to long-chain fatty acids in the ZR-75-1 human breast cancer cell line. J Biol Chem 1989;264:9335–43.
77. Martel C, Rhéaume E, Takahashi M, et al. Distribution of 17β-hydroxysteroid dehydrogenase gene expression and activity in rat and human tissues; distribution of 17β-hydroxysteroid dehydrogenase gene expression and activity in rat and human tissues. J Steroid Biochem 1992 (in press).

78. Beranek PA, Folkerd EJ, Ghilchik M, James VHT. 17β-hydroxysteroid dehydrogenase and aromatase activity in breast fat from women with benign and malignant breast tumors. Clin Endocrinol (Oxf) 1984;20:205–12.
79. Bogovich K, Payne AH. Purification of rat testicular microsomal 17-ketosteroid reductase. Evidence that 17-ketosteroid reductase and 17beta-hydroxysteroid dehydrogenase are distinct enzymes. J Biol Chem 1980; 255:5552–9.
80. Lieberman S, Greenfield NJ, Wolfson A. A heuristic proposal for understanding steroidogenic processes. Endocr Rev 1984;5:128–48.
81. Blomquist CH, Lindemann NJ, Hakanson EY. Steroid modulation of 17β-hydroxysteroid oxidoreductase activities in human placental villi in vitro. J Clin Endocrinol Metab 1987;65:647–52.

9

Structure and Regulation of Gonadal Cell Phosphodiesterases

M. Conti, J.V. Swinnen, C.S.-L. Jin, J.E. Welch,
D.A. O'Brien, and E.M. Eddy

The differentiation and function of gonadal cells are controlled by a complex array of endocrine, paracrine, and autocrine stimuli (1–4). These external stimuli are received and elaborated by gonadal cells through an equally complex array of intracellular second-messenger pathways (1, 5, 6). A constant flow of information through these pathways is necessary to maintain the endocrine and gametogenic functions of the gonads.

Gonadotropins are among the most important signals controlling gonadal cell function. These pituitary hormones bind to surface receptors coupled to a membrane signal transduction system (7, 8) that transfers the hormone signal inside the cell. With the cloning of complementary DNA encoding these plasma membrane proteins and deduction of their primary sequences (9, 10), it has been shown that the FSH and LH/hCG receptors are members of the family of opsin/adrenergic receptors that interact with guanine nucleotide binding proteins (11). A major signal resulting from the activation of the gonadotropin receptor is the *guanine nucleotide stimulatory protein* (G_s)-mediated activation of adenylate cyclase (5, 6, 12, 13). This membrane-bound enzyme catalyzes the synthesis of the second-messenger cAMP. While diffusing from the site of synthesis, cAMP either binds and activates a protein kinase or is degraded and inactivated by cyclic nucleotide *phosphodiesterases* (PDEs) (5, 6, 14, 15). The balance between these two events determines most biological effects of gonadotropins. While extensive studies have been conducted to examine the activation of the cAMP-dependent protein kinase, there has been little attention devoted to understanding how PDEs and cyclic nucleotide degradation are regulated in gonadal cells. Studies in our laboratory have focused on the role of these enzymes in the control of gonadal cell responsiveness.

Although the role of cAMP in mediating gonadotropin action in somatic cells is well established, little is known about the function of the cyclic nucleotide system in the male germ cells. Several laboratories, including ours, have shown that the enzymes involved in the cAMP-dependent pathway are expressed in germ cells during their maturation (16). Haploid germ cells express an adenylate cyclase different from somatic cells (17–20). This form is recovered in the cytosol of spermatogenic cells and is membrane bound in spermatozoa. It has been shown that this adenylate cyclase does not interact with guanine nucleotide binding proteins since it is insensitive to GTP and its analogs and is more active in the presence of Mn^{++} than Mg^{++} (17–20). This indicates that if this enzyme is under the control of external signals, the mechanisms of transduction are different from those described in somatic cells. That the signal transduction machinery is somewhat anomalous in these cells is further suggested by the observation that an *inhibitory guanine nucleotide binding protein* (G_i), but not G_s, can be detected in germ cells (20, 21). It is possible that adenylate cyclase produces cAMP in a constitutive fashion and that PDEs play a major role in regulating intracellular cAMP levels of germ cells. Changes in protein kinase and cyclic nucleotide PDEs have been described, suggesting that this pathway is modified during germ cell differentiation (23–26).

That cyclic nucleotide regulation might play a crucial role during gametogenesis is suggested by the findings in *Xenopus laevis* (27) and mammalian oocytes (28–31). It is widely accepted that a drop in oocyte intracellular cAMP signals the resumption of meiosis. This decrease in cAMP and the following *germinal vesicle breakdown* (GVBD) can be blocked by PDE inhibitors, thus indicating that PDEs play a role in this cAMP regulation (27–30). Although a blockade of the first meiotic division is not present during spermatogenesis, cAMP-dependent mechanisms similar to what has been observed in oocytes may be active in male germ cells. It is possible that a number of check points requiring external stimuli are functioning during the meiotic and postmeiotic stages of spermatogenesis. If appropriate signals regulating intracellular cAMP are not applied to maturing germ cells at these transition points, spermatogenesis may not proceed.

This chapter concentrates on the mechanisms of inactivation of cyclic nucleotides functioning in gonadal cells. We will summarize our findings on the structure and regulation of gonadal cell PDEs and discuss the possible roles of these enzymes in gonadal cell function.

Multiple Genes Encode Different cAMP PDEs

To understand the function of the cAMP degradative pathway of gonadal cells, the PDE forms expressed in these cells have been separated and characterized. The two major forms of PDE expressed belong to the families of *Ca/calmodulin-activated PDEs* (CaM PDEs) (32) and of *cAMP-specific PDEs* (cAMP PDEs) (33). In addition to CaM PDEs similar to those expressed in other tissues, a unique 67-kd CaM PDE that hydrolyzes cAMP and cGMP with similar K_m has been isolated from mouse germ cells (34, 35). Early reports also suggested that a complex number of cAMP PDEs are present in these cells (23, 24). A clearer understanding of the cAMP PDE forms expressed in the gonads comes from molecular cloning data. In *Drosophila melanogaster*, mutations in a *dunce* locus, which encodes a PDE, produce flies with altered central nervous system function and infertility (36). On the basis of this latter observation, rat testis *complementary DNA* (cDNA) libraries were screened with the dunce PDE cDNA to identify homologous PDE sequences. At least 4 genes encoding cAMP PDEs homologous to the dunce PDE are expressed in the rat testis (37, 38). The products of these genes will be referred to as *ratPDE1, ratPDE2, ratPDE3*, and *ratPDE4*. The cloning of cDNAs that correspond to ratPDE2 and ratPDE4 from rat brain libraries also indicates that multiple cAMP PDE genes are present in the rat (39, 40). These cAMP PDE genes encode proteins with molecular weights of approximately 60,000–70,000, and their structure has been in part elucidated. In the center of the protein sequence, there is a region that is highly conserved in all 4 cAMP PDEs (32, 33, 41). This region shares substantial similarities with the sequence of other PDEs (33, 41). Using deletion mutations and site-directed mutagenesis, we have demonstrated that the conserved region contains the catalytic domain of these enzymes (Jin et al., manuscript submitted). Although sequences similar to signals for subcellular localization and phosphorylation sites have been observed (42), the function of the amino and carboxy terminus of this protein is largely unknown.

cAMP PDE Genes Are Differentially Expressed in Testicular Cells

Expression of ratPDE1, ratPDE2, ratPDE3, and ratPDE4 is tissue dependent, and different genes are expressed in different organs (37, 38). In the testis, transcripts corresponding to all 4 genes are present. However, several lines of evidence indicate that ratPDE1 and ratPDE2 are preferentially expressed in the germ line, while ratPDE3 and ratPDE4 are expressed in somatic cells of the seminiferous epithelium. In utero irradiation, a treatment that causes depletion of germ cells in the gonad of the newborn animal, leads to the disappearance of ratPDE1 and ratPDE2 transcripts (37). Conversely, the abundance of ratPDE3 and ratPDE4 transcripts is increased in germ cell-depleted testes (37). Northern blot analysis of mRNA from testes at different stages of development demonstrates that the distribution of ratPDE1 and ratPDE2 transcripts is consistent with expression in germ cells. RatPDE1 transcripts appear at days 15–20 of development, while ratPDE2 transcripts become prominent after 29 days of development, suggesting sequential expression in meiotic and postmeiotic cells (Welch et al., manuscript submitted).

Studies conducted with isolated germ cells have confirmed conclusively the hypothesis that ratPDE1 and ratPDE2 are primarily expressed in these cells. Transcripts corresponding to ratPDE1 and ratPDE2 are present in RNA isolated from an enriched germ cell preparation (37). Furthermore, ratPDE1 and ratPDE2 transcripts were present in the RNA prepared from germ cells at different stages of development isolated using sedimentation at unit gravity (Welch et al., manuscript submitted). Conversely, the low level of signal obtained with ratPDE3 and ratPDE4 probes could be entirely attributed either to contamination of somatic cells in the germ cell preparation or to some crosshybridization of the probes (Swinnen et al., unpublished observation).

On the other hand, enriched Sertoli cell cultures apparently do not express ratPDE1 and ratPDE2. Under all culture conditions tested, only ratPDE3 and ratPDE4 transcripts could be detected. This is also consistent with cloning data. In fact, only ratPDE3 and ratPDE4 cDNA clones could be retrieved from a library derived from immature Sertoli cells in culture (37, 38).

Expression of a cAMP PDE Is Regulated During Spermatogenesis in the Rat

Several lines of evidence indicate that expression of ratPDE1 and ratPDE2 is regulated during meiotic and postmeiotic stages of spermatogenesis. Monn and collaborators were the first to show that a change in pattern of PDE expression occurs during sexual maturation in the rat (23). At that time, however, few biochemical tools were available to distinguish between the different PDE forms. More recently, biochemical analysis of cAMP PDEs derived from the cytosol fractions of isolated germ cells confirms and extends this initial observation. In extracts from pachytene spermatocytes, 2 peaks of cAMP PDE activity can be separated by HPLC ion exchange chromatography (Welch et al., manuscript submitted). They most likely represent the products of ratPDE1 and ratPDE2 genes. As spermatogenesis proceeds to the haploid stage of early spermatids, the second peak of activity decreases and is at the limit of detection in elongating spermatids (Welch et al., manuscript submitted). Measurement of ratPDE1 and ratPDE2 mRNA in meiotic and postmeiotic germ cells is consistent with these biochemical data. RatPDE1 transcripts are present in large amount in pachytene spermatocytes. They are markedly reduced in early spermatids and absent in elongating spermatids. Conversely, ratPDE2 transcripts increase in relative concentration from pachytene spermatocytes to round spermatids. In addition to these major changes in overall ratPDE1 and ratPDE2 mRNA levels, it was found that multiple transcripts of different sizes originate from ratPDE1 and ratPDE2 genes (Welch et al., manuscript submitted). Differences were observed in the pattern of transcript sizes both during testis development and in germ cells at different stages of spermatogenesis. In summary, these studies suggest that the ratPDE1 gene is actively transcribed only during meiotic prophase, while ratPDE2 transcription occurs mostly during the haploid phase of spermatogenesis. The biological significance of this regulation is being investigated.

Expression of cAMP PDEs in the Sertoli Cell Is Under Hormonal Control

Stringent regulation of cAMP PDE gene expression is also suggested by studies in the Sertoli cell. Northern blot analysis of mRNA from immature Sertoli cells shows that transcripts corresponding to ratPDE3 and ratPDE4 are expressed in this cell under different culture conditions (38, 42). After 4 days in culture without hormones, no ratPDE3 transcripts could be detected, while ratPDE4 transcripts were present. Incubation for 24 h with FSH or dibutyryl cAMP led to the appearance and accumulation of a large amount of ratPDE3 transcripts and to a smaller increase in ratPDE4 transcripts (38, 42). Thus, ratPDE3 is entirely FSH- and cAMP-dependent, while ratPDE4 is, to some extent, constitutively expressed. Preliminary evidence from our laboratory indicates that ratPDE3 is present in the Sertoli cell at the time of plating and disappears as the cells are maintained in culture, suggesting that continuous FSH stimulation is required for ratPDE3 expression (Swinnen, Conti, unpublished results). A similar increase in ratPDE3 mRNA was observed in vivo after FSH injection in 15-day-old rats (42).

It should be pointed out that ratPDE3 and ratPDE4 regulation is not a phenomenon unique to the Sertoli cell. Similar stimulation has been observed in C6 glioma cells (38) and in FRTL-5 cells (43), a rat thyroid cell line responsive to TSH. In addition, FSH regulates the PDE activity of cultured granulosa cells (44). These cAMP PDEs are therefore part of a ubiquitous intracellular feedback mechanism by which cAMP regulates the expression of its own degrading enzymes.

The increase in cAMP PDE activity is blocked by actinomycin D treatment of the Sertoli cell (45), suggesting that ongoing RNA synthesis is required for FSH activation. That cAMP PDE gene transcription is regulated by FSH has been confirmed by run-on experiments showing that ratPDE3 gene transcription is increased within 30 min of FSH stimulation (42). Protein synthesis inhibition does not abolish the FSH-dependent increase in ratPDE3 and ratPDE4 mRNAs (Swinnen, Conti, unpublished results). This, together with the finding that FSH can activate transcription of a reporter gene through *cAMP response elements* (CRE) (Hall et al., manuscript in preparation), would indicate that FSH regulates ratPDE3 transcription in the absence of de novo synthesis of proteins, possibly via a posttranslational modification of a *cAMP response element binding protein* (CREB) (46) or related transacting factors. Comparison of the time course of FSH activation of the protooncogene c-*fos* and ratPDE3 shows that the FSH-dependent increase in c-*fos* mRNA occurs in 15 min, while ratPDE transcripts appear with a time lag of approximately 1 h (42). Nuclear run-off experiments show that FSH increases c-*fos* and ratPDE3 gene transcription with different time courses (42). The significance of this is under investigation, but it might indicate differences in the mechanisms involved in the transcriptional activation of c-*fos* and ratPDE3 genes. The increase in PDE mRNA is followed by increased accumulation of the PDE protein as determined by immunoblot analysis of Sertoli cell extracts (42). The increase in PDE protein is consistent with the increase in cAMP PDE activity observed (45, 47–49). It is, however, possible that posttranslational modifications of the newly synthesized PDE protein are also necessary to express the full catalytic capacity of the enzyme.

Although FSH treatment of the Sertoli cell causes an increase in ratPDE4 mRNA, no significant changes in ratPDE4 transcriptions were observed under these conditions (42). This is an indication that the increase in ratPDE4 mRNA observed after FSH stimulation might be the result of mRNA stabilization.

Hormonal Activation of the cAMP PDE Causes a Decreased Response of the Target Cell

Studies from several laboratories, including ours, have shown that an increase in cAMP PDE activity coincides with the onset of the refractoriness of the Sertoli cell (47–50). Sertoli cell refractoriness is reversed if cells are exposed to different PDE inhibitors (50, 51), indicating that the observed increase in PDE activity has a major role in regulating Sertoli cell sensitivity to hormones. Thus, FSH stimulation of estrogen accumulation in desensitized cells in the presence of *methyl-isobutyl xanthine* (IBMX) or RO-201724 is comparable to fully responsive cells (51). Furthermore, the cAMP analog dibutyryl cAMP, which is poorly hydrolyzed by the cAMP PDEs, stimulates steroidogenesis of the desensitized cells to levels comparable to control cells (51). Also, Sertoli cell responses to heterologous stimuli are dependent on the state of activation of the cAMP PDE. In fact, the decrease in beta-adrenergic agonist, forskolin, or cholera toxin stimulation that follows FSH treatment can be completely reverted by including PDE inhibitors in the incubation medium (50, 51).

The loss of responsiveness of the Sertoli cell is the result of several different desensitizing mechanisms. Exposure of these cells to FSH produces a rapid decrease in the adenylate cyclase response to FSH (47, 50), and after approximately 1 h, an increase in the PDE activity (45, 47, 48). At later times, the number of FSH receptors exposed on the surface of the Sertoli cell decreases (52, 53). These three mechanisms contribute to the overall decrease in responsiveness of the Sertoli cell to repeated stimulation. The relative contribution of these three mechanisms, which operate at different steps of the cAMP-dependent cascade, to the overall decrease in response is dependent both on time and on the concentration of FSH that induces the refractory state (50). Early on, after exposure to FSH (30–60 min), adenylate cyclase desensitization plays a major role in reducing FSH responsiveness. At later times (2–24 h), the induction of the cAMP PDE is a major factor in limiting the response. Receptor down-regulation also has delayed effects. Furthermore, low concentrations of FSH (1–10 ng/mL) produce little desensitization of the adenylate cyclase and receptor down-regulation, but maximally stimulate the PDE activity (50). This is due to the fact that PDE induction is a highly amplified phenomenon requiring activation of many steps in the hormone-dependent cascade, including activation of gene expression. Conversely, receptor down-regulation and adenylate cyclase desensitization require activation of few steps of the cAMP-dependent cascade and are less amplified (33).

The role of an increase in PDE activity in the control of cell responsiveness to gonadotropins has been recently confirmed in the MA-10 Leydig cell tumor line (54). These cells respond to the gonadotropin LH/hCG with an increase both in cAMP and in steroid production (progesterone in this case) (55). The PDE activity present in these cells can be manipulated by transfection of expression vectors containing the full-length cDNA encoding a cAMP PDE. Stable transfected clones have been isolated that display a 2- to 7-fold increase in recombinant cAMP PDE activity when compared to wild-type cells. When the responsiveness of these cells containing an elevated PDE activity was tested by challenging the cells with hCG, it was found that cAMP response to hCG is decreased up to 95% and progesterone production up to 75%. This loss of response was present despite the fact that hCG is still able to fully activate adenylate cyclase in these transfected cells. Furthermore, the responsiveness to hCG in these clones could be restored either by incubating the cells with cAMP PDE inhibitors or by stimulating the cells with cAMP analogs that are poor substrates for the cAMP PDE. These latter findings confirm that the increase in PDE activity is the primary cause of the reduction in MA-10 response. Taken together, the above findings support the concept that a change in PDE activity has a major impact on cell responsiveness to gonadotropin. It is worth noting that a

5-fold increase in MA-10 cell PDE activity is sufficient to reduce cAMP accumulation by more than 90%. If it is considered that FSH produces more than a 10-fold increase in PDE activity in the immature Sertoli cells in culture, it can be concluded that the cAMP PDE induction can be by itself a major cause of loss of response.

In addition to a role during the acute FSH regulation of response, an increase in PDE activity might cause the changes in responsiveness of the Sertoli cell that occur during testicular maturation. It has been shown that the FSH-dependent cAMP accumulation in the Sertoli cell decreases progressively during testicular maturation (55). Although a reduction in adenylate cyclase activation by FSH plays a role (56), the following findings are consistent with a role of cAMP PDEs. An increase in cAMP PDE activity occurs in the Sertoli cell during testicular development (5, 24). Furthermore, the response to FSH can be, at least in part, restored by PDE inhibitors (5). Therefore, a change in adenylate cyclase and PDE are probably responsible for this loss in response during maturation.

Summary

The data reviewed above demonstrate that cyclic nucleotide degradation in gonadal cells is carried out by a complex array of phosphodiesterases. This function of second-messenger inactivation is not constitutively expressed in gonadal cells. The activity of PDEs is regulated by gonadotropins, and their activation plays an important role in regulation of cell responsiveness to different external stimuli. The observation that these enzymes are regulated during germ cell development suggests an important role of these enzymes in the control intracellular cAMP levels in male germ cells. It is then possible that regulation of intracellular cAMP plays a regulatory role during spermatogenesis in a manner similar to what has been shown for oocyte maturation.

Acknowledgments. The authors are indebted to Frank French for the continuous support during the completion of the studies described. The work from the authors' laboratories was supported by NIH Grants HD-20788 (M. Conti), HD-26485 (D. O'Brien), and the Center for Population Research Grant P-30-HD-18968. The support of grants from the Andrew W. Mellon Foundation and Glaxo, Inc., is also acknowledged.

References

1. Hsueh AJW, Adashi EY, Jones PBC, Welsh TH Jr. Hormonal regulation of the differentiation of cultured ovarian granulosa cells. Endocr Rev 1984;5:76–126.

2. Parvinen M. Regulation of the seminiferous epithelium. Endocr Rev 1982;3:404–17.
3. Stefanini M, Conti M, Geremia R, Ziparo E. Regulatory mechanisms of mammalian spermatogenesis. In: Metz, Monroy A, eds. Biology of fertilization. Academic Press, 1984:3–46.
4. Skinner MK. Cell-cell interaction in the testis. Endocr Rev 1991;12:45–77.
5. Means AR, Dedman JR, Tash JS, Tindall DJ, van Sickle M, Welsh MJ. Regulation of the testis Sertoli cell by follicle stimulating hormone. Annu Rev Physiol 1980;42:59–70.
6. Conti M, Monaco L, Hall S, Joseph D, French F. Modulatory mechanisms of FSH and LH action in gonadal cells. In: Parvinen M, Huhtaniemi, Pelliniemi, LJ, eds. Development and function of the reproductive tract. Serono Symposia Rev 1988;14:139–51.
7. Bhalla VK, Reichert LE. Properties of follicle-stimulating hormone receptor interaction. J Biol Chem 1974;249:43–51.
8. Catt KJ, Dufau ML. Spare gonadotropin receptors in rat testis. Nature 1973;244:219–21.
9. McFarland KC, Sprengel R, Phillips HS, et al. Lutotropin-choriogonadotropin receptor: an unusual member of the G protein coupled receptor family. Science 1989;245:494.
10. Loosfelt H, Misrahi M, Atger M, et al. Cloning and sequencing of porcine LH-hCG receptor cDNA: variant lacking transmembrane domain. Science 1989;245:525.
11. O'Dowd BF, Lefkowitz RJ, Caron MG. Structure of the adrenergic and related receptors. Annu Rev Neurosci 1989;12:67–83.
12. Birnbaumer L, Kirchick HJ. In: Grenwald GS, Terranova PF, eds. Factors regulating ovarian function. New York: Raven Press, 1983:287–310.
13. Abou-Issa H, Reichert LEJ. Modulation of follicle-stimulating hormone-sensitive rat testicular adenylate cyclase activity by guanyl nucleotides. Endocrinology 1979;104:189–93.
14. Fakunding JL, Tindall DJ, Dedman JR, Mena CR, Means AR. Biochemical actions of follicle-stimulating hormone in the Sertoli cell of the rat testis. Endocrinology 1976;98:392–402.
15. Cooke BA, Dix CJ, Magee-Brown R, Jansezn FHA, Van Der Molen HJ. Adv Cyclic Nucleotide Protein Phosphorylation Res 1981;14:593–609.
16. Conti M, Monaco L. In: Stefanini M, Conti M, Geremia R, Ziparo E, eds. Molecular and cellular endocrinology of the testis. Excerpta Medica Amsterdam, 1986:89–100.
17. Braun T, Dods RF. Development of Mn^{2+}-sensitive "soluble" adenylate cyclase in rat testis. Proc Natl Acad Sci USA 1975;72:1097–101.
18. Neer EJ. Physical and functional properties of adenylate cyclase from mature rat testis. J Biol Chem 1978;253:5808–12.
19. Adamo S, Conti M, Geremia R, Monesi V. Particulate and soluble adenylate cyclase activities of mouse male germ cells. Biochem Biophys Res Commun 1980;97:607–13.
20. Gordeladze JO, Hansson V. Purification and kinetic properties of the soluble Mn^{2+}-dependent adenylyl cyclase of the rat testis. Mol Cell Endocrinol 23:125.

21. Kopf GS. Mechanisms of signal transduction in mouse spermatozoa. Ann NY Acad Sci 1989;564:289–302.
22. Strathmann M, Wilkie TM, Simon MI. Diversity of the G-protein family: sequences from five additional alpha subunits in the mouse. Proc Natl Acad Sci USA 1989;86:7407–9.
23. Monn E, Desatel M, and Christiansen RO. Highly specific testicular adenosine 3'-5'-monophosphate phosphodiesterases associated with sexual maturation. Endocrinology 1972;91:716–20.
24. Geremia R, Rossi P, Pezzotti R, Conti M. Cyclic nucleotide phosphodiesterase in developing rat testis. Identification of somatic and germ cell forms. Mol Cell Endocrinol 1982;28:37–53.
25. Conti M, Adamo S, Geremia R, Monesi V. Developmental changes of cAMP-dependent protein kinase activity in mouse male germ cells. Biol Reprod 1983;28:860–9.
26. Oyen O, Scott JD, Cadd GG, et al. A unique mRNA species for regulatory subunit of cAMP-dependent protein kinase is specifically induced in haploid germ cells. FEBS Lett 1988;229:391–4.
27. Maller JL. Interaction of steroids with the cyclic nucleotide system in amphibian oocytes. Adv Cyclic Nucleotide Protein Phosphorylation Res 1983;15:295–337.
28. Cho WK, Stern S, Biggers JD. Inhibitory effect of dibutyryl cAMP on mouse oocyte maturation in vitro. J Exp Zool 1974;187:383–6.
29. Dekel N, Lawrence TS, Gilula NB, Beers W. Modulation of cell-to-cell communication in the cumulus-oocyte complex and the regulation of oocyte maturation. Dev Biol 1981;80:356–62.
30. Bornslaeger EA, Wilde MW, Schultz RM. Regulation of mouse oocyte maturation: involvement of cyclic AMP phosphodiesterase and calmodulin. Dev Biol 1984;105:448–99.
31. Salustri A, Petrungaro S, De Felici M, Conti M, Siracusa G. Effect of follicle-stimulating hormone on cyclic adenosine monophosphate level and meiotic maturation in mouse cumulus cell-enclosed oocytes cultured in vitro. Biol Reprod 1985;33:797–802.
32. Beavo JA. Multiple isozymes of cyclic nucleotide phosphodiesterase. Adv Second Messenger and Phosphoprotein Res 1988;22:1–38.
33. Conti M, Swinnen JV. Structure and function of the rolipram-sensitive low-Km cyclic AMP phosphodiesterase: a family of highly related enzymes. In: Beavo J, Houslay MD, eds. Cyclic nucleotide phosphodiesterases: structure, regulation and drug action. Chichester, England: Wiley & Sons, 1990:243–66.
34. Geremia R, Rossi P, Mocini D, Pezzotti R, Conti M. Characterization of a calmodulin-dependent high affinity cyclic AMP and cyclic GMP phosphodiesterase from male mouse germ cells. Biochem J 1984; 217:693–700.
35. Rossi P, Giorgi M, Geremia R, Kincaid RL. Testis-specific calmodulin-dependent phosphodiesterase. J Biol Chem 1988;263:15521.
36. Chen CN, Denome S, Davis RL. Molecular analysis of cDNA clones and the corresponding genomic coding sequences of the *Drosophila* dunce + gene, the structural gene for cAMP phosphodiesterase. Proc Natl Acad Sci USA 1986;83:9313–7.

37. Swinnen VM, Joseph DR, Conti M. Molecular cloning of rat homologues of the *Drosophila melanogaster* dunce cAMP phosphodiesterase: evidence for a family of genes. Proc Natl Acad Sci USA 1989;86:5325–9.
38. Swinnen VM, Joseph DR, Conti M. The mRNA encoding a high affinity cAMP phosphodiesterase is regulated by hormones and cAMP. Proc Natl Acad Sci USA 1989;86:8197–201.
39. Davis RL, Takaysasu H, Eberwine M, Myres J. Cloning and characterization of mammalian homologs of the *Drosophila* dunce + gene. Proc Natl Acad Sci USA 1989;86:3604–8.
40. Colicelli J, Birchmeier C, Michaeli T, O'Neill K, Riggs M, Wigler M. Isolation and characterization of a mammalian gene encoding a high-affinity cAMP phosphodiesterase. Proc Natl Acad Sci USA 1989;86:3599–603.
41. Charbonneau H. Structure-function relationships among cyclic nucleotide phosphodiesterase. In: Beavo JA, Houslay MD, eds. Cyclic nucleotide phosphodiesterases: structure, regulation and drug action. Chichester, England: Wiley and Sons, 1990;2:267–96.
42. Swinnen JV, Tsikalas KE, Contin M. Properties and hormonal regulation of two structurally related cAMP-phosphodiesterases from the rat Sertoli cell. J Biol Chem 1991.
43. Takahashi S, Swinnen JV, Van Wyk JD, Conti M. TSH regulation of phosphodiesterase in FRTL-5 cells [Abstract 1035]. Endocr Soc 1991:292.
44. Conti M, Kasson BG, Hsueh AJW. Hormonal regulation of 3',5'-adenosine monophosphate phosphodiesterases in cultured granulosa cells. Endocrinology 1984;114:2361–8.
45. Conti M, Toscano MV, Petrelli L, Geremia R, Stefanini M. Regulation by follicle-stimulating hormone and dibutyryl adenosine 3',5'-monophosphate of a phosphodiesterase isoenzyme of the Sertoli cell. Endocrinology 1982;110:1189–96.
46. Gonzales GA, Montminy MR. Cyclic AMP stimulates somatostatin gene transcription by phosphorylation of CREB at serine 133. Cell 1989;59:675.
47. Conti M, Toscano MV, Geremia R, Stefanini M. Follicle-stimulating hormone regulates in vivo testicular phosphodiesterase. Mol Cell Endocrinol 1983;29:73–89.
48. Verhoeven G, Cailleau J, de Moor P. Hormonal control of phosphodiesterase activity in cultured rat Sertoli cells. Mol Cell Endocrinol 1981;24:41–51.
49. Verhoeven G, Cailleau J, de Moor P. Desensitization of cultured rat Sertoli cells by follicle stimulating hormone and by L-isoproterenol. Mol Cell Endocrinol 1980;20:113–26.
50. Conti M, Toscano MV, Petrelli L, Geremia R, Stefanini M. Involvement of phosphodiesterase in the refractoriness of the Sertoli cell. Endocrinology 1983;113:1845–53.
51. Conti M, Monaco L, Geremia R, Stefanini M. Effect of phosphodiesterase inhibitors on Sertoli cell refractoriness: reversal of the impaired androgen aromatization. Endocrinology 1986;118:901–8.
52. O'Shaghnessy PJ. FSH receptor autoregulation and cyclic AMP production in the immature rat testis. Biol Reprod 1980;23:810–4.
53. Francis GL, Brown TJ, Bercu BB. Control of Sertoli cell response to FSH: regulation by homologous hormone exposure. Biol Reprod 1981;24:955–61.

54. Swinnen JV, D'Souza B, Conti M, Ascoli M. Attenuation of cAMP-mediated responses in MA-10 Leydig tumor cells by genetic manipulation of a cAMP phosphodiesterase. J Biol Chem 1991;266:14383–9.
55. Ascoli M. Characterization of several clonal lines of cultured Leydig tumor cells: gonadotropin receptors and steroidogenic responses. Endocrinology 1981;108:88–95.
55. Steinberger A, Hintz M, Heindel JJ. Changes in cAMP responses to FSH in isolated rat Sertoli cells during sexual maturation. Biol Reprod 1978;19:566.
56. Van Sickle M, Oberwetter JM, Birnbaumer L, Means AR. Developmental changes in the hormonal regulation of rat testis Sertoli cell adenylyl cyclase. Endocrinology 1981;109:1270–80.

Part III

Placenta and Fetus

10

Differentiation of Human Trophoblasts: Structure-Function Relationships

LEE-CHUAN KAO, GBOLAGADE O. BABALOLA, GREGORY S. KOPF, CHRISTOS COUTIFARIS, AND JEROME F. STRAUSS III

The human placenta and the chorion laeve are derived from the trophectoderm of the implanting blastocyst (1). During the process of implantation, the trophoblast cells replicate and invade into the uterine endometrium, initiating the formation of a hemochorial placenta. The trophoblast cells differentiate along several different pathways, becoming extravillous trophoblasts (sometimes called intermediate trophoblasts), extravillous multinucleated giant cells, columns of cytotrophoblasts that anchor the conceptus to the uterus, and floating chorionic villi. The chorionic villi form from avascular buds of cytotrophoblasts that develop into multilayered ramifications. The villi comprise an outer layer of syncytiotrophoblast overlying mononucleate cytotrophoblasts that are connected to each other and the syncytiotrophoblast by desmosomes. The cytotrophoblasts sit upon a basement membrane that encapsulates the villus core that contains blood vessels, macrophages, and mesenchymal elements. Each of the trophoblast phenotypes noted above displays characteristic functional properties that have been elucidated by immuno-cytochemical studies, in situ hybridization histochemistry, and analysis of isolated tissues and cells in vitro.

The endocrine functions of the human placenta are largely, but not exclusively, carried out by the syncytiotrophoblast of the chorionic villi. The syncytiotrophoblast is a polarized cell type with a dense microvillous system on its apical surface. The cytoplasm of the syncytial cells contains abundant stacks of dilated endoplasmic reticulum tubules, a prominent Golgi apparatus, lipid droplets, numerous vesicles, and many small mitochondria. These ultrastructural features are indicative of intense secretory activity. Indeed, the syncytiotrophoblast elaborates a diverse repertoire of glycoprotein, protein, and steroidal hormones, including *chorionic gonadotropin* (CG), *chorionic somatomammotropin* (CS), variant growth hormone, progesterone, and estrogens (2, 3). An impressive number of growth factors and cytokines are also produced, including interleukins, hepatocyte growth factor, insulin growth factors, and colony stimulatory factor I. These hormones and growth factors are produced in different patterns during the course of gestation. The trophoblast cells (cytotrophoblasts and syncytiotrophoblasts) not only synthesize hormones, growth factors, cytokines, and extracellular matrix proteins, they express receptors for a number of these and other hormones and extracellular matrix proteins (integrins).

The syncytiotrophoblast is a terminally differentiated cell type. It is formed by the fusion of postmitotic mononucleate cytotrophoblasts (4, 5). The precursor cytotrophoblasts are the replicating trophoblast cells; they are believed to multiply under the direction of growth factors derived from the embryo and possibly from endometrial cells at the implantation site (6, 7). Cytotrophoblasts have a distinctive structure and pattern of endocrine activity that differs from that of the syncytiotrophoblast. Compared to the syncytiotrophoblast, the cytotrophoblast cytoplasm is relatively simple; it contains large mitochondria, scattered irregular segments of rough endoplasmic reticulum, and few lipid droplets. Cytotrophoblasts produce a variety of hormones, including gonadotropin releasing hormone, somatostatin, and inhibin/activin subunits. As cytotrophoblasts differentiate and form syncytiotrophoblasts, their functional properties change as well as their morphology and ultrastructure. The dynamics of this process in vivo are intricate and involve interactions between the trophoblast cells through paracrine and autocrine signaling, interactions with the extracellular matrix, and paracrine substances elaborated by the uterus. Given the potential complexity, it is unreasonable to expect that existing in vitro systems could fully mimic these events. However, in vitro systems do offer opportunities to examine some of the fundamental aspects of the differentiation process. This chapter reviews results obtained from in vitro models that address aspects of structure-function relationships during trophoblast differentiation.

In Vitro Systems for the Study of Trophoblast Differentiation

Term human placental tissue can be enzymatically dispersed and the isolated cells fractionated by the use of density gradients and various immunoselection methods, with the resulting isolation of highly purified populations of cytotrophoblasts (8). The syncytium is sensitive to enzymatic treatment and is usually destroyed. Cytotrophoblasts isolated from term placenta are mostly postmitotic cells; less than 10% of the cells incorporate ^{3}H-thymidine, and mitotic figures are rare (5). By 24 h of culture, there is virtually no DNA synthesis as assessed by ^{3}H-thymidine incorporation. In contrast, dispersion of placentae obtained in the first trimester of pregnancy, when the placenta is growing most rapidly, yields replicating cytotrophoblast that can be passed in culture through several generations (8).

Cytotrophoblasts isolated from term placentae are particularly useful for studying the final stages of trophoblast differentiation. They can be cultured under a variety of conditions, including suspension culture and standard tissue culture systems on plastic or on extracellular matrix in serum-supplemented and serum-free medium (8–11).

Cell lines derived from choriocarcinomas are an alternative to the use of primary trophoblast cell cultures (8). The various choriocarcinoma lines that have been established are generally cytotrophoblast-like. Although some cells in some lines will fuse to form syncytia, these cells are rarely capable of expressing genes characteristic of terminally differentiated syncytiotrophoblast (e.g, CS). Moreover, they oftentimes respond to stimuli, including cAMP analogs, in ways opposite to what is observed in primary cultures of trophoblast cells. Thus, choriocarcinoma cell lines have not proved to be optimal model systems for the exploration of certain aspects of trophoblast differentiation.

Morphological and Functional Differentiation of Cytotrophoblasts In Vitro

Cytotrophoblasts freshly isolated from term placental tissue and cultured in serum-supplemented medium or in serum-free medium on surfaces coated with extracellular matrix proteins undergo morphological differentiation to form syncytia (5, 10). The mononucleate cells are seen to move in a random fashion by time-lapse video microscopy in the initial hours after plating. When the cells make contact, they adhere and may continue to move, but in a more circumscribed area. The motile cells are rounded, whereas cells that have made contact with other cells tend to spread on the culture substrate. With time in culture, cell aggregates increase in number and size. The extent and kinetics of aggregation are dependent on plating density. The aggregated cells ultimately fuse to form multinucleated giant cells that show relatively little motility compared to the freshly isolated cytotrophoblasts. The kinetics of cell fusion and the size of the syncytia formed are again related to plating density.

The morphological changes in the cultured cells are associated with significant changes in endocrine function as evidenced by an increase in CG secretion (Fig. 10.1), CS, and an increase in the level of the mRNA encoding cytochrome $P450_{scc}$, the rate-determining enzyme in steroidogenesis. These biochemical changes seem to reflect those occurring in the chorionic villi as cytotrophoblasts form syncytiotrophoblasts. In situ CG gene expression and steroidogenic activity are primarily

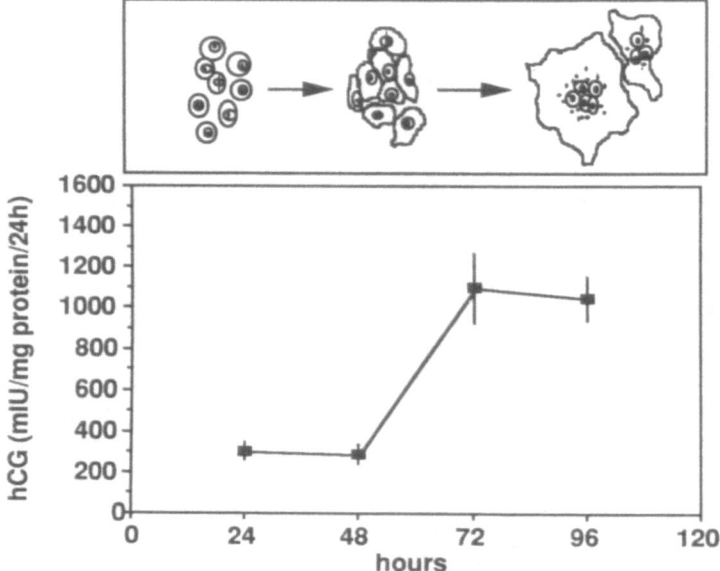

FIGURE 10.1. Morphological differentiation of human trophoblasts CG secretion. The stages of morphological differentiation are shown schematically. The temporal pattern of CG secretion during trophoblast differentiation is also plotted. Modified with permission from Kliman, Nestler, Sermasi, Sanger, and Strauss III (5).

localized in the syncytiotrophoblast, with little or no expression of these genes in cytotrophoblasts (11). The concomitant changes in morphology and function suggest that the morphological differentiation of cytotrophoblasts into syncytiotrophoblasts is coupled to endocrine differentiation.

cAMP and Trophoblast Differentiation

In most endocrine tissues, cAMP plays a central role as a second messenger governing hormone production. cAMP also appears to have a key function in trophoblast differentiation. Cytotrophoblasts possess adenylate cyclase, G-proteins, and cAMP-dependent protein kinase (12). Interestingly, adenylate cyclase is localized to the basal membranes of the syncytiotrophoblast that abut against those of the underlying cytotrophoblasts (13), suggesting that interactions of these two trophoblast cell types regulate cAMP production.

   Treatment of isolated cytotrophoblasts with agents that activate endogenous adenylate cyclase, such as forskolin and cholera toxin, and cAMP analogs increases the production of progesterone and CG within 24 h of exposure (14, 15). Continuous exposure of trophoblast cells to cAMP analogs appears to be required for maximal responses (16). The augmented CG and progesterone secretion is due to increases in the expression of the genes encoding the CG subunits and enzymes involved in progesterone synthesis. This response to cAMP analogs (8-Br-cAMP) occurs under conditions in which cytotrophoblast aggregation and fusion are not possible (i.e., culture in serum-free medium in the absence of extracellular matrix), and cAMP treatment does not promote cell fusion under any culture condition. However, the cAMP analog does stimulate cytoplasmic differentiation of the cytotrophoblasts (10) with proliferation of endoplasmic reticulum and dilation of its tubules and increased prominence of the Golgi. Thus, cytotrophoblasts isolated from term placentae have the capacity to express the endocrine functions of syncytiotrophoblasts although they remain mononucleated. Hence, gross morphological differentiation—that is, syncytium formation—can be dissociated from endocrine differentiation. It appears that the cytotrophoblasts of term placentae have the machinery to respond to cAMP and await a stimulus to activate the system.

The mRNAs encoding the CG subunit genes and the components of the cholesterol side-chain cleavage system respond to 8-Br-cAMP in different temporal patterns and show a different sensitivity to the protein synthesis inhibitor cycloheximide (15). The αCG mRNA responds most rapidly, followed by increases in adrenodoxin, the βCG subunit, and $P450_{scc}$ mRNAs. The cAMP-induced increase in αCG mRNA is not inhibited by cycloheximide, whereas the increase in the other mRNAs is blocked. The response of trophoblastic cells to cAMP analogs can also be modulated by other hormones. Glucocorticoids augment the secretion of both CG and progesterone during the spontaneous differentiation of trophoblasts in culture and enhance the response to 8-Br-cAMP (17). The different kinetics of response, the differential sensitivity to protein synthesis inhibitors, and effects of glucocorticoids indicate that a complex regulatory system governs trophoblast gene expression.

Mechanisms of Cytotrophoblast Aggregation and Fusion

The process of cytotrophoblast aggregation is cell specific in that cytotrophoblasts aggregate preferentially with other cytotrophoblasts and not nontrophoblast cell types. The aggregation is dependent upon trypsin-sensitive cell-surface molecules and requires calcium (11). The calcium dependence of the aggregation process suggested the possibility that a calcium-dependent cell adhesion molecule mediates the aggregation. *E-cadherin*, a 120-kd glycoprotein that is the same as or homologous to uvomorulin, cell CAM 120/80, L-CAM, and Arc-1, was considered to be a good candidate for the key cell adhesion molecule since it is localized on the surface of cytotrophoblasts in chorionic villi in situ (18).

E-cadherin is detected by immunocytochemistry at areas of contact between trophoblast cells in vitro (19). Moreover, antibodies to the functional extracellular domain of cadherins block cytotrophoblast fusion in a dose-dependent manner, whereas antibodies generated against the cadherin cytoplasmic domain do not. These observations implicate E-cadherin in cytotrophoblast aggregation and fusion. However, it is likely that other molecules also participate in this process, including desmosomal proteins, as desmosomes form at sites of adherence of the aggregated cells. Analysis of major desmosomal proteins—desmoplakins I and II—by immunocytochemistry and Western blotting reveals that these proteins accumulate at sites of cell contact after E-cadherin, suggesting a sequential establishment of the cell adhesion complexes. The signals resulting from cell contact that trigger the development of the adhesion system remain to be elucidated.

The pattern of E-cadherin expression in trophoblast cells is of interest given recent observations suggesting that a reduced level of E-cadherin is associated with tumor cell invasion (20). Cytotrophoblasts at the tips of cell columns that intrude into the uterus do not stain for E-cadherin, whereas cytotrophoblasts in the interior of the villi do. This pattern of trophoblast E-cadherin expression is consistent with the reported inverse relationship between E-cadherin levels and invasive behavior. The loss of E-cadherin from syncytiotrophoblasts that form in vitro from fusing cyto-trophoblasts would seem contrary to this notion since the syncytial structures display reduced motility and invasiveness in in vitro assays. However, in situ, the syncytiotrophoblast is linked to the underlying layer of cytotrophoblasts, and E-cadherin is localized at these junctions. Thus, the syncytiotrophoblast remains firmly anchored by a specialized dis-tribution of the E-cadherin that is not disclosed in the standard tissue culture system.

Cytotrophoblast Fusion: Reorganization of the Cell Adhesion Complex

The aggregation of cytotrophoblasts is followed by cell fusion. The cell fusion process is cell specific in that cytotrophoblasts only fuse with other trophoblastic elements. Fusion also appears to be dependent on calcium in the extracellular medium. Cell fusion can take place between tropho-blasts isolated from different placentae and between cytotrophoblasts and JEG-3 cells, a clonal cell line derived from a choriocarcinoma. These cytotrophoblast-like cells do not normally undergo fusion in culture. These findings suggest that the specificity of cytotrophoblast interaction does not extend to major histocompatability antigens and that fusion competence can be conferred on cytotrophoblast-like cells that do not normally fuse by a fusion-competent partner.

In association with cell fusion, there is a dramatic reorganization of the adhesion complexes that bind aggregated trophoblast cells together (19). Immunocytochemistry reveals that E-cadherin is lost from the cell surface, and desmosomal proteins (desmoplakins) are apparently internalized and subsequently degraded. Western blot analysis documents a decline in cellular E-cadherin with time in culture as cytotrophoblasts fuse, and Northern hybridization analysis reveals that E-cadherin mRNA levels also decline. These findings suggest that the loss of E-cadherin as syncytia form is due, at least in part, to diminished gene expression, as well as turnover of preexisting cell adhesion molecules. In contrast, choriocarcinoma cell lines, like JEG-3, that do not fuse maintain levels of E-cadherin protein and mRNA during culture. It is not yet known whether the dramatic change in the cell adhesion system precedes, is concomitant with, or follows cell fusion. Moreover, the mechanisms by which changes in cell morphology impact on E-cadherin gene expression remain obscure.

A Relationship Between Fusion and Functional Differentiation

As noted above, when cytotrophoblasts spontaneously differentiate in vitro, there is a correlation between the formation of syncytial structures and the increase in CG secretion. Since cAMP has been implicated in the

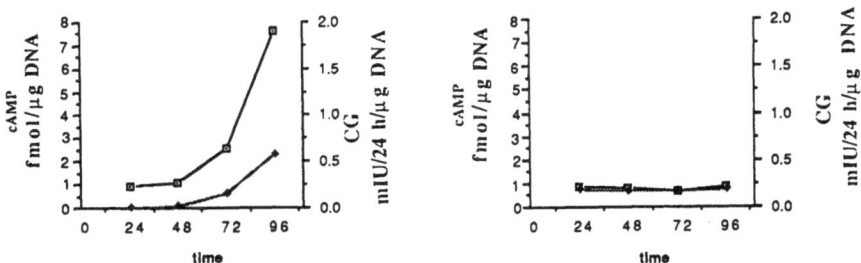

FIGURE 10.2. Temporal changes of intracellular cAMP and secreted CG levels in cytotrophoblasts during culture. Trophoblasts were cultured for the indicated periods of time. Culture media were collected for assay of CG (solid squares), and cells were terminated for cAMP (open squares) extraction and radioimmunoassay. The left panel presents results of an experiment in which cells were plated at a density of 2×10^6 cells/30-mm culture well. The right panel presents results of an experiment in which cells were plated at a density of 0.25×10^6 cells/30-mm culture well. At this plating, density cell aggregation and fusion are markedly reduced.

regulation of CG subunit genes, we wondered whether the aggregation and cell fusion events were associated with an increase in cellular cAMP levels that in turn might activate expression of the genes encoding CG subunits. Measurement of cellular cAMP levels revealed a temporal association between the appearance of multinucleated syncytia and a rise in cellular cAMP and CG secretion (Fig. 10.2). Moreover, cAMP levels appeared to rise prior to or concomitant with CG secretion. Since CG has been reported to activate placental adenylate cyclase (21) and a correlation existed between cAMP and CG secretion, we also examined the effects of exogenous CG on cellular cAMP levels. We found no effect of exogenous CG at concentrations (5000 mIU/mL) that stimulate human ovarian granulosa cell steroidogenesis. Thus, it seems likely that the rise in CG does not cause the rise in cAMP trophoblast levels, leaving the more likely scenario to be that a rise in cAMP promotes CG production.

These findings suggest a possible mechanism whereby trophoblast structural and functional differentiation can be coordinated: An aggregation or fusion-associated rise in cellular cAMP triggers CG gene activation. The localization of adenylate cyclase to the basal aspect of syncytiotrophoblast cells, where they contact cytotrophoblasts, is consistent with a role for intimate contact between these cells and regulation of cellular cAMP levels. By coupling fusion with cAMP increases, cytotrophoblast nuclei, newly recruited into syncytiotrophoblasts, could be activated so that they would be capable of expressing the same repertoire of genes as the resident nuclei. It is not yet known whether the rise in cAMP as cytotrophoblasts differentiate precedes or follows cell fusion. Because the cell fusion event cannot yet be synchronized in culture, the precise kinetics and temporal relationships to CG gene expression have not been defined.

Antibodies that block cytotrophoblast aggregation prevent the spontaneous rise in CG secretion during cytotrophoblast differentiation in vitro. This suggests that paracrine interactions between cells do not trigger endocrine differentiation and that intimate cell contact is required. Additional evidence supporting this notion includes the direct relationship between plating density and cellular cAMP levels (Fig. 10.3) and the absence of a rise in cellular cAMP and CG secretion when cells are plated at densities that discourage cell interaction (Fig. 10.2). Moreover, Hochberg et al. (22) recently reported that cocultures of trophoblasts isolated from normal placentae with choriocarcinoma cells stimulated trophoblast CG gene expression. Given the potential of the choriocarcinoma cells to aggregate and fuse with the placental trophoblasts, the activation of trophoblast gene expression that these authors observed could be accounted for by increased cellular interaction and, hence, increased cAMP formation. Notably, nontrophoblast cell types could not substitute for choriocarcinoma cells in the latter experiments, confirming the specificity of the cellular interactions.

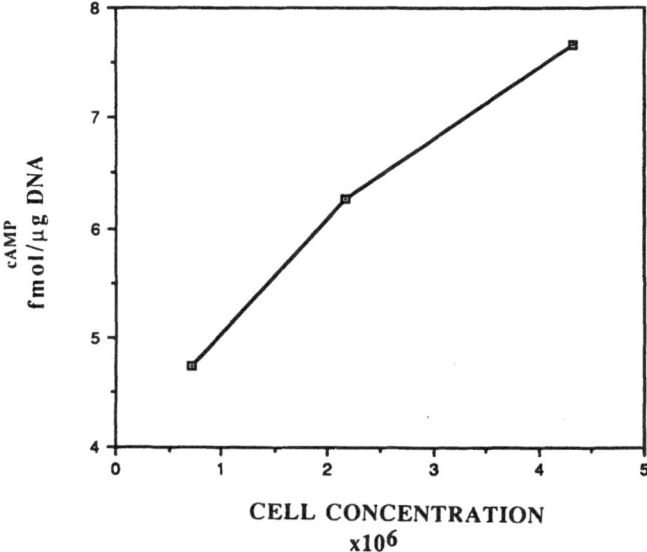

FIGURE 10.3. Intracellular cAMP levels in cultured cytotrophoblasts at various plating densities. Trophoblasts were plated at 3 different densities—0.72×10^6, 2.16×10^6, and 4.32×10^6 cells per 50-mm culture well—and cultured for 24 h. Cells were terminated for assay of cAMP.

The relationship between the aggregation-fusion of cytotrophoblasts and the cAMP increase is reminiscent of the changes in cAMP secretion associated with the aggregation of the slime mold *Dictyostelyium discoideum* (23). As amoeba aggregate, cAMP is secreted, and cAMP promotes the differentiation of the aggregated cells into a multicellular organism. A rise in cellular cAMP is also observed as myoblasts fuse to form myotubes (24). Thus, the formation of multicellular and multinucleated structures is associated with an increase in cAMP in other systems.

Cell fusion is associated with other functional changes in the trophoblasts, including changes in cell motility. The freshly isolated cytotrophoblasts display motility in culture, but the extent of their excursions declines substantially as they aggregate and fuse to form syncytia. During differentiation an elaborate cytoskeleton of intermediate filaments and other components is constructed, and adhesion plaques containing integrins form (25) that probably anchor the large cells to their substrate. It is not known if the endocrine differentiation of the trophoblasts is linked to the alterations in the cytoskeleton and changes in motility or whether these are independent processes.

Caveats and Questions

Besides controlling genes involved in endocrine activities, cAMP may regulate cytotrophoblast replication. Treatment of BEWO choriocarcinoma cells with forskolin results in a reduction in mitotic index, enlargement of the nuclei, and expansion of the cytoplasmic volume (26). Similar changes are observed when cells are exposed to methotrexate. The common feature of the action of cAMP and methotrexate on choriocarcinoma cells appears to be inhibition of cell replication, which is associated with morphologic and biochemical changes indicative of cellular differentiation. Thus, cAMP may have effects on mitotic activity that subsequently entrain differentiation events.

The data reviewed above should not lead the reader to believe that cAMP is the only regulator of trophoblast gene expression. Indeed, this cyclic nucleotide cannot account for the diverse changes in gene expression that are associated with trophoblast differentiation, including CS expression. Other second-messenger systems, including protein kinase C, undoubtedly contribute to the integrated effects on gene expression. Moreover, it is evident that trophoblast function in different anatomical sites differs. Thus, syncytial cells in chorionic villi produce hormones in different patterns than giant wandering cells in extravillous sites. These differences may be due to modifying influences that include the nature of the extracellular matrix and the influence of factors produced by other cells in the vicinity of the trophoblasts, rather than to an intrinsic differentiation program.

Why does spontaneous differentiation of cytotrophoblasts into syncytiotrophoblasts take place in culture, whereas it is suppressed for various periods of time in situ? The fact that removal of the cytotrophoblasts from their villous environment in term placentae is associated with spontaneous differentiation suggests that some factors in the villi hold this process in check (paracrine factors or extracellular matrix?). This phenomenon is not unlike the spontaneous differentiation of granulosa cells in vitro following their removal from preovulatory follicles. Morrish et al. recently reported that transforming growth factor β1 (TGFβ1) inhibited the morphological differentiation of cytotrophoblasts in culture as well as CG and CS secretion (27), whereas these authors found that EGF enhanced the transformation of cytotrophoblasts into syncytia. These observations indicate that trophoblast differentiation can be modified by substances acting in a paracrine mode. Whether TGFβ1 or EGF have such roles in situ remains to be established.

The observations discussed in this review were derived mainly from a simple in vitro system and do not address the influence of matrix or paracrine factors produced by other placental or uterine cell types on trophoblast differentiation and gene expression. Studies using culture systems in which a differentiating chorionic villous is reconstituted will be needed to evaluate these interactions.

Acknowledgments. Research described in this article was supported by NIH Grant HD-06274 and the Rockefeller and Mellon Foundations. The authors thank Mrs. Barbara McKenna for help in preparing this manuscript.

References

1. Boyd JD, Hamilton WJ. The human placenta. London: MacMillan, 1970.
2. Simpson ER, MacDonald PC. Endocrine physiology of the placenta. Annu Rev Physiol 1981;43:163–88.
3. Albrecht ED, Pepe GJ. Placental steroid hormone biosynthesis in primate pregnancy. Endocr Rev 1990;11:124–50.
4. Midgley AR, Pierce GB, Deneau GA, Gosling JRG. Morphogenesis of syncytiotrophoblast in vivo: an autoradiographic demonstration. Science 1963;141:350–1.
5. Kliman HJ, Nestler JE, Sermasi E, Sanger JM, Strauss JF III. Purification, characterization, and in vitro differentiation of cytotrophoblasts from human term placentae. Endocrinology 1986;118:1567–81.
6. Ohlsson R. Growth factors, protooncogenes and human placental development. Cell Differ Dev 1989;28:1–16.
7. Mercola M, Stiles CD. Growth factor super families and mammalian embryogenesis. Development 1988;102:451–60.
8. Ringler GE, Strauss JF. In vitro systems for the study of human placental endocrine function. Endocr Rev 1990;11:105–23.
9. Nelson DM, Crouch EC, Curran EM, Farmer DR. Trophoblast interaction with fibrin matrix; epithelialization of perivillous fibrin deposits as a mechanism for villous repair in the human placenta. Am J Pathol 1990;136:855–65.
10. Kao L-C, Caltabiano S, Wu S, Strauss JF III, Kliman HJ. The human villous cytotrophoblast: interactions with extracellular matrix proteins, endocrine function, and cytoplasmic differentiation in the absence of syncytium formation. Dev Biol 1988;130:693–702.
11. Babalola GO, Coutifaris C, Soto EA, Kliman HJ, Shuman H, Strauss III JF. Aggregation of dispersed human cytotrophoblastic cells: lessons relevant to the morphogenesis of the placenta. Dev Biol 1990;137:100–8.
12. Nulsen JC, Woolkalis MJ, Kopf GS, Strauss JF III. Adenylate cyclase in human cytotrophoblasts: characterization and its role in modulating human chorionic gonadotropin secretion. J Clin Endocrinol Metab 1988;66:258–65.
13. Matsubara S, Tamada T, Saito T. Ultracytochemical localizations of adenylate cyclase, guanylate cyclase and cyclic $3',5'$-nucleotide phosphodiesterase activity on the trophoblast of human placenta; direct histochemical evidences. Histochemistry 1987;87:505–9.
14. Feinman MA, Kliman HJ, Caltabiano S, Strauss JF III. 8-Bromo-$3'$-$5'$-adenosine monophosphate stimulates the endocrine activity of human cytotrophoblasts in culture. J Clin Endocrinol Metab 1986;63:1211–7.
15. Ringler GE, Kao L-C, Miller WL, Strauss JF III. Effects of 8-bromo-cAMP on expression of endocrine functions by cultured human trophoblast cells; regulation of specific mRNAs. Mol Cell Endocrinol 1989;61:13–21.

16. Deutsch PJ, Sun Y, Kroog GS. Vasoactive intestinal peptide increases intracellular cAMP and gonadotropin-a gene activity in JEG-3 syncytial trophoblasts. J Biol Chem 1990;265:10274–81.
17. Ringler GE, Kallen CB, Strauss JF III. Regulation of human trophoblast function by glucocorticoids: dexamethasone promotes increased secretion of chorionic gonadotropin. Endocrinology 1989;124:1625–31.
18. Eidelman S, Damsky CH, Wheelock MJ, Damjanov I. Expression of the cell-cell adhesion glycoprotein cell-CAM 120/80 in normal human tissues and tumors. Am J Pathol 1989;135:101–10.
19. Coutifaris C, Kao L-C, Sehdev HM, et al. E-cadherin expression during the differentiation of human trophoblasts. Development (in press).
20. Vleminckx K, Vakaet L Jr, Mareel M, Fiers W, Van Roy F. Genetic manipulation of E-cadherin expression by epithelial tumor cells reveals an invasion suppressor role. Cell 1991;66:107–19.
21. Menon KM, Jaffe RB. Chorionic gonadotropin sensitive adenylate cyclase in human term placenta. J Clin Endocrinol Metab 1973;36:1104–9.
22. Hochberg A, Sibley C, Pixley M, Sadovsky Y, Strauss B, Boime I. Choriocarcinoma cells increase the number of differentiating human cytotrophoblasts through an in vitro interaction. J Biol Chem 1991;266:8517–22.
23. Devreotes P. *Dictyostelium discoideum:* a model system for cell-cell interactions in development. Science 1989;245:1054–8.
24. Moryama Y, Hajegawa S, Murayama K. cAMP and cGMP changes associated with the differentiation of cultured chick embryo muscle cells. Exp Cell Res 1976;101:159–63.
25. Coutifaris C, Babalola GO, Abisogun AO, et al. In vitro systems for the study of human trophoblast implantation. Ann New York Acad Sci 1991; 662:191–201.
26. Sekiya S, Kaiho T, Shirotake S, Iwasawa H, Takeda B, Takamizawa H. Effect of methotrexate on the growth and human chorionic gonadotropin secretion of human choriocarcinoma cell lines in vitro. Am J Obstet Gynecol 1983;146:57–64.
27. Morrish DW, Bhardwaj D, Paras MT. Transforming growth factor β1 inhibits placental differentiation and human chorionic gonadotropin and human placental lactogen secretion. J Clin Endocrinol Metab 1991;129:22–6.

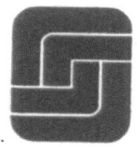

11

Placental Lactogen/Growth Hormone Gene Family

HENRY G. FRIESEN, MAY C. ROBERTSON, MARY LYNN DUCKWORTH, INGO SCHROEDTER, NI QUAN, AND JEAN-CLAUDE VUILLE

Before focusing on selective aspects of current research on *placental lactogens* (PLs), it is useful to look at advances in this area from a historical perspective. Several landmark studies can be recognized that were instrumental in providing new directions to the field (Table 11.1). Four major periods, each ushered in by the applications of new technologies or methods, can readily be identified, underscoring the importance of technology to scientific advance.

The first period, the classical period, was the time when hormones were discovered as the result of ablation experiments frequently combined with the administration of crude extracts from putative endocrine glands or from serum obtained at a propitious time. The end point was the induction or maintenance of a biological effect often confirmed by histological examination. The studies of Astwood and Greep (1) in 1938 on the identification of "A corpus luteum stimulating substance in the rat placenta" appeared following a number of studies indicating that the rodent placenta functioned as an endocrine gland. These investigators showed convincingly that placental tissue extracts around midpregnancy contained a potent luteotropic principle. This report was followed by others that demonstrated as well that serum and placental extracts obtained at day 12 of gestation exerted a mammotropic effect (2).

The next major impetus to the field was the application of a disarmingly simple immunological procedure: the double-immunodiffusion technique of Ouchterlony. Antibodies to human *growth hormone* (GH) crossreacted with a line of partial identity with crude human placental extracts. This seminal report by Josimovich and Maclaren (3) ushered in a burst of research activity on placental lactogens, especially as the initial studies suggested that *human placental lactogen* (hPL) might be a useful substitute for human growth hormone in the treatment of growth hormone-deficient children.

TABLE 11.1. Phases in research on placental lactogens.

1. Classical period	Astwood and Greep, 1938
2. Immunological techniques—protein purification	Josimovich and MacLaren, 1962 (hPL)
3. Radioreceptor and Nb2 cell bioassays	Shiu, 1973; Tanaka, 1982
4. Molecular biology—cloning	Linzer; Duckworth; 1984–86

The third period was triggered by the introduction of two sensitive assay techniques that allowed for the detection of a variety of prolactins and placental lactogens. These assays provided the essential tools for the identification and, ultimately, the isolation of a rapidly expanding number of placental lactogens from different species. The two assays were the *prolactin radioreceptor assay* (PRL-RRA) (4) and the Nb2 node lymphoma assay (5) that were able to detect PRL-like activity in serum and placental tissue extracts of many different species. The list included PLs from sheep, cow, goats, mice, guinea pig, and so forth (6). In a recent review, Forsyth (7) reported that PLs had been demonstrated in 10 species of rodents, 20 ruminants, and 5 primates. As shown below, in several species multiple forms and types of PLs are synthesized and secreted at specific stages of pregnancy, making the repertoire of PLs even in one species astonishingly rich. The first hint of this complexity was provided by the initial studies using an RRA that identified a second PL peak late in rat pregnancy (4), whereas the report of Astwood and Greep (1) had suggested only a single midpregnancy peak. Subsequently, chemical and immunological studies (8, 9) indicated that there were likely to be significant structural differences between the two PLs, a conclusion that proved entirely accurate when the cDNAs for each were cloned (10, 11).

The fourth period, which takes us to the present, is the era when molecular and cell biology techniques have been applied to studies on PL. As in so many fields, the application of these techniques has led to a major expansion of the number and knowledge of members of the PL family (12–18). This has been especially true for, but by no means confined to, members of the rodent PL family. This chapter focuses particularly on our studies on rat placental lactogens.

The initial application of the PRL-RRA demonstrated two major peaks of lactogenic activity during pregnancy in the rat. The first peak from days 8 to 13 was identified as due to the secretion of *rat placental lactogen I* (rPL-I), while the second peak, beginning at day 12 to term, was due to *rat placental lactogen II* (rPL-II) (8, 9). Studies in the mouse showed a similar pattern. The task of purifying each species of PL from rats or mice and developing antibodies to each proved a daunting task, particularly given the small amount of placental tissue available at midpregnancy (19).

With the application of techniques for the isolation and characterization of cDNA clones, there has been a veritable explosion in the number

TABLE 11.2. Comparison of members of the rat PL/GH family.

PRL-like protein	Appearance (day)	mRNA (kb)	Protein (kd)	Glycosylated	Amino acid sequence	Cell type
1. Luteotropin	8–11			Yes	?	Decidua
2. rPL-I	8–12	1.0	36–40	Yes	230 (30)*	Giant cells
3. rPL-I$_v$	15 to term	1.0	29 and 33	Yes	223 (29)	Cytotrophoblast
4. rPL-II	12 to term	1.0	25	No	191 (30)	Giant cell (basal and labyrinth)
5. rPLP-A	14 to term	1.0	25 and 27	Yes	196 (31)	Cytotrophoblast
6. rPLP-B	8–11	1.2	26	Yes	201 (32)	Cytotrophoblast
	14 to term	0.9				Decidua
7. rPLP-C	16 to term	1.0	25 and 29	Yes	partial	Cytotrophoblast (spongiotrophoblast)
8. GH-related protein	15+	?	25–30	?	partial	?

* Number in parentheses indicates length of signal peptide.

of members of the PL/GH family expressed in the placenta during pregnancy. The list in the rat continues to expand. As many as 7 PRL-related members and 4 GH-related proteins have been identified, as shown in Table 11.2. The details regarding the GH-related placental proteins remain sketchy, but undoubtedly these, too, will be more fully characterized in the next several years. As will become apparent, there is a much greater amount of structural information on each member of the PL family than knowledge about the factors that regulate the expression and secretion of each. As well, there is only limited knowledge about the biological function of each member of the PL family. Although there is a high degree of homology among several members of the PL gene family (e.g., rPL-I and *mouse PL-I* [mPL-I] and rPL-II and *mouse PL-II* [mPL-II]), there are also significant differences. In the mouse 2 genes related to PRL, proliferin and proliferin-related proteins are expressed in the placenta at midpregnancy (13, 20), but they do not appear to have rat homologs. Conversely, no counterpart to *rat prolactin-like proteins A* and *B* (rPLP-A and rPLP-B) have been identified in the mouse.

The expression of members of the PL gene family can be divided into 3 gestational periods: days 7–10, days 8–13, and days 12 to term. The first to appear are two proteins—decidual luteotropin and rPLP-B—that are synthesized by decidual tissue and are of maternal origin. Decidual luteotropin (21) and rPLP-B (22) are expressed by the antimesometrial uterine decidual tissue both in pseudopregnancy and pregnancy. Circumstantial evidence, including the temporal and tissue-specific expression of both decidual luteotropin and rPLP-B, raises the question whether the two may in fact be identical. The former was identified on the basis of an RRA using ovarian tissue membrane PRL receptors (23), but so far, no information is available on the binding of rPLP-B to tissue receptors. As growth and maintenance of the deciduoma depends on estrogen and progesterone, one might speculate that either or both hormones may regulate the tissue-specific expression of the two decidual hormones. With the availability of the rPLP-B gene, studies of progesterone response elements in the 5' flanking region are planned.

The second period of intense endocrine activity of the placenta occurs in the interval surrounding midpregnancy. It is during this stage that rPL-I is synthesized and secreted, beginning as early as day 8 and reaching its apogee at day 12 (8). There are now several reports on the characterization of rPL-I (9, 11). Its mouse counterpart has been purified from placental extracts and cultured trophoblasts (19, 24, 25), while rPL-I from serum or placental tissue has only been partially purified. However, a full-length cDNA has been isolated, permitting the cloning of rPL-I and the assignment of a number of structural features (11). These are summarized in an abbreviated fashion in Table 11.2. The nucleotide and amino acid sequence of rPL-I shows extensive homology (73%) with mPL-I (11) and shares 41% identity at the amino acid level with rPL-II, 24% with rPLP-A, 26% with rPLP-B, and 31% with rat PRL. Maximal expression of rPL-I mRNA transcript was observed on days 11 and 12. However, with in situ hybridization studies, rPL-I mRNA was detected in a few rapidly dividing cells as early as day 8 of gestation. Subsequently, the dominant location of rPL-I was in giant cells.

To generate recombinant rPL-I, Chinese hamster ovary cells were transfected with an expression construct with an rPL-I cDNA insert, and the transfected cells were selected upon culturing in the appropriate antibiotic. Resistant colonies were amplified by stepwise increases in methotrexate, and recombinant rPL-I from the culture medium was purified using monoclonal antibodies to rPL-I. The recombinant rPL-I in many respects was similar to placental rPL-I. Of particular interest was the biological activity of recombinant rPL-I. In the Nb2 lymphoma cell assay, its potency was substantially greater than that of other lactogens. The availability of recombinant rPL-I should allow a number of additional biological studies to be done that previously could not be considered because of the limited amounts of rPL-I that could be obtained.

The second major lactogen identified by the initial RRA studies was rPL-II (8). It was the first of the PLs to be cloned in our laboratory (10). It is the only nonglycosylated PL and is composed of 191 amino acids plus a 30-amino acid signal peptide. As in the case of rPL-I, there is striking structural homology between rPL-II and mPL-II. Using in situ hybridization, rPL-II mRNA can be detected as early as day 12 in giant cells of the basal zone (26). As pregnancy continues, there is a progressive increase in rPL-II expression, particularly in the trophoblast cells of the labyrinth.

During the course of our studies on rPL-II, we isolated and characterized a novel PRL-like cDNA clone from developing rat placenta (15). Because of its homology with prolactin, the 1-kb transcript that codes for a 25- to 27-kd protein was named *rPLP-A*. It is first expressed at day 14, 2 days later than rPL-II, and then increases and remains elevated until term. The amino acid sequence deduced from the cDNA suggests that rPLP-A is a secreted glycoprotein of 196 amino acids with 2 potential glycosylation signals. Its sequence is 40% homologous to rat and human prolactin. In the circulation, rPLP-A circulates primarily as a high-molecular-weight complex (27).

The cDNA for another abundant placental protein, *rPLP-B*, was identified because of its homology with members of the PL gene family (16). The predicted amino acid sequence showed a 44% homology to rPRL and 40% to rPL-II and rPLP-A. Two rPLP-B mRNA transcripts of 0.9 and 1.2 kb are strongly expressed in almost equal amounts from day 14 of pregnancy to term. The difference in transcript size appears to be due to differences in their 5' untranslated region. All intron-exon boundaries of the clone are of the same splice class and occur in the identical locations within the coding region as in the rPRL gene.

As mentioned above, rPLP-B is expressed at an early stage of pregnancy by decidual cells (22). At day 13 of pregnancy, the transcripts for rPLP-B are detected concurrently both in decidual cells of maternal origin and in the cytotrophoblasts (spongiotrophoblasts), which are of fetal origin. As yet, no estimates of the concentration of rPLP-B during pregnancy have been reported.

During the course of purification of rPLP-A, a major contaminating protein with similar molecular weight but different isoelectric point was identified by Deb et al. (28). This major secretory protein of the placenta in the second half of pregnancy was isolated and characterized and found to have significant homology to other members of the PL gene family; it was named *rat prolactin-like protein C* (rPLP-C). The protein consists of 2 major species with molecular weights of 25,000 and 29,000. It bears striking similarity with one of the GH-related peptides reported by Ogilvie et al. (29). Until the complete structure of each species is defined by molecular cloning, the relationship between rPLP-C and the GH-related placental proteins cannot be resolved with certainty. The expression of rPLP-C is confined to spongiotrophoblast cells in the basal zone of the rat chorioallantoic placenta beginning at day 14, reaching peak concentrations at day 18, but continuing until term.

The last member of the rat PL family to be considered is *rat placental lactogen I variant* (rPL-I$_v$). Robertson et al. (30) identified this species when they unexpectedly observed hybridization upon Northern analysis of mRNA isolated from day 18 placentas with an rPL-I cDNA probe. The result was surprising because all previous studies suggested rPL-I was only expressed until days 12 to 13. When a day-18 placental cDNA library was probed, positive clones were identified and sequenced. Further analysis revealed that the cDNA insert had an open reading frame that coded for a protein of 223 amino acids that was closely related to rPL-I

(85% homology). Independently, Deb et al. (31) characterized a protein identified in conditioned medium from late-gestation placental explants. They noted specific differences between the newly identified protein rPL-I$_v$ and rPL-I in terms of affinity for concanavalin A (major) temporal expression (late in pregnancy) and biological activity in the Nb2 cell assay (weak lactogen).

When comparing the cell-specific and temporal pattern of expression, 3 members of the PL family—rPLP-A, rPLP-C, and rPL-I$_v$—show a pattern that is quite similar. In contrast, rPL-I, rPL-II, and rPLP-B exhibit a different pattern. The expression of rPL-I is restricted to the early period postimplantation, while rPL-II expression shifts from the basal zone trophoblast giant cells to the labyrinth giant cells as pregnancy advances. The pattern of rPLP-B is unique in that it is the only member of the PL family that is expressed in both decidual cells and, later, cytotrophoblast cells. Finally, the nature and distribution of members of GH-related proteins and their relationship to rPLP-C remain to be defined.

At present, at least 8 members of the PL/GH gene family have been isolated and characterized (Table 11.2). The primary structural information on most has been reported. Limited studies on the patterns of secretion during pregnancy have been carried out on rPL-I and rPL-II. Several studies on factors that influence the secretion of rPL-II have been carried out. The studies have shown that the pituitary and ovary inhibit, and fetal factors appear to stimulate, rPL-II secretion (32, 33). It has been suggested that GH may be at least one of the factors involved (34). As yet, however, the molecular mechanisms involved are completely unknown. A promising approach to exploring the cellular factors that regulate rPL gene expression may be possible using a transplantable rat choriocarcinoma cell line (35, 36). This cell line is composed of trophoblast cells that undergo differentiation in culture, which should allow a search for the nuclear factors that regulate rPL gene expression to be carried out.

While there has been a remarkable increase in the number of members of the PL gene family that have been recognized, progress in the understanding of the role and biological significance of each in most cases remains rudimentary. A list of some of the biological effects and possible functions of the PLs, particularly of rPL-I and rPL-II, are outlined in Table 11.3. For the other members, our knowledge of functions or effects is almost nonexistent. However, one can expect that with the ability to produce recombinant proteins using expression vectors, sufficient amounts of each of the PLs will be generated in order to explore possible effects.

In the case of recombinant rPL-I and rPL-I$_v$, we have initiated several studies and have found in preliminary experiments that rPL-I appears to be especially potent in stimulating Nb2 cell growth, while rPL-I$_v$ is only a

TABLE 11.3. Functions of placental lactogen.

Mammary gland	Growth
Ovary	Luteotropic
Growth	↑ IGF-I
Metabolism	↑ islet cell growth
	↑ insulin secretion
Neuroendocrine effects	Maternal behavior
	Pituitary PRL cycles

weak lactogen. Sorensen et al. (37) reported that PRL stimulates islet cell proliferation and insulin secretion, and one might anticipate that recombinant rPL-I will do the same. It is also planned to examine directly whether recombinant rPL-I can inhibit the twice-daily PRL surges that abruptly cease at midpregnancy at a time when rPL-I levels peak (38). The pituitary hormone PRL has been shown to have a key role in inducing maternal behavior (39). Given the large amount of PLs in the circulation, especially as pregnancy advances, it will be of interest to examine under what conditions and at what locus central administration of members of the PL family induces maternal behavior.

While many of these potential effects of PLs are similar to known actions of PRL, an entirely unexpected biological effect of one of the new members of the mouse PL family has been reported. Proliferin was shown to inhibit myogenic differentiation (40) and to repress myogenic-specific transcription by the suppression of an essential serum response factor-DNA binding activity (41). It seems likely that as more of the novel PL proteins become available, a host of new, and at times surprising, results will emerge. For a detailed review of the expanding field of placental lactogens, interested readers may wish to consult several excellent reviews (34, 42).

Acknowledgments. This research was supported by grants from the MRC Canada and PHS HD-07843-16.

References

1. Astwood EB, Greep RO. A corpus luteum stimulating substance in the rat placenta. Proc Soc Exp Biol Med 1938;38:713–6.
2. Matthies DL. Studies of the luteotropic and mammotropic factor found in trophoblast and maternal peripheral blood of the rat at mid pregnancy. Anat Rec 1967;159:55–68.
3. Josimovich JB, Maclaren JA. Presence in the human placenta and term serum of a highly lactogenic substance immunologically related to pituitary growth hormone. Endocrinology 1962;71:209–20.

4. Shiu RPC, Kelly PA, Friesen HG. Radioreceptor assay for prolactin and other lactogenic hormones. Science 1973;180:968–71.
5. Tanaka T, Shiu RPC, Gout PW, Beer CT, Nobel RL, Friesen HG. A new sensitive and specific bioassay for lactogenic hormones: measurement of prolactin and growth hormone in human serum. J Clin Endocrinol Metab 1980;51:1058–63.
6. Kelly PA, Tsushima T, Shiu RPC, Friesen HG. Lactogenic and growth hormone-like activities in pregnancy determined by radioreceptor assays. Endocrinology 1976;99:765–74.
7. Forsyth IA. The roles of placental lactogen and prolactin in mammogenesis: an overview of in vivo and in vitro studies. In: Hoshino K, ed. Prolactin gene family and its receptors. Amsterdam: Elsevier, 1988:137–44.
8. Robertson MC, Friesen HG. Two forms of rat placental lactogen revealed by radioimmunoassay. Endocrinology 1981;108:2388–90.
9. Robertson MC, Gillespie B, Friesen HG. Characterization of the two forms of rat placental lactogen (rPL): rPL-I and rPL-II. Endocrinology 1982; 111:1862–6.
10. Duckworth ML, Kirk KL, Friesen HG. Isolation and identification of a cDNA clone of rat placental lactogen II. J Biol Chem 1986;261:10871–8.
11. Robertson MC, Croze F, Schroedter IC, Friesen HG. Molecular cloning and expression of rat placental lactogen-I complementary deoxyribonucleic acid. Endocrinology 1990;127:702–10.
12. Linzer DIH, Nathans D. Nucleotide sequence of a growth related mRNA encoding a member of the prolactin-growth hormone family. Proc Natl Acad Sci USA 1984;4255–9.
13. Linzer DIH, Lee S-J, Ogren L, Talamantes F, Nathans D. Identification of proliferin mRNA and protein in mouse placenta. Proc Natl Acad Sci USA 1985;82:4356–9.
14. Jackson LL, Colosi P, Talamantes F, Linzer DIH. Molecular cloning of mouse placental lactogen cDNA. Proc Natl Acad Sci USA 1986;83:8496–500.
15. Duckworth ML, Peden LM, Friesen HG. Isolation of a novel prolactin-like cDNA clone from developing rat placenta. J Biol Chem 1986;261:10879–84.
16. Duckworth ML, Peden LM, Friesen HG. A third prolactin-like protein expressed by the developing rat placenta: complementary deoxyribonucleic acid sequence and partial structure of the gene. Mol Endocrinol 1988;2: 912–20.
17. Schuler LA, Shimomura K, Kessler MA, Zieler CG, Bremel RD. Bovine placental lactogen: molecular cloning and protein structure. Biochemistry 1988;27:8443–8.
18. Colosi P, Thordarson G, Hellmiss R, et al. Cloning and expression of ovine placental lactogen. Mol Endocrinol 1989;3:1462–9.
19. Colosi P, Marr G, Lopez J, Haro L, Ogren L, Talamantes F. Isolation, purification and characterization of mouse placental lactogen. Proc Natl Acad Sci USA 1982;79:771–5.
20. Linzer DIH, Nathans D. A new member of the prolactin-growth hormone gene family expressed in mouse placenta. EMBO J: 4:1419–23.
21. Jayatilak PG, Puryear TK, Herz Z, Fazleabas A, Gibori G. Protein secretion by mesometrial and antimesometrial rat decidual tissue: evidence for differential gene expression. Endocrinology 1989:659–66.

22. Croze F, Kennedy TG, Schroedter IC, Friesen HG. Expression of rat prolactin-like protein-B in deciduoma of pseudopregnant rat and in decidua during early pregnancy. Endocrinology 1990;2665–72.
23. Jayatilak PG, Glaser LA, Basuray R, Kelly PA, Gibori G. Identification and characterization of a prolactin-like hormone produced by the rat decidual tissue. Proc Natl Acad Sci USA 1985;82:217–21.
24. Colosi P, Talamantes F, Linzer DIH. Molecular cloning and expression of mouse placental lactogen I complementary deoxyribonucleic acid. Mol Endocrinol 1987;1:767–76.
25. Colosi P, Ogren L, Southard JN, Thordarson G, Linzer DIH, Talamantes F. Biological immunological and binding properties of recombinant mouse placental lactogen-I. Endocrinology 1988;123:2662–7.
26. Duckworth ML, Schroedter IC, Friesen HG. Cellular localization of rat placental lactogen II and rat prolactin-like proteins A and B by in situ hybridization. Placenta 1990;2:143–55.
27. Southard JN, Talamantes F. High molecular weight forms of placental lactogen: evidence for lactogen-macroglobulin complexes in rodents and humans. Endocrinology 1989;125:791–800.
28. Deb S, Roby KF, Faria TN, Larsen D, Soares MJ. Identification and immunochemical characterization of a major placental secretory protein related to the prolactin-growth hormone family prolactin-like protein-C. Endocrinology 1991;128:3066–72.
29. Ogilvie S, Buhi WC, Olson JA, Shiverick KT. Identification of a novel family of growth hormone-related proteins secreted by rat placenta. Endocrinology 1990;126:3271–3.
30. Robertson MC, Schroedter IC, Friesen HG. Molecular cloning and expression of rat placental lactogen-I_v, a variant of rPL-I present in late pregnant rat placenta. Endocrinology 1991.
31. Deb S, Faria TN, Roby KF, et al. Identification and characterization of a new member of the prolactin family: placental lactogen-I variant. J Bio Chem 1991;266:1605–10.
32. Robertson MC, Owens RE, Klindt J, Friesen HG. Ovariectomy leads to a rapid increase in rat placental lactogen secretion. Endocrinology 1984;114:1805–11.
33. Robertson MC, Owens RE, McCoshen JA, Friesen HG. Ovarian factors inhibit and fetal factors stimulate the secretion of rat placental lactogen. Endocrinology 1984;114:22–30.
34. Ogren L, Talamantes F. Prolactins of pregnancy and their cellular source. Int Rev Cytol 1988;112:1–65.
35. Verstuyf A, Sobis H, Goebels J, Fonteyn E, Cassiman JJ, Vandeputte M. Establishment and characterization of a continuous in vitro line from a rat choriocarcinoma. Int J Cancer 1990;45:752–6.
36. Faria TN, Deb S, Kwok SCM, Vandeputte M, Talamantes F, Soares MJ. Transplantable rat choriocarcinoma cells express placental lactogen: identification of placental lactogen-I immunoreactive protein and messenger ribonucleic acid. Endocrinology 1990;127:3131–7.
37. Brelje TC, Sorenson RL. Role of prolactin versus growth hormone on islet B cell proliferation in vitro: implications for pregnancy. Endocrinology 1991;128:45–57.

38. Tonkowicz P, Robertson M, Voogt J. Secretion of rat placental lactogen by the fetal placenta and its inhibitory effect on prolactin surges. Biol Reprod 1983;28:707–16.
39. Bridges RS, Numan M, Ronsheim PM, Mann PE, Lupini CE. Central prolactin infusions stimulate maternal behavior in steroid-treated nulliparous female rats. Proc Natl Acad Sci 1990;87:8003–7.
40. Wilder EL, Linzer DIH. Participation of multiple factors, including proliferin, in the inhibition of myogenic differentiation. Mol Cell Biol 1989;9:430–41.
41. Muscat GEO, Gobius K, Emery J. Proliferin, a prolactin/growth hormone-like peptide represses myogenic-specific transcription by the suppression of an essential serum response factor-like DNA-binding activity. Mol Endocrinol 1991;5:802–14.
42. Soares MJ, Faria FN, Raby KF, Deb S. Pregnancy and the prolactin family of hormones: coordination of anterior pituitary, uterine and placental expression. Endocr Rev 1992.

12

Human Relaxin Genes in the Decidua and Placenta

Gillian D. Bryant-Greenwood

It is now generally accepted that the endocrine events of human parturition are different from those in animal models of parturition. The lack of obvious systemic endocrine changes in late pregnancy in the human led to the hypothesis that the events involved are predominantly localized within the uterus (1, 2). The research focused on the prostaglandins that fulfilled the criteria of local production and action and were clearly involved in parturition as judged by exogenous administration. Although of major value in the clinical setting, it became clear that the stimulation of prostaglandin production was not the starting point of human peripartal changes. Interestingly, a similar conclusion has recently been reached for another group of locally produced and acting intrauterine bioactive agents, the cytokines: "The accumulation of bioactive agents characteristic of the inflammatory response in amniotic fluid during term and preterm labor is an accompaniment of parturition and not its cause (3)." This does not mean that these substances have no role in cell:cell signaling in the early stages of parturition since the end points in each case have been the macrodetermination of high levels as "overspill" into amniotic fluid.

Human Relaxin and Parturition

The interest in our laboratory has been the role of relaxin in human pregnancy and parturition, especially as an autocrine-paracrine hormone (4). Although the corpus luteum is the major systemic source of relaxin in pregnancy (5), it is not essential for the maintenance and cessation of human pregnancy. We have demonstrated extraluteal sources of human relaxin by molecular techniques (6) and shown that the corpus luteum is not the sole source of relaxin in human pregnancy (5).

Relaxin has two biological activities in animal models: a rapid effect on myometrial contractility to cause quiescence and a longer-term effect on connective tissues, such as the cervix and interpubic ligament (5). The evidence that human relaxin is a myometrial inhibitor in the human is now questionable since it has recently been shown that it has only a limited effect on the contraction of human myometrium in vitro (7). However, it does cause the inhibition of contraction of pig myometrium in vitro (7), but porcine relaxin, while being very active on the homologous tissue, has no activity on the human myometrium (8). This suggests that the role of human relaxin in the control of myometrial activity in the human may be different from that in other species. Its role in cervical dilatation, however, was demonstrated in the human by topical application of porcine relaxin (9, 10) and formed the basis of the current phase II clinical trials with human relaxin, now taking place in Australia (Genentech, Inc.). The endogenous relaxin acting on the cervix is likely to be from a paracrine source and not from the corpus luteum since an oophorectomized woman in an in vitro fertilization program had a normal labor despite having no detectable circulating relaxin (11).

Almost a decade ago we proposed that relaxin is a paracrine hormone involved in localized changes in collagen preceding follicular and fetal membrane rupture (4). Preliminary work in our laboratory suggested that a local source of relaxin regulates collagenase production within the fetal membranes (12). We have not been alone in proposing that the proteolytic processes that cause local remodeling of the connective tissue in the fetal membranes are under hormonal control from within the uterus and that these changes may be part of the same process that occurs in the cervix (13). These authors independently suggest that the dynamic interplay of proteases and their inhibitors is important for the structural integrity of the membranes, and we have obtained preliminary evidence for this at the mRNA level.

The timing of membrane rupture is of some importance. A recent multicenter case-control study of risk factors for preterm premature rupture of the fetal membranes showed that antepartum vaginal bleeding, history of preterm delivery, and cigarette smoking were related to an increased risk of preterm premature rupture of membranes (14). This study did not attempt to explain the action of the risk factors or possible common denominators; however, infants surviving preterm delivery often have moderate to severe neurodevelopmental impairment (15).

The amniotic collagen is the thinnest but strongest of the fetal membranes (16) and, therefore, has a major mechanical function, especially in the last weeks of pregnancy. However, at this time when there is maximal growth of the uterine contents, there is a significant decrease (20%) in the amount of collagen in the amnion (17). This is a gradual loss of collagen that may be important for the membranes to be able to stretch rather than rupture as they come under increasing tension. This controlled loss of collagen is probably a result of the production of the enzyme collagenase, which in itself is under very tight regulation by an array of enzymes and inhibitors. The production of the inhibitors is very important as part of the control process and to prevent the fetus's exposure to these very active proteases in the amniotic fluid.

In addition, the decidua itself should be considered to be a membrane (18) since each cell is surrounded by a basket of major basement membrane components (19). The increased awareness of the interaction of endocrine cells with their surrounding extracellular matrix (20) suggests that to fully understand the endocrine activities of the decidual cell, its interaction with its basement membrane is important. The additional association of decidual cells with resident macrophages and their "activation," which takes place peripartum, suggests a microsystem of decidual cell:macrophage:extracellular matrix (21).

However, we still know relatively little about the expression of the human relaxin genes, the control of relaxin secretion from the human decidua, or the expression and control of the relaxin receptor. The task also involves linking relaxin to the other hormones involved in the complex system regulating collagenase production within the fetal membranes.

Transcription of the Human Relaxin Gene(s)

The presence of two relaxin genes, called H1 and H2, in the human genome appears to be unique to this species (22). It has been suggested that these two genes evolved by a second gene duplication subsequent to the separation from the insulin gene (23). The structure of the coding region of the H1 preprorelaxin gene was obtained from a genomic clone using porcine relaxin cDNA for its identification (24). Its structure showed similarities with the insulin gene with a single large intron in similar position, thereby strengthening the view that these hormones arose by gene duplication from a common ancestor. During this work there was some indication of a possible second relaxin gene that was subsequently successfully identified using RNA from the corpus luteum of pregnancy and specific probes to exon 1 and 2 of the human relaxin H1 gene. This was called the H2 relaxin gene and differed significantly in sequence from the H1 structure (25). The evidence to date is that the H2 gene, but not the H1 gene, is expressed in the corpus luteum (26, 27). The product of this H2 gene appears to be the only systemic relaxin present in the serum of pregnant women (28). We have focused on characterizing mRNAs for relaxin in a number of intrauterine tissues as sources of paracrine rather than endocrine relaxins: the decidua parietalis, basalis, the chorionic cytotrophoblast of the fetal membranes, and the placental trophoblast (6). The mRNA isolated from these tissues varied slightly in size (Table 12.1). When the poly(A)$^+$ tails were removed from the decidua and trophoblast mRNAs by digestion with RNase H and

TABLE 12.1. Species of relaxin mRNA and the human relaxin genes expressed in the corpus luteum and intrauterine tissues.

Tissue	mRNA (kb)	Relaxin gene(s) identified
Corpus luteum	2.00	H2
	1.05	
Decidua parietalis	1.05	H1 + H2
Chorion laeve	1.05	Not available
Placental trophoblast	1.00	H1 + H2
Decidua basalis	1.05	Not available
	1.00	

subjected to Northern analysis, the differences were maintained. Hence, the truncation of the trophoblast relaxin mRNA is located in the 5' or 3' untranslated regions, suggesting different transcription initiation sites or different polyadenylation sites in the 3' region.

The detection of mRNA for relaxin in these tissues has been difficult in the past, reflecting transcription at low copy number. This has been circumvented by the use of reverse transcription and the *polymerase chain reaction* (PCR), followed by selective amplification of the specific cDNA and its detection on Southern analyses (27). Using this methodology, we have been able to demonstrate the coexpression of both the H1 and H2 relaxin genes in the decidua and trophoblast (27). The identification of the gene products was made possible by the use of specific restriction enzyme analysis since *Hpa*I cuts only the H2 gene product and *Hpa*II cuts only the H1 gene product. Specific H1 and H2 PCR primers that will selectively amplify H1 and H2 cDNAs can be made, but it has not yet been possible to quantitate precisely the relative expressions of the H1 and H2 genes in these tissues.

Translation of the Human Relaxins in Intrauterine Tissues

Early work suggested that the cyclic human endometrium produces relaxin (29), detected immunohistochemically with an antibody to porcine relaxin. Monoclonal antibodies to human relaxin, kindly supplied by Genentech, Inc., have been used to reevaluate this claim and follow the reproductive cycle of this tissue into pregnancy. We have found that relaxin immunostaining predominates in the glandular cells of both proliferative- and secretory-phase endometria. The predecidual cells begin to immunostain in the late secretory phase, whereas the decidualized stromal cells stain heavily in early pregnancy (to be published). There is no information to date on relaxin gene expression in the human cyclic endometrium.

It is not known whether the monoclonal antibodies used for these and other studies can distinguish between the H1 and H2 relaxin gene product, but it seems likely since they were made to a synthetic H2 relaxin and were screened with this hormone. Differences found in staining patterns in the placenta and chorionic cytotrophoblast (6) nonetheless suggest that they detect different conformations of endogenously produced and/or receptor-bound relaxin. Studies on relaxin gene expression have confirmed potential extraluteal sites of relaxin synthesis (30), but antibodies specific for the H1 and H2 gene products are clearly needed in order to show translation of both genes concomitantly in intrauterine tissues.

Hormone:Receptor Linkage

The human decidua is maternal tissue, whereas the placenta is derived from the fetal trophoblast. There is an increasing list of hormones being detected at low levels in tissues other than their primary site of production. It is likely that decidual hormones may target the placenta and that placental hormones target the decidua. However, the decidua and the fetal membranes are themselves a complex paracrine system, just as the placental syncytiotrophoblast and cytotrophoblast form another such system. The placental basal plate, or decidua basalis, is a region of exceptional complexity where both maternal and fetal cells are mixed. An understanding of the autocrine-paracrine roles of hormones produced by these tissues is central to an understanding of the events of human parturition.

We have used four criteria developed several years ago in our laboratory for the study of relaxins as autocrine-paracrine hormones (31). These are (1) the demonstration of relaxin production in the tissues adjacent to target sites; (2) the detection of specific receptors for relaxin in the same tissues; (3) the demonstration of a biological consequence of exogenous relaxin added to the putative target tissues; and (4) the demonstration of changes in concentration of relaxin production, receptors, and, hence, biological effects in pregnancy up to parturition. In order to satisfy the classification as an autocrine-paracrine hormone, all intrauterine hormones should be studied using these criteria, and molecular techniques make a significant contribution.

The synchronization of endocrine events in reproduction involves the linkage between hormone production and the concentration and affinity of its receptor. It is evident that the increasing rate of secretion and local concentration can up-regulate or down-regulate receptor concentrations on adjacent cells. Similarly, a hormone can up-regulate the receptors for the next hormone in a cascade; for example, the effect of estrogen upon the oxytocin receptor (32). However, one hormone may down-regulate its own receptor and, at the same time, down-regulate that of another hormone, an effect shown recently for prolactin on LH receptor concentration in the ovary (33). These linkages suggest that a complex interplay of locally produced hormones may orchestrate key changes both within and between intrauterine tissues. The advent of molecular technology applicable to these problems now makes possible the sensitivity and precision necessary to be able to further define these interactions both temporally and spatially and to relate the changes to the events of parturition.

Acknowledgments. This work has been supported by Grants HD-24314 and RCMI-RR-03061. The dedicated work of students in my laboratory, past and present, is gratefully acknowledged.

References

1. Liggins GC, Forster CS, Grieves SA, Schwartz AL. Control of parturition in man. Biol Reprod 1977;16:39–56.
2. MacDonald PC, Porter JC, Schwarz BE, Johnston JM. Initiation of parturition in the human female. Semin Perinatol 1978;2:273–86.
3. MacDonald PC, Casey ML. Decidual activation in parturition [Abstract]. 1st conf on the Primate Endometrium, New York, May, 1990.
4. Bryant-Greenwood. Relaxin as a new hormone. Endocr Rev 1982;3:62–90.
5. Sherwood OD. Relaxin. In: Knobil E, Neill J, eds. The physiology of reproduction. New York: Raven Press, 1988:585–673.
6. Sakbun V, Ali SM, Greenwood FC, Bryant-Greenwood GD. Human relaxin in the amnion, chorion, decidua-parietalis, basal plate and placental trophoblast by immunocytochemistry and Northern analysis. J Clin Endocrinol Metab 1990;70:508–14.
7. MacLennan AH, Grant P. The response of human and pig myometrium in vitro to human relaxin. J Reprod Med 1991.
8. MacLennan AH, Grant P, Ness D, Down A. Effect of porcine relaxin and progesterone on rat, pig and human myometrial activity in vitro. J Reprod Med 1986;31:43–9.
9. MacLennan AH, Green R, Bryant-Greenwood GD, Greenwood FC, Seamark RF. Ripening of the human cervix and induction of labor with purified porcine relaxin. Lancet 1980;1:220–3.
10. Evans MI, Dougan MB, Moawad AH, Evans WJ, Bryant-Greenwood GD, Greenwood, FC. Ripening of the human cervix with porcine ovarian relaxin. Am J Obstet Gynecol 1983;147:410–4.
11. Eddie LW, Cameron IT, Leeton JF, Healy DL, Renon P. Ovarian relaxin is not essential for dilatation of the cervix. Lancet 1989;336:243–4.

12. Koay ESC, Bryant-Greenwood GD, Yamamoto SY, Greenwood FC. The human fetal membranes: a target tissue for relaxin. J Clin Endocrinol Metab 1986;62:513–21.
13. Lonky NM, Hayashi RH. A proposed mechanism for premature rupture of membranes. Obstet Gynecol Surv 1988;41:22–7.
14. Harger JH, Hsing AW, Tuomala RE, et al. Risk factors for preterm premature rupture of fetal membranes: a multicenter case-control study. Am J Obstet Gynecol 1990;163:130–7.
15. Hack M, Fanaroff AA. Outcomes of extremely low-birth-weight infants between 1982 and 1988. New Engl J Med 1989;321:1642–7.
16. Polishuk WZ, Kohane S, Peranio A. The physical properties of fetal membranes. Obstet Gynecol 1962;20:204–10.
17. Skinner SJM, Campos GA, Liggins GC. Collagen content of human amniotic membranes: effect of gestation length and premature rupture. Obstet Gynecol 1981;57:487–9.
18. Damjanov I. Vesalius and Hunter were right: decidua is a membrane! Lab Invest 1985;53:597–8.
19. Wewer UM, Faber M, Liotta LA, Albrechtsen R. Immunochemical and ultrastructural assessment of the nature of the pericellular basement membrane of human decidual cells. Lab Invest 1985;53:624–33.
20. Getzenberg RH, Pienta KJ, Coffrey DS. The tissue matrix: cell dynamics and hormone action. Endocr Rev 1990;11:399–417.
21. Bryant-Greenwood GD. Decidual and placental relaxins. Reprod Fertil Dev (in press).
22. Crawford RJ, Hammond VE, Roche PJ, Johnston PD, Tregear GW. Structure of rhesus monkey relaxin predicted by analysis of the single-copy rhesus monkey relaxin gene. J Mol Endocrinol 1989;3:169–74.
23. Crawford RJ, Hudson P, Shine J, Niall HD, Eddy RL, Shows TB. Two relaxin genes are on chromosome 9. EMBO J 1984;3:2341–5.
24. Hudson P, Haley J, John M, et al. Structure of a genomic clone encoding biologically active human relaxin. Nature 1983;301:628–31.
25. Hudson P, John M, Crawford R, et al. Relaxin gene expression in the human ovaries and the predicted structure of a human preprorelaxin by analysis of cDNA clones. EMBO J 1984;3:2333–9.
26. Ivell R, Hund N, Khan-Dawood F, Dawood MY. Expression of the human relaxin gene in the corpus luteum of the menstrual cycle and in the prostate. Mol Cell Endocrinol 1989;66:251–5.
27. Hansell DJ, Bryant-Greenwood GD, Greenwood FC. Expression of the human relaxin H1 gene in the decidua, trophoblast and prostate. J Clin Endocrinol Metab 1991;72:899–904.
28. Winslow J, Shih A, Laramee G, Bourell J, Stults J, Johnston P [Abstract 889]. Proc 71st meet Endocr Soc 1989.
29. Yki-Jarvinen H, Wahlstrom T, Seppala M. Immunohistochemical demonstration of relaxin in gynecologic tumors. Cancer 1983;52:2077–80.
30. Bryant-Greenwood GD. The human relaxins: concensus and dissent. Mol Cell Endocrinol (in press).
31. Koay ESC, Greenwood FC, Bryant-Greenwood GD. Relaxin: a local hormone in human parturition. Serono Symposia, USA, 1985;21:247–53.

32. Alexandrova M, Soloff MS. Oxytocin receptors and parturition, I. Control of oxytocin receptor concentration in the rat myometrium at term. Endocrinology 1980;106:730−5.

33. Lane TA, Chen TT. Heterologous down-modulation of luteinizing hormone receptors by prolactin: a flow cytometry study. Endocrinology 1991;128: 1833−40.

13

Molecular Mechanisms Regulating Oxytocin Gene Expression

HANS H. ZINGG, STÉPHANE RICHARD, AND DIANA L. LEFEBVRE

The hypothalamic nonapeptide *oxytocin* (OT) acts as a circulating hormone and as a neurotransmitter. Moreover, peripherally produced OT may function as a paracrine mediator. Despite the manifold sites of OT secretion and OT action, all its diverse effects are intimately related to the physiology of reproduction. During parturition, circulating OT regulates the contraction of uterine smooth muscle and, during lactation, OT triggers milk ejection by its action on myoepithelial cells of the mammary gland (1). At the level of the pituitary, OT stimulates prolactin release (2). In addition, parvicellular OT neurons project to distinct areas in the brain and the spinal cord, where OT functions as a neurotransmitter. In the female rat, OT's central actions include facilitation of specific sexual behavior, as well as induction of maternal behavior (nesting and pup-gathering) (3, 4). Also, in the male specific central actions have been assigned to OT. These include the strange combination of penile erection and yawning (5). Whereas the close relationship of the former phenomenon to reproduction has long been established, the significance of the latter reflex remains less clear in the present context. OT-like immunoreactivity has also been demonstrated in the ovary, testis, placenta, and adrenal. Ovarian OT is thought to act on uterine prostaglandin production and can thus promote luteal regression (6). Phylogenetically, OT is derived from more ancient molecules present in nonmammalian species, where OT-related peptides assume functions related to salt and water homeostasis. Perhaps as a relic of these more ancient functions, OT exerts a natriuretic effect also in mammals (7).

Although much is known of the physiology of the OT systems, the molecular mechanisms involved in cell-specific and hormone-induced regulation of the OT gene are less well understood. The following section summarizes available data on the regulation of OT mRNA accumulation, and the remainder of the chapter describes recent attempts to unveil the molecular mechanisms involved in OT gene regulation.

Regulation of Oxytocin Messenger RNA Abundance

OT is biosynthetically derived from a 16-kd precursor molecule that consists of the nonapeptide OT and, in addition, the OT-associated neurophysin. In humans, cow, rat, and mouse, the structure of the gene encoding the OT precursor molecule has been determined and shown to consist of 3 exons, with OT being encoded by the first exon (8–11). The gene encoding the related peptide vasopressin has a very similar organization and shows significant sequence homology with the OT gene, most notably in exon 2. However, the 5′ promoter regions of the two genes show very little similarity, suggesting that different mechanisms are involved in the regulation of each gene. Using Northern blot analysis in conjunction with an oligonucleotide probe complementary to exon 3, we have investigated the dynamics of hypothalamic OT mRNA accumulation during pregnancy and lactation, two periods in which OT plays major roles. As illustrated in Figure 13.1, we found that, in the rat, pregnancy induced a gradual rise in hypothalamic OT mRNA accumulation. By day 18, OT mRNA levels exceeded control levels by a factor of 2.5, and throughout the ensuing lactation period, OT mRNA levels remained elevated at levels corresponding to 3 times that of control (12). Similar observations have been made by Van Tol et al. (13). As shown by Caldwell et al. (14), this rise in mRNA is paralleled by a concomitant increase in OT-like immunoreactivity in hypothalamic neurons. By contrast, during this same period, the secretion of OT into the blood stream is not stimulated, inasmuch as OT serum levels remain unchanged throughout pregnancy, with the exception of the hours immediately preceding delivery (15). Therefore, the increased biosynthesis of hypothalamic OT does not seem to be a direct consequence of increased release. Interestingly, the rise of OT mRNA levels occurs in parallel with an increase in serum estrogen levels (16). This raises the possibility that estrogens may be involved in stimulating OT gene transcription (see below). Although estrogens have a rapid and dramatic effect on OT secretion (17), we were unable to detect a global change in hypothalamic OT mRNA in response to systemic estrogen application. The only other stimulus that led to a dramatic overall rise in hypothalamic OT mRNA was osmotic stimulation by salt loading (Fig. 13.1).

The effect of estrogen on hypothalamic OT mRNA is more subtle and could only be demonstrated by in situ hybridization since only a subclass of OT neurons are affected by this stimulus (18). Van Tol et al. were able to demonstrate subtle changes in hypothalamic OT mRNA during the cycle with a maximum at estrus (13). Additional indications that estrogens are involved in OT gene regulation stem from developmental studies. We have shown earlier that by Northern blot analysis, hypothalamic OT mRNA becomes detectable in the rat brain around birth and rises dramatically in the early postnatal period (19). More

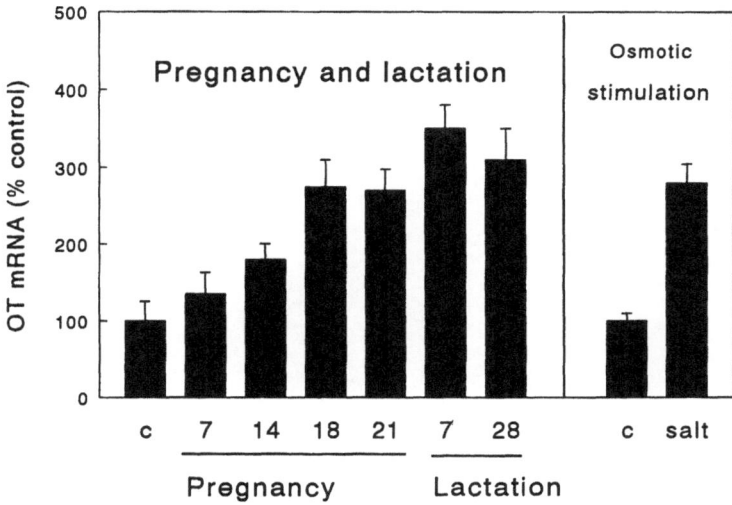

FIGURE 13.1. Oxytocin mRNA accumulation in rat hypothalamus during pregnancy and lactation as well as in response to osmotic stimulation. mRNA levels were estimated by densitometric scanning of Northern blots. The probe used was a 27-bp oligonucleotide, complementary to a region in the third exon of the oxytocin gene (12). Each bar corresponds to the mean ± SE of at least 3 independent determinations.

recently, Miller et al. demonstrated that the onset of puberty is associated with an additional major rise in hypothalamic OT mRNA and that this rise can be prevented by ovariectomy (20). In a follow-up study, it was further demonstrated that exogenously added estrogens abolished the effect of ovariectomy (21). As described in the following section, we therefore investigated the possibility that estrogens exert their effect on OT gene expression via a direct action on the OT gene promoter.

Analysis of Molecular Mechanisms Controlling Oxytocin Gene Promoter Activity

For the study of mechanisms regulating gene expression at a transcriptional level, one takes advantage of the fact that the organization of most genes is modular in nature. Regions mediating regulation of gene transcription can function in relative independence from regions encoding the structure of the gene product. It is thus possible to link regulatory regions of a given gene to another structural gene that acts as a reporter gene. Following transfection of such a chimeric gene construct into a host cell line, the expression of the reporter gene serves as an indicator for the transcriptional activity of the linked regulatory region.

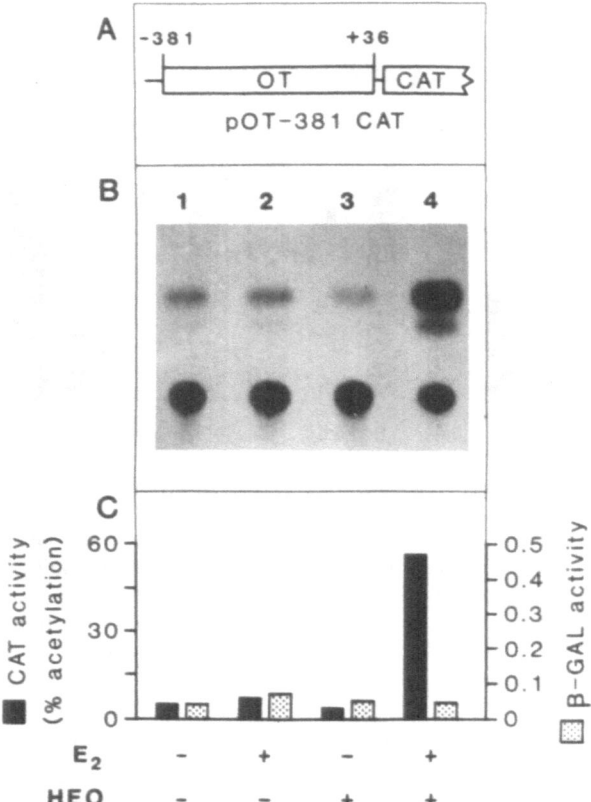

FIGURE 13.2. Estrogen responsiveness of the OT gene promoter. *A:* Shown is a schematic illustration of the chimeric gene used for transfection where 381 bp of the 5' flanking region of the human OT gene were linked to the structural gene for CAT. *B:* Shown are the CAT activities resulting from transient transfections of pOT/381CAT in Neuro-2a cells. Lanes 1 and 2 represent pOT-381CAT cotransfected with carrier DNA, whereas lanes 3 and 4 represent cotransfection with the expression vector containing the estrogen receptor cDNA (HE0) as described in reference 26. As an external control for transfection efficiency, a plasmid containing the structural gene for β-galactosidase under the control of the SV40 early promoter was also cotransfected. Cell extracts from either control (lanes 1 and 3) or 17β-estradiol-treated cultures (10^{-7} M, lanes 2 and 4) were assayed for CAT activity by thin-layer chromatography. *C:* Shown is the quantitation of CAT activity (filled bars) and β-galactosidase activity (stippled bars) in extracts from control (E_2^-) or estradiol-treated (E_2^+) cultures transfected with either carrier DNA (HE0$^-$) or with an estrogen receptor expression vector (HE0$^+$). Reprinted with permission from Richard and Zingg (23).

As a first step, we have linked 381 bp of the 5' flanking region of the human OT gene to the bacterial reporter gene *chloramphenicol acetyltransferase* (CAT) (22) (Fig. 13.2A). In the absence of an established cell line that highly expresses the OT gene, we have used the mouse neuroblastoma-derived cell line Neuro-2a, a cell line in which the transfected OT promoter elements displayed a readily detectable baseline activity in the absence of external stimulation (Fig. 13.2B). When these cells were endowed with estrogen receptors by cotransfection with expression vector containing the estrogen receptor cDNA, addition of 10^{-7} M estradiol elicited a 12-fold increase in expression of the chimeric gene pOT-381CAT (Fig. 13.2B) (23). This effect was specific since the expression of a cotransfected plasmid containing the LacZ gene placed under the control of the SV40 early promoter remained unaffected (Fig. 13.2C). Moreover, this effect was estrogen receptor dependent since estrogen induction could only be observed in host cells cotransfected with the estrogen receptor expression plasmid (Fig. 13.2).

In order to delineate more precisely the area necessary for estrogen responsiveness, we constructed 5' and 3' deletion mutants and tested their estrogen inducibility. As a result of these studies, we identified a near-perfect palindrome situated at −164 that was both necessary and sufficient for conferring estrogen responsiveness to the OT gene promoter. As shown in Figure 13.3, this sequence element bears significant sequence similarity with other *estrogen response elements* (EREs). The OT ERE differs from other EREs by containing a G in position 4 instead of a C. Data from methylation interference analysis suggest that position 4 in the palindrome is indeed of functional importance since the G that base-pairs with the C in this position forms a contact site with the estrogen receptor molecule (24). Thus, the present ERE represents a novel version of a fully functional ERE that differs at a relevant position from the consensus palindrome. We have mutated the G in position 4 into a C, thus rendering this sequence element perfectly palindromic. This nucleotide change induced only a very modest (21%) increase in estrogen responsiveness of the OT promoter (unpublished results).

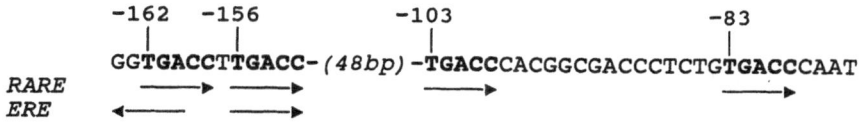

ERE consensus sequence: GGTCAnnnTGACC

FIGURE 13.3. Hormone response elements in the 5' flanking region of the human OT gene. The ERE consists of an imperfect palindrome. The RARE consists of 4 consecutive repeats of the pentanucleotide sequence TGACC. Note the overlap of the first 2 TGACC repeats with the ERE.

When we investigated the response of the OT gene promoter fragment to ligands interacting with other members of the steroid receptor superfamily, we found that it was unresponsive to glucocorticoids, thyroid hormone, and vitamin D (unpublished data), even in cells that were cotransfected with expression plasmids encoding the cognate receptors. In contrast, we found that the OT promoter fragment was highly responsive to *retinoic acid* (RA), another ligand that exerts its effects via interaction with a specific subclass of nuclear receptors. 5′ and 3′ deletion analysis revealed that the regions conferring RA responsiveness extended over a much longer area than the region conferring estrogen responsiveness. By site-directed mutagenesis, we were able to determine that the presence of 4 pentanucleotide repeats with the sequence TGACC was necessary for full RA responsiveness. As shown in Figure 13.3, the first 2 TGACC repeats are contained within the previously characterized ERE. The third and fourth TGACC repeats, which are both necessary for retinoic acid inducibility, lay 53 bp downstream. To what extent each of these pentamer pairs serves as a binding site for a retinoic acid receptor dimer remains to be investigated. Since our deletion and mutation experiments indicated that all 4 pentamer repeats are necessary for RA action, the 2 pentamer pairs are likely to act in synergism over the 53-bp distance.

The primary role of RA is generally thought to be linked to its action as a morphogen during embryogenesis. The presence of retinoic acid receptors in various adult tissues is, however, an indication that the role of RA extends beyond the developmental period. Specifically, the high amounts of retinoic acid receptor α and β expression in adult brain (25) may suggest hitherto unrecognized roles of RA in brain function. Our present data demonstrating RA responsiveness of the OT gene promoter may therefore serve as an additional impetus for widening our current concepts of RA physiology and point towards a possible role of RA in the modulation of neuropeptide gene expression and brain function.

We were next interested in determining the sequence element necessary for maintaining the high baseline expression of the OT gene promoter observed in Neuro-2a cells. The rationale for these studies was that elucidation of the mechanisms that induced OT promoter activity in Neuro-2a cells may help to identify sequence elements and transcription factors that play a role in mediating the cell-specific expression of the OT gene in vivo. By 5′ deletion analysis, we determined that a segment from −49 to +36 was capable of mediating cell-specific promoter activity in Neuro-2a cells, but was inactive in NIH3T3 fibroblasts or JEG-3 placental cells. Within this segment, we identified 3 *proximal promoter elements* (PPE-1, PPE-2, and PPE-3) that are each required for promoter activity (Fig. 13.4). Gel mobility shift analysis with 3 different double-stranded oligonucleotides demonstrated that each proximal promoter element binds distinct nuclear factors (unpublished data). In each case, only the homologous oligonucleotide, but neither of the oligonucleotides

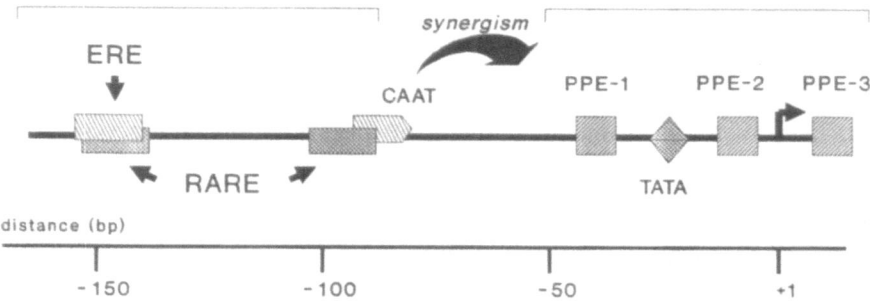

FIGURE 13.4. Schematic diagram of regulatory sequences in the 5' flanking region of the human OT gene. Distance from initiation site of transcription is indicated in bp. (ERE = estrogen response element; RARE = retinoic acid response element; CAAT = CAAT box; PPE-1, PPE-2, and PPE-3 = proximal promoter elements 1, 2, and 3; TATA = TATA box.)

corresponding to adjacent elements, was able to act as a competitor. Thus, a different set of factors appears to bind independently to each element. Proteins binding to these sites form a mutually synergistic complex since occupation of all 3 sites is essential for maximal activity, and deletion or mutation of any of these sites resulted in a dramatic reduction of promoter activity. Although the factors binding to the 3 sites identified seem to act synergistically to stimulate transcription, it is unlikely that this synergism results from cooperative binding of these proteins to DNA since each is capable of binding independently of the others. Rather, it appears that the synergism is the result of direct interaction of these proteins with the transcriptional machinery or via an indirect interaction involving additional non-DNA binding proteins, referred to as co-activators. The involvement of multiple proteins binding to the OT gene proximal promoter region could participate in the generation of a versatile transcriptional control mechanism required for the diversity of OT gene expression in different neuronal and nonneuronal cell types as well as in response to different extracellular stimulators.

The emerging overall picture of the factors interacting with the OT gene promoter is illustrated in Figure 13.4. A hormone regulatory region can be distinguished from a region that mediates baseline expression of the OT gene. The hormone regulatory elements we have identified comprise an ERE as well as a *retinoic acid response element* (RARE). These sequence elements serve, most likely, as a binding site for the corresponding receptor-ligand complexes. In vitro DNA binding studies are currently in progress in order to define the precise stoichiometry of receptor-DNA interactions. Interestingly, the EREs and RAREs overlap partially. However, the combined application of estrogens and RA to cells transfected with both receptor types leads to an additive effect, indicating that there is no negative interaction between the two receptor types. Most probably, estrogen and RA receptors are capable of interacting synergistically.

Moreover, the hormone regulatory elements do not act in isolation, but are, to some extent, dependent on the synergistic interaction with elements mediating the basal expression of the gene. We have identified 3 of these elements and termed them PPE-1, PPE-2, and PPE-3. The factors interacting with these elements are currently unknown. In addition, the OT gene promoter contains a TATA box as well as a CCAAT box, both binding sites for well-characterized, ubiquitous transcription factors. Identification and characterization of the factors interacting with the additional proximal promoter elements will significantly advance our understanding of the mechanisms involved in cell-specific expression of the OT gene and, perhaps, other neuropeptide genes.

We have most recently determined that the OT gene is highly expressed in the pregnant rat uterus (unpublished results). It is conceivable that the identified hormone response elements play a major role in the regulation of OT gene expression in the periphery. Especially, since the uterus has long been known to be a prime target site for estrogen action, the identified ERE might well be of major importance in mediating OT expression in the pregnant rat uterus. The significance of the RARE for the peripheral OT gene expression is currently an unknown territory. Our finding of a RARE in the OT gene promoter has now prompted investigations into the possible role of RA in gene expression in the rat brain and uterus. Whereas, in many instances, molecular biology helps to explain phenomena observed at a physiological level, there might be moments where discoveries at a molecular level lead to the definition of novel physiological mechanisms. Our identification of an ERE in the OT gene promoter is a paradigm of the former case. Our characterization of a RARE in the same promoter may emerge as an example of the latter strategy.

Acknowledgments. We thank Ms. C. Younge for excellent secretarial help. During execution of this present work, Stéphane Richard and Diana L. Lefebvre were supported by Studentships from the Medical Research Council (MRC) of Canada, and Hans H. Zingg was Scholar of the MRC, Canada. The work was supported by operating grants from the MRC, Canada.

References

1. Forsling ML. Regulation of oxytocin release. In: Ganten D, Pfaff D, eds. Neurobiology of oxytocin; current Topics in Neuroendocrinology, vol. 6. Berlin: Springer, 1986:19–54.

2. Mori M, Vigh S, Miyata A, Yoshihara T, Oka S, Aimutra A. Oxytocin is the major prolactin releasing factor in the posterior pituitary. Endocrinology 1990;126:1009–13.

3. Pederson CA, Ascher JA, Munroe YL, Prange AJ. Oxytocin induces maternal behavior in virgin female rats. Science 1982;216:648–50.

4. Caldwell JD, Prange AJ, Pederson CA. Oxytocin facilitates sexual receptivity of estrogen-treated female rats. New Peptides 1986;7:175–89.

5. Argiolas A, Melis MR, Gessa GL. Yawning and penile erection: central dopamine oxytocin-adrenocorticotropin connection. Ann NY Acad Sci 1988;525:330–7.

6. Flint APE, Sheldrick EL. Ovarian oxytocin. In: Amico JA, Robinson AG, eds. Oxytocin, clinical and laboratory studies. Excerpta Medica (Amsterdam) 1985:335–50.

7. Verbalis JG, Mangione MP, Stricker EM. Oxytocin produces natriuresis in rats at physiological plasma concentrations. Endocrinolology 1991;128:1317–22.

8. Sausville E, Carney D, Battey J. The human vasopressin gene is linked to the oxytocin gene and is selectively expressed in a cultured lung cancer cell line. J Biol Chem 1985;260:10236–41.

9. Ruppert S, Scherer G, Schutz G. Recent gene conversion involving bovine vasopressin and oxytocin precursor genes suggested by nucleotide sequence. Nature 1984;308:554–7.

10. Ivell R, Richter D. Structure and comparison of the oxytocin and vasopressin genes from rat. Proc Natl Acad Sci USA 1984;81:2006–10.

11. Hara Y, Battey J, Gainer H. Structure of mouse vasopressin and oxytocin genes. Mol Brain Res 1990;8:319–24.

12. Zingg HH, Lefebvre DL. Oxytocin and vasopressin gene expressing during gestation and lactation. Mol Brain Res 1988;4:1–6.

13. Van Tol HHM, Bolwerk ELM, Burbach JBH. Oxytocin and vasopressin gene expression in the hypothalamo-neurohypophysial system of the rat during the estrous cycle, pregnancy, and lactation. Endocrinology 1988;122:945–51.

14. Caldwell JD, Grenn RA, Johnson MF, Prange AJ, Pederson CA. Oxytocin and vasopressin immunoreactivity in hypothalamic and extrahypothalamic sites in late pregnancy and postpartum rats. Neuroendocrinology 1987;46:39–47.

15. Fuchs AR, Daywood MY. Oxytocin release and uterine activation during parturition in rabbits. Endocrinology 1980;107:1117–26.

16. Yoshinaga K, Hawkins RA, Stocker JF. Estrogen secretion by the rat ovary in vivo during the estrous cycle and pregnancy. Endocrinology 1969;85:103–12.

17. Amico JA, Seif SM, Robinson AG. Oxytocin in human plasma: correlation with neurophysin and stimulation with estrogen. J Clin Endocrinol Metab 1981;52:988–93.

18. Caldwell JD, Brooks PJ, Jirikowski GF, Barakat AS, Lund PK, Pederson CA. Estrogen alters oxytocin mRNA levels in the preoptic area. J Neuroendocrinol 1989;1:273–8.

19. Almazan G, Lefebvre DL, Zingg HH. Ontogeny of hypothalamic vasopressin, oxytocin and somatostatin gene expression. Devel Brain Res 1989;45:69–75.

20. Miller FD, Ozimek G, Milner RJ, Bloom FE. Regulation of neuronal oxytocin mRNA by ovarian steroids in the mature and developing hypothalamus. Proc Natl Acad Sci USA 1989;86:2125–36.
21. Chibbar R, Toma JG, Mitchell BF, Miller FD. Regulation of neural oxytocin gene expression by gonadal steroids in pubertal rats. Mol Endocrinol 1990;4:2030–8.
22. Gorman CM, Moffat LF, Howard BH. Recombinant genomes which express chloramphenicol acetyltransferase in mammalian cells. Mol Cell Biol 1982;2:1044–51.
23. Richard S, Zingg HH. The human ocytocin gene promoter is regulated by estrogens. J Biol Chem 1990;265:6098–103.
24. Klein Hitpass L, Tsai SY, Greene BL, Clarke GH, Tsai M-J, O'Malley BW. Mol Cell Biol 1989;9:43–9.
25. Giguere V, Ong ES, Segui P, Evans RM. Identification of a receptor for the morphogen retinoic acid. Nature 1987;330:624–9.
26. Green S, Walter P, Kumar V, et al. Human estrogen receptor cDNA: sequence, expression and homology to v-exb A. Nature 1986;320:134–9.

Part IV

Poster Presentation Manuscripts

14

Structure-Function Relationships of Multiple Rat Members of the 3β-Hydroxysteroid Dehydrogenase Family

YVAN DE LAUNOIT, JACQUES SIMARD, HUI-FEN ZHAO,
PATRICK COUTURE, AND FERNAND LABRIE

The conversion of 3β-hydroxy-5-ene steroids by the membrane-bound enzyme 3β-hydroxysteroid dehydrogenase/Δ^5-Δ^4-isomerase, hereafter called 3βHSD, is an essential step in the biosynthesis of all classes of hormonal steroids. We have recently characterized 3 types of cDNAs encoding rat 3βHSD (1, 2). The predicted rat type I and type II 3βHSD-expressed proteins share 94% homology (1), while they share only 80% similarity with the rat type III 3βHSD, which is also a 372-amino acid protein (2) as observed for the human (3–5), macaque (6), and bovine (7) 3βHSD predicted proteins. Using the highly sensitive RNase protection method, we have shown that the type I and type II 3βHSD mRNAs are present in several rat tissues, including the ovary, testis, adrenal, and adipose tissue (1, 2), whereas the type III was only found in liver (2). In addition, we have demonstrated by computer analysis that the type I 3βHSD and the type III 3βHSD-encoded proteins possess 2 predicted trans*membrane-spanning domains* (MSD) (1, 2, 8), whereas type II 3βHSD is devoid of one of the 2 MSDs (1, 8). Moreover, transient expression of rat type I and type II 3βHSD cDNAs in non-steroidogenic cells revealed that these two 42-kd proteins catalyze both the oxidation and isomerization of Δ^5-3β-hydroxysteroid precursors into Δ^4-3-ketosteroids, as well as the interconversion of 3β-hydroxy and 3-keto-5α-androstane steroids (1, 8). This chapter investigates the structure-function relationships of the 3 rat 3βHSD isoenzymes so far isolated.

Materials and Methods

Construction of Wild-Type and Chimeric Rat 3βHSD cDNAs by Site-Directed Mutagenesis

We used the full-length cDNA inserts corresponding to the rat type I (ro3β-HSD56) and type II (ro3β-HSD112) 3βHSD clones (1). We have also constructed the full-length rat liver 3βHSD clone by ligating the *Eco*RI/*Sau*I DNA fragment corresponding to nucleotides −84 to +205 of clone rl 3β-HSD 20, the *Sau*I/*Taq*I DNA fragment corresponding to nucleotides +206 to 720 of clone rl 3β-HSD 33 (2), and the *Taq*I/*Eco*RI DNA fragment corresponding to nucleotides +721 to +1871 of clone rl 3β-HSD 35 (2). Those full-length rat type I, type II, and type III 3βHSD clones were then cloned into the unique *Eco*RI site of the pCMV vector, downstream from the *cytomegalovirus* (CMV) promoter, to produce the recombinant plasmids pCMV type I 3βHSD, pCMV type II 3βHSD, and pCMV type III 3βHSD, respectively. Those plasmids were sequenced in both orientations, amplified, and subsequently purified by 2 cesium chloride-ethidium bromide density gradient ultracentrifugations. We then constructed by site-directed mutagenesis 2 chimeric cDNAs in which the 4 codons for Ala[83], Ile[85], Val[87], and His[89] potentially involved in an MSD predicted between residues 75 and 91 in the type I 3βHSD protein using the SOAP program (9) of the PC/GENE software (release 6.01; Intelli Genetics, Inc./Genofit SA: Mountain View, CA) were substituted by the codons Ser[83], Met[85], Phe[87], and Arg[89] present in the type II 3βHSD protein and vice versa, thus leading to cDNA inserts encoding a type I 3βHSD protein without the potential MSD (I − MSD) or a type II 3βHSD protein containing type I MSD (II + MSD) as represented schematically in Figure 14.1. Briefly, the full-length cDNA inserts ro3β-HSD56 (type I) and ro3β-HSD112 (type II) were excised from the respective pCMV plasmids by partial *Eco*RI digestion, and the DNA fragments released were size-fractionated on a 1.2% low-melting-temperature agarose gel. The purified full-length cDNA inserts were then digested with *Kpn*I, and the generated *Eco*RI-*Kpn*I and *Kpn*I-*Eco*RI fragments were purified. The small *Eco*RI-*Kpn*I DNA fragment from type I (ro3β-HSD56) was ligated with the long *Kpn*I-*Eco*RI DNA fragment from type II (ro3β-HSD112) and then subcloned into pCMV vector to produce the pCMV-I − MSD plasmid. Similarly, the small *Eco*RI-*Kpn*I DNA fragment from type II (ro3β-HSD112) was ligated with the long *Kpn*I-*Eco*RI DNA fragment from type I (ro3β-HSD56) and then subcloned into the pCMV vector to produce the pCMV-II + MSD plasmid. The only differences between the deduced protein sequences encoded by the corresponding 2 small *Eco*RI-*Kpn*I fragments are amino acid positions 83, 85, 87, and 89.

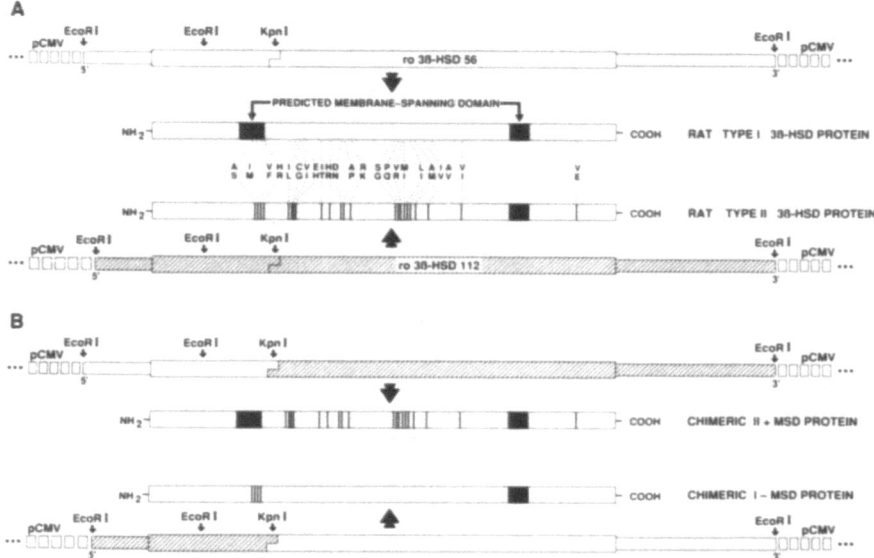

FIGURE 14.1. Structures of wild-type rat types I and II 3βHSD proteins and chimeric 3βHSD proteins. *A:* Shown are a schematic representation and comparison of the structures of wild-type rat type I and type II 3βHSD proteins encoded by pCMV type I 3βHSD containing the ro3β-HSD56 cDNA insert and pCMV type II 3βHSD containing the ro3β-HSD112 cDNA insert, respectively. The 23 nonidentical amino acid residues between the two rat 3βHSD types are indicated by vertical lines in the schema illustrating the deduced type II 3βHSD protein. The predicted MSDs are represented by black boxes. Only the *Eco*RI and *Kpn*I restriction sites that are of interest for the construction of chimeric cDNAs are illustrated. *B:* Shown is a schematic representation of the chimeric 3βHSD proteins encoded by site-directed mutated rat 3βHSD cDNAs constructed as described in "Materials and Methods" and "Results" sections in the pCMV expression vector. The chimeric protein II + MSD was generated by introducing a putative MSD, while the chimeric protein I − MSD, on the other hand, was generated removing the same putative MSD. Reprinted with permission from Simard, de Launoit, and Labrie (8), © by The American Society for Biochemistry and Molecular Biology.

Transient Expression of Multiple Rat 3βHSD cDNAs

Expression of the plasmids was carried out in the HeLa human cervical carcinoma cells by the transfection method previously described (1, 4, 5, 8). Briefly, the plasmids were introduced into HeLa cells by the calcium phosphate precipitation procedure. Mock transfections were carried out with the pCMV alone, while transfection efficiency was monitored by cotransfecting the tested plasmids with the control pXGH5 plasmid that expresses growth hormone. In order to determine 3βHSD activity, cells were incubated for the indicated time periods at 37°C in the presence of ^{3}H-labeled steroid substrates in 50 mM Tris buffer (pH 7.5) containing 1 mM of the appropriate cofactor, namely NAD$^+$ for *dehydroepiandrosterone* (DHEA, Steraloids), *5α-androstane-3β,17β-diol* (3β-diol, Steraloids) and *pregnenolone* (PREG, Steraloids) or NADH for *5α-dihydrotestosterone* (DHT, Steraloids) and *5α-androstane-dione* (A-dione, Steraloids). The enzymatic reaction was stopped by chilling the incubation mixture in an ice-water slurry and adding 4 volumes of ether/acetone (9/1, v/v). Cell protein content was measured by the method of Bradford (1, 8) using bovine serum albumin as standard, while the relative amounts of translated type I, type II, and type III, as well as the 2 chimeric 3βHSD 42-kd proteins, were evaluated by immunoblot analysis and quantification with an image analyzer of the integrated optical intensity of the corresponding 42-kd bands, followed by the correction of transfection efficiency with GH expression (8). Values of K_m as well as relative V_{max} and relative specificity (V_{max}/K_m ratio) values were calculated by the Lineweaver-Burk method, as previously described (8).

Thin-Layer Chromatography and High-Performance Liquid Chromatography Analyses

The organic phase was then evaporated and separated either on TLC plates using a 4:1 mixture of benzene and acetone or by HPLC. For TLC, substrates and formed steroids were identified by comigration on each TLC plate of the nonlabeled steroid(s). The corresponding area was cut and transferred to scintillation vials containing 0.1-mL ethanol to which 10-mL scintillation fluid was added for measurement of radio-activity (1, 4, 5, 8).

Steroids were also identified by HPLC analysis using a System Gold (Beckman) unit consisting of a model 126 pump, a 507 automatic injector, a Radial-Pak NovaPak C_{18} column (8 mm × 10 cm) and a model Beckman 168 photodiode array detector. The mobile phase for A-dione, DHT, and 3β-diol was H_2O/methanol/tetrahydrofuran/acetonitrile (50/35/10/5, v/v/v/v) at 1.5 mL/min flow rate over a 30-min period. Radioactivity was monitored in the eluent using a Beckman 171 HPLC Radioactivity Monitoring System using Formula 963 (NEN) as a scintillation mixture at a flow rate of 4.5 mL/min.

Results and Discussion

As mentioned above, the deduced amino acid sequences of rat type I and type II 3βHSDs display 93.8% similarity with only 23 nonidentical residues (1). Following transient expression, we have recently observed, however, that rat type I 3βHSD possesses much higher activity than the type II 3βHSD protein for all substrates tested (1). As predicted by computer analysis (9), there is a potential MSD common to the deduced rat type I, type II, and type III, as well as to human, macaque, and bovine 3βHSD protein sequences, corresponding to a membrane-associated alpha helical segment present in all proteins between residues 287 and 303. Furthermore, such analysis of rat 3βHSDs indicates that the change of residues 83, 85, 87, and 89 observed in the wild-type rat type II 3βHSD protein prevents the formation of a second potential MSD present in the rat type I enzyme between residues 75 and 91 as well as in human and macaque 3βHSD proteins. In order to characterize the functional significance of MSD 75–91 in type I 3βHSD, we have compared the catalytic properties of wild-type rat types I and II 3βHSD proteins with those of a chimeric type I protein lacking this MSD (I − MSD) and of a chimeric type II protein having gained this putative MSD (II + MSD) by site-directed mutagenesis.

As illustrated in Figure 14.2, pCMV type I 3βHSD encodes a protein having a 3βHSD/Δ^5-Δ^4 isomerase relative specificity, as determined by the relative V_{max}/K_m ratio (relative V_{max} values were calculated assuming

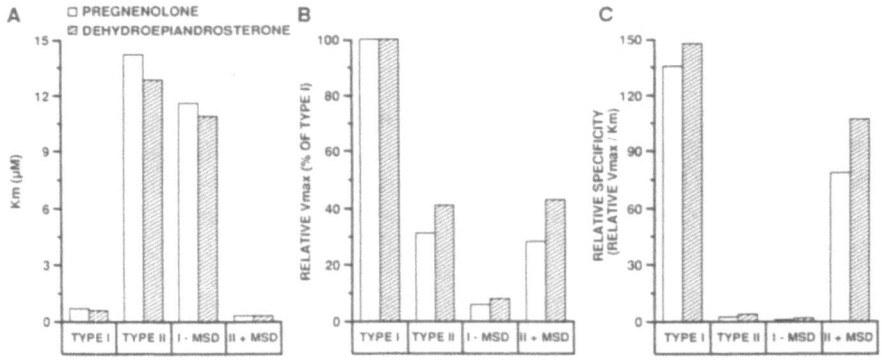

FIGURE 14.2. Catalytic properties of the expressed rat type I and type II isoenzymes of the 3βHSD family, as well as of the chimeric pCMV I − MSD- and pCMV II + MSD-encoded proteins. The procedure for transient expression of wild-type and chimeric cDNAs in HeLa cells and the 3βHSD assay are as described in "Materials and Methods." Kinetic parameters were determined using Lineweaver-Burk plot (1/v vs 1/[S]) analysis of the catalytic activity of expressed proteins. The enzymatic reaction was performed during 30 min for expressed type I 3βHSD and II + MSD proteins, while the reaction was stopped after 60 min with homogenates from cells transfected with pCMV type II 3βHSD and pCMV I − MSD cDNAs. Relative V_{max} values were calculated assuming the V_{max} for the wild-type type I 3βHSD enzyme equal to 100. The specific activity values for the type I 3βHSD enzyme using PREG and DHEA as substrates were 5.1 and 4.1 nmol/min/mg protein, respectively. Reprinted with permission from Simard, de Launoit, and Labrie (8), © by The American Society for Biochemistry and Molecular Biology.

that the rate for the pCMV type I 3βHSD encoded protein is equal to 100), that is 64 times higher than that of the pCMV type II 3βHSD-encoded protein with corresponding K_m values of 0.74 µM (type I) and 14.3 µM (type II), and 100 and 30.8 relative V_{max} values, respectively (Fig. 14.2). The much higher relative specificity of rat type I 3βHSD compared to type II 3βHSD was confirmed using DHEA as labeled substrate (Fig. 14.2). The present data thus indicate that the lower activity and lower relative specificity of type II 3βHSD result primarily from an approximately 95% decrease in affinity for both substrates, while the V_{max} value differs by only 60%–70%.

As illustrated in Figure 14.2, the chimeric I − MSD protein, which lacks MSD at position 75–91, shows a markedly reduced affinity for PREG and DHEA, with K_m values of 11.7 and 11 µM, respectively, compared to 0.74 and 0.68 µM, while its relative specificity was dramatically decreased to 0.35% and 0.47% compared to wild-type type I when PREG and DHEA were used as substrates, respectively, thus demonstrating that removal of the putative MSD in the type I 3βHSD markedly affects the specific activity as well as the affinity of the enzyme (Fig. 14.2). In an opposite fashion, the chimeric II + MSD protein gained an affinity for PREG and DHEA comparable to that of the wild-type type I 3βHSD protein, with K_m values of 0.36 and 0.40 µM, respectively. The present data show that the introduction of a putative MSD in the type II 3βHSD protein increased the relative specific activity of the type II protein to 58% (for PREG) and 73% (for DHEA) compared to the wild-type I 3βHSD protein. The present data provide strong evidence supporting the crucial role of the predicted MSD between residues 75 and 91 for the high level of enzymatic specificity of rat type I 3βHSD.

Surprisingly, we have observed that the expressed type I 3βHSD protein can also convert DHT into its 17β-oxidative form (i.e., A-dione), thus demonstrating that this enzyme possesses a secondary 17βHSD activity. In fact, as illustrated in Figure 14.3B, the endogenous 17βHSD activity in HeLa cells is very low (Fig. 14.3B), whereas homogenates from cells transfected with pCMV type I 3βHSD convert about 65% of DHT into A-dione after a 10-h incubation period in the presence of 1 mM NAD^+ (Fig. 14.3C). In contrast, type II 3βHSD isoenzyme, as well as expressed chimeric I − MSD or II + MSD protein did not have such 17βHSD activity (data not shown). The fact that the chimeric II + MSD 3βHSD protein is devoid of 17βHSD activity strongly suggests that lack of such 17βHSD enzymatic activity of expressed type II 3βHSD protein is not only due to the absence between residues 75 and 91, but should rather be due to other specific differences between the type I and type II isoenzymes. Although dual enzymatic activity has been previously demonstrated with purified steroid enzymes (10, 11), this represents the first demonstration of such dual catalytic activity after transient expression of a specific cDNA.

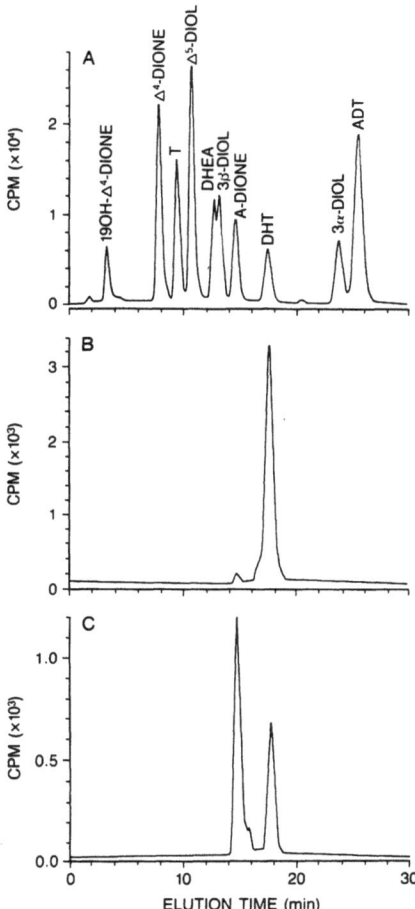

FIGURE 14.3. Secondary 17βHSD activity of the rat type I 3βHSD enzyme. Twenty μg of protein from homogenates of cells transfected with the pCMV alone or pCMV type I 3βHSD plasmid were incubated in the presence of 1 mM NAD⁺ and 300 nM tritiated DHT for 12 h. Steroids were extracted, and the organic phase was then evaporated and steroids were identified by HPLC, as described in "Materials and Methods." DHT and A-dione were separated with a mobile phase of H₂O-methanol-tetrahydrofuran/acetonitrile (50/35/10/5, v/v/v/v) at a flow rate 1.5 mL/min over a 30-min period.

We then made the unexpected observation that the expressed liver-specific type III isoform of the 3βHSD family is unable to catalyze the oxidation and isomerization of Δ^5-3β-hydroxysteroid precursors, such as PREG, DHEA, or Δ^5-androstenediol, into their Δ^4-3-keto forms nor to convert 3β-hydroxysteroids, such as 3β-diol, into their corresponding 3-

TABLE 14.1. Lineweaver-Burk plot analyses of the catalytic properties of the expressed rat type I and liver-specific type III isoforms of the 3βHSD family using increasing concentrations of DHT as substrate and the indicated cofactor.

Expressed protein	K_m (μM)	Relative V_{max}	Relative specificity (relative V_{max}/K_m)
Rat type I			
NADH	4.9	100.0	20.4
NADPH	2.9	47.8	16.5
Rat type III			
NADH	5.5	27.3	5.0
NADPH	0.7	15.4	22.0

Incubations were performed for 20 min at 37°C as described in "Materials and Methods."

keto-5α-androstane form (data not shown). However, time course studies with homogenates from cells transfected with pCMV type III 3βHSD plasmid in the presence of A-dione or DHT supplemented with 1 mM NADH clearly showed a potent reductase activity, this reductase activity being shared with type I 3βHSD. As indicated in Table 14.1, Lineweaver-Burk plot analysis shows that the K_m values for DHT using homogenate from cells transfected with pCMV type I 3βHSD or pCMV type III 3βHSD are in the same range with respective values of 4.9 μM and 5.5 μM using NADH as cofactor. The type I 3βHSD protein possesses, however, a relative V_{max} and a relative specificity 3.7- and 4.1-fold higher than those of expressed type III isoform, respectively (Table 14.1). The K_m value for the type I 3βHSD protein in the presence of DHT supplemented with 1 mM NADPH is 2.9 μM, a value that is in the same range as the value obtained when NADH is used as cofactor. However, the affinity of the expressed type III isoform for DHT is much higher with NADPH, with a K_m value of 0.69 μM compared to 5.51 μM with NADH. Using NADPH as cofactor, the relative V_{max} value is 3.1-fold higher in type I than in type III, whereas the relative specificity shows comparable values (16.5 vs 22.0). The present data indicate that NADPH is the preferred cofactor for the reductase activity of the expressed type III isoform (Table 14.1). Detailed investigation of the structure-function characteristics of the catalytic sites of the multiple 3βHSD isoenzymes by site-directed mutagenesis should provide crucial information concerning the molecular basis for their differential enzymatic activity.

References

1. Zhao HF, Labrie C, Simard J, et al. Characterization of 3β-hydroxysteroid dehydrogenase Δ^5-Δ^4 isomerase cDNA and differential tissue-specific expression of the corresponding mRNAs in steroidogenic and peripheral tissues. J Biol Chem 1991;266:583–93.

2. Zhao HF, Rhéaume E, Trudel C, Couet J, Labrie F, Simard J. Structure and sexual dimorphic expression of a liver-specific rat 3β-hydroxysteroid dehydrogenase/isomerase. Endocrinology 1990;127:3237–9.
3. Luu-The V, Lachance Y, Labrie C, et al. Full length cDNA structure and deduced amino acid sequence of human 3β-hydroxy-5-ene steroid dehydrogenase. Mol Endocrinol 1989;3:1310–2.
4. Rhéaume E, Lachance Y, Zhao HF, et al. Structure of a new cDNA encoding the almost exclusive 3β-hydroxysteroid dehydrogenase/Δ^5-Δ^4 isomerase in human adrenals and gonads. Mol Endocrinol 1991;5.
5. Lachance Y, Luu-The V, Labrie C, et al. Characterization of human 3β-hydroxysteroid dehydrogenase/Δ^5-Δ^4 isomerase gene and its expression in mammalian cells. J Biol Chem 1990;265:20469–75.
6. Simard J, Melner MH, Breton N, et al. Characterization of macaque 3β-hydroxy-5-ene steroid dehydrogenase/Δ^5-Δ^4 isomerase: structure and expression in steroidogenic and peripheral tissues in primates. Mol Cell Endocrinol 1991;75:101–10.
7. Zhao HF, Simard J, Labrie C, et al. Molecular cloning, cDNA structure and predicted amino acid sequence of bovine 3β-hydroxy-5-ene steroid dehydrogenase/Δ^5-Δ^4 isomerase. FEBS Lett 1989;259:153–7.
8. Simard J, de Launoit Y, Labrie F. Characterization of the structure-activity relationships of rat type I and type II 3β-hydroxysteroid dehydrogenase/Δ^5-Δ^4 isomerase by site-directed mutagenesis and expression in HeLa cells. J Biol Chem 1991;266.
9. Klein P, Kanehisa M, De Lisi C. The detection and classification of membrane-spanning proteins. Biochem Biophys Acta 1985;815:468–76.
10. Smirnov AN. Estrophilic 3α,3β,17β,20α-hydroxysteroid dehydrogenase from rabbit liver, II. Mechanisms of enzyme-steroid interaction. J Steroid Biochem 1990;36:617–29.
11. Chen Q, Kosik LO, Nancarrow CD, Sweet F. Fetal lamb 3β,20α-hydroxysteroid oxidoreductase: dual activity at the same active site examined by affinity labeling with 16α-(bromo[2'-^{14}C]acetoxy)progesterone. Biochemistry 1989;28:8856–63.

15

Preovulatory Collagenase Activity in the Follicle of the Chicken

JANE A. JACKSON AND JANICE M. BAHR

Follicles in the hen ovary are arranged in a hierarchy that allows identification of a *preovulatory follicle* (F1) that will ovulate next and a *less-mature follicle* (F2) that will ovulate the next day. Prior to ovulation, collagen fibers are dissociated and/or degraded specifically at the stigma, site of follicular rupture, in both mammals (1, 2) and hens (3). Collagenase activity in preovulatory ovaries increased in PMSG-primed rats following hCG treatment (4, 5). However, it is unknown whether collagenase has a role in ovulation in the hen. Our objective was to determine active and total (active + latent) collagenase activities in the granulosa layer and *stigma* (S) and *nonstigma* (NS) regions of the theca layer of F1 and F2 that would have ovulated within ~30 min and 26 h, respectively. Collagenase activity was measured using a specific [^{3}H]-collagen substrate and is expressed as percent collagen digested ± SE. Collagenase activity was undetectable in granulosa. Active collagenase was greater (P < 0.05) in F1S (18.8 ± 2.1) compared to F1NS (8.8 ± 4.5) and greater (P < 0.01) in the F2S (17.5 ± 4.2) compared to nearly undetectable levels in the F2NS (3.0 ± 2.1). Total activity was greater (P < 0.05) in the theca layer of the F1S (28.0 ± 1.8) compared to F1NS (20.8 ± 1.6) and greater (P < 0.001) in F2S (35.4 ± 2.7) compared to F2NS (14.6 ± 2.2). Active and total collagenase activity was completely inhibited by the metalloprotease inhibitor phenanthroline. This is the first demonstration of increased active and latent collagenase enzyme activity specifically in the stigma of the follicle. The preponderance of active enzyme in the stigma supports a role for collagenase in ovulation in the chicken and may explain, in part, why follicles rupture only in the stigma region.

210

Methods

Tissue Collection

Ovulation in the hen occurs 5–30 min after oviposition. The F1 and F2 follicles were collected from white Leghorn hens at the time of oviposition. The S and NS regions of the theca layer were isolated. One NS and 8 S regions from 8 follicles were used for each sample (~150-mg tissue).

Extraction of Collagenase

Stigma and NS regions of the theca layer were extracted by the method of Curry et al. (4) with slight modification. For the Triton extraction, samples were homogenized with a Brinkman Polytron in 10 vol of buffer (10 mM $CaCl_2$, 0.25% Triton X-100 and 2 mM dithiothreitol to destroy the putative collagenase inhibitors) and centrifuged. The Triton supernatant was saved, and the pellet was resuspended in 10 vol of high-calcium Tris buffer (4) containing 2 mM dithiothreitol, heated at 60°C for 6 min, and centrifuged. The Triton and heat extracts were treated with 5 mM iodoacetamide and dialyzed into assay buffer (4).

Measurement of Collagenase Activity

Collagenase was measured by the method of Dean and Woessner (6). Active collagenase was determined by incubation of 20 µL [^{3}H]-collagen substrate (0.2-µg rat type I; 1×10^5 cpm) with 100 µL of the Triton or heat (collagen-associated) extracts for 60 h at 30°C. To measure total (active + latent) collagenase activity, Triton and heat extracts were treated with 0.5 mM aminophenylmercuric acetate, diluted 3-fold with assay buffer, and combined in a 1:1 ratio. A 100-µL aliquot of the combined Triton and heat extracts was incubated with 20-µL [^{3}H]-collagen for 36 h at 30°C. Active and total collagenase activity are expressed as percent collagen digested.

Results

Total collagenase activity was measurable only in the theca layer of the preovulatory F1 follicle, with nearly undetectable activity in the granulosa layer (results not shown). There was no difference in active collagenase activity in the Triton extract (readily extractable) between the S and NS regions of the F1 or F2 follicles (Figs. 15.1A and 15.1B). However, active collagenase in the heat extracts (collagen-associated enzyme activity) was higher ($P < 0.05$) in the S regions compared to the NS regions of the F1 and F2 follicles (Figs. 15.1C and 15.1D). There was no further increase in

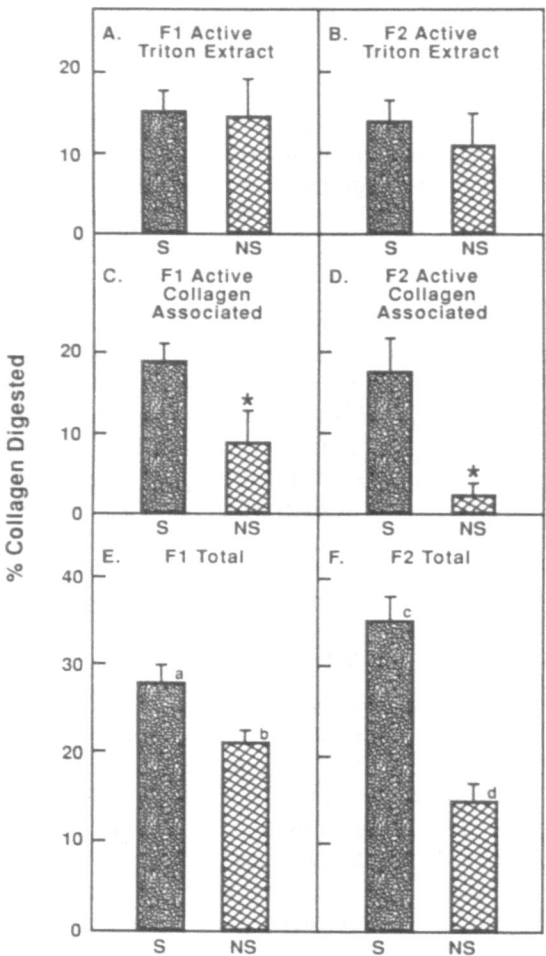

FIGURE 15.1. Stigma (S) and nonstigma (NS) regions of the theca layer of the preovulatory (F1) and less-mature (F2) follicles collected 5–30 min and 26 h before ovulation, respectively. Active collagenase was measured in the Triton extract of the F1 (A) and F2 (B) follicles (n = 3) and in the heat extract (collagen-associated) of the F1 (C) and F2 (D) follicles (n = 4). The Triton and heat extracts were combined in a 1:1 ratio and treated with aminophenylmercuric acetate to measure total collagenase (active + latent) in the F1 (E) and F2 (F) follicles (n = 4). (* = $P < 0.05$, S vs NS within the same follicle; different lower case letters = $P < 0.05$.)

the amount of active enzyme in the S region of the F1 compared to the F2 follicle. Total (active + latent) collagenase activity was also higher (P < 0.05) in the S regions compared to NS regions of the F1 and F2 follicles (Figs. 15.1E and 15.1F). The F1NS region had increased (P < 0.05) total collagenase activity compared to the F2NS region. In contrast, the F2S region had increased (P < 0.001) levels of total collagenase activity compared to the F1S region. Active and total collagenase activity in both the Triton and heat extracts were completely inhibited by the metalloprotease inhibitor phenanthroline (results not shown).

Discussion

We have demonstrated the presence of active and latent collagenase enzyme activity that is increased in the S compared to NS region of the theca layer of the hen follicle. The higher amounts of active enzyme specifically in the S area support a role for collagenase in ovulation in the hen and may explain, in part, why follicles rupture only in the S region.

There was no further increase in active collagenase, associated with the collagen matrix, in the S region of the F1 compared to the S region of the less-mature F2 follicle. This suggests that the regulation of the amount of active enzyme in the S region in vivo may depend on the activity of collagenase inhibitor(s). A role for collagenase inhibitors in the regulation of ovulation is suggested by recent studies that demonstrated that granulosa cells cultured in the presence of LH produced a dose-dependent increase in metalloproteinase inhibitor activity (7). Moreover, the preovulatory increase in *tissue inhibitor of metalloproteinase* (TIMP) mRNA from PMSG-primed rat ovaries following hCG treatment (7) coincided with the increase in ovarian collagenase activity (4).

In the present study, total collagenase activity decreased prior to ovulation in the S region of the F1 compared to the S region of the F2, which is consistent with previous work (4, 8). Collagenase activity at the apex of human preovulatory follicles decreased prior to ovulation compared to less-mature follicles (8). Total collagenase activity also declined in rat preovulatory ovaries 12 h after hCG administration (time of ovulation) compared to high levels of enzyme activity at 8 h after hCG treatment (4). The collagenase enzyme may be inactivated or degraded by other proteases immediately before ovulation.

In conclusion, active and total collagenase activities are present predominantly in the S region of the hen follicle prior to ovulation. Similar amounts of active collagenase in the S region of the immature F2 and S region of the preovulatory F1 follicle suggest a possible role for collagenase inhibitor(s) in the ovulatory process.

Acknowledgments. This work was supported in part by USDA-Ag 89-37240-4769.

References

1. Espey LL. Ultrastructure of the apex of the rabbit graafian follicle during the ovulatory process. Endocrinology 1967;81:267–76.
2. Bjersing L, Cajander S. Ovulation and the mechanism of follicle rupture, V. Ultrastructure of tunica albuginea and theca externa of rabbit graafian follicles prior to induced ovulation. Cell Tissue Res 1974;153:15–30.
3. Yoshimura Y, Koga O. Ultrastructural changes of the stigma of the follicle during the process of ovulation in the hen. Cell Tissue Res 1982;224:349–59.
4. Curry TE, Dean DD, Woessner JF, LeMaire WJ. The extraction of a tissue collagenase associated with ovulation in the rat. Biol Reprod 1985;33:981–91.
5. Reich R, Tsafriri A, Mechanic GL. The involvement of collagenolysis in ovulation in the rat. Endocrinology 1985;116:522–7.
6. Dean DD, Woessner JF. A sensitive, specific assay for tissue collagenase using telopeptide-free [^{3}H]acetylated collagen. Anal Biochem 1985;148:174–81.
7. Mann JS, Kindy MS, Edwards DR, Curry TE. Hormonal regulation of matrix metalloproteinase inhibitors in rat granulosa cells and ovaries. Endocrinology 1991;128:1825–32.
8. Fukumoto M, Yajima Y, Okamura H, Midorikawa O. Collagenolytic enzyme activity in human ovary: an ovulatory enzyme system. Fertil Steril 1981; 36:746–50.

16

Transforming Growth Factor β_1 Stimulation of Progesterone Production in Cultured Granulosa Cells from Gonadotropin-Primed Adult Rats

Hiroshi Ohmura, William Y. Chang, Mehmet Uzumcu, Serdar Coskun, Shigeo Akira, Tsutomu Araki, and Young C. Lin

Transforming growth factor β (TGFβ) is a homodimeric polypeptide that enhances the aromatase activity in rat granulosa cells. So far, the effect of TGFβ on granulosa cells has been primarily studied in estrogen-primed immature rats. Our study is designed to investigate the effect of TGFβ_1 on steroidogenesis in cultured mature rat granulosa cells. Mature rat granulosa cells were obtained from ovarian follicles of adult rats that have demonstrated 3 consecutive 4-day estrous cycles. The rats were treated by subcutaneous injections of 20-IU *pregnant mare serum gonadotropin* (PMSG) at 2000 h on the day of estrus followed by 15-IU PMSG at 0800 and 2000 h the next day. In each experiment, a total of 10 ovaries were collected for granulosa cell preparation 40 h after the last PMSG injections. The average viability of mature rat granulosa cells was about 80%. The culture wells were precoated with Dulbecco's Modified Eagle Medium/Ham's F-12 nutrient mixture (50:50) (DME/F-12) supplemented with 10% *fetal calf serum* (FCS) under 5% CO_2, 95% air at 37°C for 24 h, and the media in the culture wells were removed by washing 3 times with serum-free DME/F-12. Then, each culture well was seeded with approximately 15,000 live granulosa cells/mL. The mature rat granulosa cells were cultured in DME/F-12 supplemented with 1-µg insulin, 1-IU thrombin and 10-ng low-density lipoprotein per mL (DME/F-12/S) for 48 h. After the 48-h incubation period, fresh DME/F-12/S containing 1-ng/mL TGFβ_1 was added to each culture well and incubated for another 48 h, at which time, media were collected for progesterone *radio-*

immunoassay (RIA). Protein contents of the mature rat granulosa cells were measured by the dye binding method.

Results showed that 1-ng/mL TGFβ$_1$ significantly induced higher progesterone secretion than controls (4.0 ± 0.2 vs 2.2 ± 0.1 pg/μg cell protein, mean ± SD). Our data suggests that TGFβ$_1$ is capable of inducing progesterone secretion in cultured mature rat granulosa cells originated from PMSG-primed adult rats. However, this effect of TGFβ$_1$ on progesterone secretion appears to be species specific as compared to cultured porcine granulosa cells (1). It is unknown at this time by what mechanism TGFβ$_1$ acts within granulosa cells to account for this species difference. Further investigation to elucidate such differential actions of TGFβ is warranted.

In the mammalian ovary, many factors are known to regulate follicular maturation (2). Many of these factors have different sites and mechanisms of action and regulate the interaction between thecal and granulosa cells. Thecal and granulosa cells are two somatic cell types within the ovary. Granulosa cells line the inner layer of the ovarian follicle and will proliferate and differentiate during follicular development. The granulosa cells play an important role during follicular development. Therefore, in order to understand the regulation of follicular maturation, we must also understand the factors that control granulosa cell functions.

A factor currently under intensive investigation is TGFβ. TGFβ is a homodimeric polypeptide which has been shown to modulate functions in many different tissues (3). TGFβ has already been shown to be produced in thecal (4) and granulosa cells (5) of both immature and mature female rats. In experiments performed on *diethylstilbestrol* (DES)-induced immature rat granulosa cells, TGFβ generally stimulates proliferation and differentiation (6–10).

However, the immature rat granulosa cell cultures used in most experiments do not represent normal cycling ovarian function. We contend that granulosa cells from mature female rats are a more appropriate model to study the regulation of granulosa cell function during follicular development. The purpose of our study is (1) to evaluate the effect of TGFβ$_1$ on steroidogenesis in cultured mature rat granulosa cells without exposing the granulosa cells to FSH during the culture period and (2) to establish a model culture system for granulosa cells from mature rats previously stimulated with PMSG.

Materials and Methods

All media used contained 100-IU/mL penicillin, 100-μg/mL streptomycin, and 250-μg/L amphotericin B (all supplements were obtained from Gibco Laboratories, Grand Island, NY). Granulosa cells were obtained from 60- to 80-day-old mature female Sprague-Dawley rats. The rats were housed in our environmentally controlled animal facility (22°C, 55% humidity) and allowed free access to food and water. Vaginal smears were taken every morning to determine their estrous stages. Mature rats demonstrating 3 consecutive 4-day estrous cycles were treated with subcutaneous injections of 20-IU PMSG (Sigma Chemical Co., St. Louis, MO) at 2000 h on the day of estrus, followed by two 15 IU PMSG injections at 0800 h and 2000 h the next day (12 and 24 h later, respectively). In each experiment, 5 adult female rats were sacrificed by overdosing with ether (J.T. Baker, Inc., Phillipsburg, NJ) 40 h after the last PMSG injection. Ten ovaries were collected by laparotomies. Ovaries were put into 20-mL DME/F-12, and follicular fluids were collected in media by puncturing follicles with a cluster of 10 sterile fine sewing needles. After follicular fluids were filtered with a 140 micron filter, the fluids were centrifuged 3 times at approximately 150 × g for 5 min at 4°C to remove unnecessary tissues and cells. The average viability of mature rat granulosa cells was about 80% as determined with trypan blue staining. The culture wells (#25820 24-well plates: Corning Glass Works, Corning, NY) were precoated by adding 1.0-mL DME/F-12 supplemented with 10% fetal calf serum (Hyclone, Logan, UT) to each culture well 12 h prior to granulosa cell seeding. The culture wells were then incubated under 5% CO_2, 95% air at 37°C to enhance the granulosa cell attachment. The fetal calf serum was completely removed by washing 3 times with FCS-free DME/F-12 before granulosa cell seeding. Then, each culture well was seeded with approximately 15,000 cells/well in 1.0-mL DME/F-12 supplemented with 1-µg/mL insulin, 1-IU/mL thrombin, and 10-ng/mL low-density lipoprotein (DME/F-12/S) (all supplements were obtained from Sigma Chemical Co.). In this culture media, the granulosa cells were incubated for 48 h to allow for cell attachment and growth. After this 48-h incubation period, culture media were replenished with fresh DME/F-12/S media containing 1.0-ng/mL of TGFβ_1 (R&D Systems, Minneapolis, MN) and allowed to incubate for another 48 h. Then, the culture media and granulosa cells were collected and extracted to measure progesterone and cell protein contents by RIA and dye binding method, respectively. RIA procedures are described in our previous publication (11). Granulosa cells were collected after removal of culture media by adding 0.2 mL of 0.5N NaOH to each culture well. Then, the cell-containing NaOH solution was collected from each culture well and mixed with a dye reagent for the Bio-Rad protein microassay (Bio-Rad Laboratories, Richmond, CA). The optical densities were measured with a Beckman DU-70 Spectrophotometer using 595-nm visible light.

Statistical Analysis

Statistical analyses were performed with the Student's t-test. The data are presented as the mean ± SD. Values of P less than 0.05 are considered statistically significant.

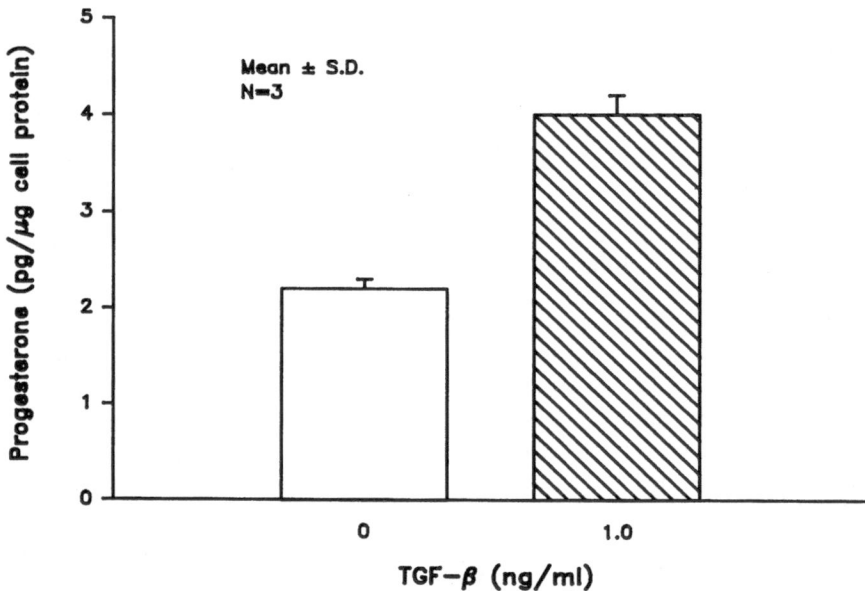

FIGURE 16.1. Progesterone production (pg/μg cell protein) by mature rat granulosa cells after 48-h treatment with 0- and 1.0-ng/mL TGFβ₁ (mean ± SD). Treatment with 1.0-ng/mL TGFβ₁ significantly (P < 0.05) stimulated progesterone production as compared to control (0 ng/mL).

Results

Our results showed that the PMSG-primed mature rat granulosa cell culture used in this study can produce a significant amount of progesterone even without FSH stimulation of the cultured granulosa cells. Also, the PMSG injections are sufficient to consistently produce 75,000 viable granulosa cells/rat (2 ovaries/rat) from 5 adult rats (10 ovaries, 3.75×10^6 cells) for each experiment and consistently attain high viability (80%). We were only able to attain 40,000 live granulosa cells/rat from mature rats not primed with PMSG.

The results showed that 1-ng/mL TGFβ₁ induced significantly higher progesterone secretion than controls (see Fig. 16.1). These data suggest that TGFβ₁ is capable of inducing progesterone secretion in cultured mature rat granulosa cells from PMSG-primed adult rats.

Discussion

It is known that rat granulosa cells (12, 13) and thecal cells (4–6) produce TGFβ-like activity. It is also known that rat granulosa cells respond well to TGFβ stimulation of differentiation and growth (6–10). In general, TGFβ produces remarkable differentiation and proliferation of rat granulosa cells, especially in the FSH-induced cell culture system. However, in almost all the articles, DES-primed immature rats (20–30 days) were used as sources of granulosa cells.

In the current study, we used PMSG-stimulated mature female rats to harvest granulosa cells. We expect that mature female rats that have complete 4-day estrous cycles would be a better physiological representation of normal reproductive ovarian function than immature female rats. Therefore, we believe this mature granulosa cell model is more appropriate to study the regulation of follicular development and granulosa cell function. Few studies have been done on mature rat granulosa cells to date. Recently, one study showed that mature rat granulosa cells produce TGFβ at a level similar to DES-primed and normal immature rat granulosa cells (12). However, it is not known whether the response of the mature rat granulosa cells to TGFβ treatment will also be similar to immature rat granulosa cells.

Our results show that 1.0-ng/mL TGFβ_1 stimulates PMSG-primed mature rat granulosa cells to secrete progesterone. This observation seems to indicate that mature rat granulosa cells respond similarly to TGFβ_1, as do immature rat granulosa cells. Interestingly, this response differs from that of porcine granulosa cells as reported by our laboratory previously (1). Perhaps the effect of TGFβ_1 is species specific. More research is necessary before the mechanism behind this apparent species difference is understood.

Three injections of PMSG (20, 15, and 15 IU) were used in the current study to stimulate follicular growth. PMSG has been reported to stimulate follicle growth in rats by rescuing atretic follicles and allowing them to reach ovulation (14). PMSG has FSH-like as well as LH-like activities (15) and a long biological half-life (54–60 h) (16). It has been shown that a single injection of 10-IU PMSG followed by *human chorionic gonadotropin* (hCG) does not maximally induce ovulation (17). In our laboratory, we have successfully superovulated adult female rats using a regimen of 3 injections of PMSG at 12-h intervals to stimulate follicular growth followed by 1 dose of hCG to induce ovulation (18). The precise mechanism of action for the difference in response between single and multiple injections of PMSG in mature rats is not understood at this time.

We can conclude from our data that (1) 1.0-ng/mL TGFβ_1 significantly stimulates progesterone secretion in mature rat granulosa cells; (2) our PMSG-primed mature rat granulosa cell model can produce a fair amount of progesterone even without FSH-stimulation of the cultured granulosa cells; and (3) PMSG successfully stimulated granulosa cell growth inside the ovarian follicles since we were able to consistently acquire 75,000 functional granulosa cells/rat with a relatively high viability (80%), as compared with porcine granulosa cells (55%–60%) (1).

Acknowledgments. This work was supported by grants from the March of Dimes Birth Defects Foundation and the Food and Drug Administration.

References

1. Chang WY, Ohmura H, Coskun S, Lin YC. Transforming growth factor-β_1 inhibits progesterone production in cultured porcine granulosa cells. In: Leung CK, Hsueh AJW, Friesen HG, eds. Molecular basis of reproductive endocrinology. New York: Springer-Verlag, 1992. (*See* Chapter 19, this volume.)

2. Hsueh AJ, Adashi EY, Jones PBC, Welsh TH. Hormonal regulation of differentiation of cultured ovarian granulosa cells. Endocr Rev 1984;5: 76–127.

3. Sporn MB, Roberts AB. The transforming growth factor-betas: past, present, and future. In: Piez KA, Sporn MB, eds. Transforming growth factor-βs. New York: NY Acad Sci, 1990:1–6.

4. Hernandez ER, Hurwitz A, Payne DW, Dharmarajan AM, Purchio AF, Adashi EY. Transforming growth factor-β_1 inhibits ovarian androgen production: gene expression, cellular localization, mechanism(s), and site(s) of action. Endocrinology 1990;127:2804–11.

5. Skinner MK, Keski-Oja J, Osteen KG, Moses HL. Ovarian thecal cells produce transforming growth factor-β which can regulate granulosa cell growth. Endocrinology 1987;121:786–92.

6. Bendell JJ, Dorrington J. Rat theca/interstitial cells secrete a transforming growth factor-β-like factor that promotes growth and differentiation in rat granulosa cells. Endocrinology 1988;123:941–8.

7. Knecht M, Feng P, Catt K. Bifunctional role of transforming growth factor-β during granulosa cell development. Endocrinology 1987;120:1243–9.

8. Dorrington J, Chuma AV, Bendell JJ. Transforming growth factor β and follicle-stimulating hormone promote rat granulosa cell proliferation. Endocrinology 1988;123:353–9.

9. Dodson WC, Schomberg DW. The effect of transforming growth factor-β on follicle-stimulating hormone-induced differentiation of cultured rat granulosa cells. Endocrinology 1987;120:512–6.

10. Feng P, Catt KJ, Knecht M. Transforming growth factor β regulates the inhibitory actions of epidermal growth factor during granulosa cell differentiation. J Biol Chem 1986;261:14167–70.

11. Gu Y, Lin YC, Rikihisa Y. Inhibitory effect of gossypol on steroidogenic pathways in cultured bovine luteal cells. Biochem Biophys Res Commun 1990;169:455–61.

12. Kim I-C, Schomberg DW. The production of transforming growth factor-β activity by rat granulosa cell cultures. Endocrinology 1989;124:1345–51.

13. Mulheron GW, Schomberg DW. Rat granulosa cells express transforming growth factor-β type 2 messenger ribonucleic acid which is regulatable in follicle-stimulating hormone in vitro. Endocrinology 1990;126:1777–9.

14. Braw RH, Tsafriri A. Effect of PMSG on follicular atresia in the immature rat ovary. J Reprod Fertil 1980;59:267–72.

15. Schams D, Menzer C, Schallenberger E, Hoffmann B, Hahn J, Hahn R. Some studies on pregnant mare serum gonadotrophin (PMSG) and on

endocrine responses after application for superovulation in cattle. In: Sreenan JM, ed. Control of reproduction in the cow. Boston: Martinus Nijhoff, 1978:122–43.

16. Sasamoto S, Sato K, Naito H. Biological active life of PMSG in mice with special reference to follicular ability to ovulate. J Reprod Fertil 1972;30: 371–9.

17. Husain SM, Saucier R. Induction of superovulation in mature rats with gonadotrophins. Can J Physiol Pharmacol 1970;48:196–9.

18. Akira S, Sanbuissho A, Lin YC, Araki T. Acceleration of embryo transport in superovulated adult rats. J Reprod Fertil 1991.

17

Induction of Steroidogenesis in a Retrovirus-Transformed Rat Granulosa Cell Line

I.M. Rao, P.J. Hornsby, and V.B. Mahesh

Ovarian follicular function is determined mainly by granulosa cells whose primary secretory products are estradiol and progesterone. To study the molecular mechanisms of the regulation of expression of key enzymes in the steroidogenic pathway, it would be advantageous to have continuously proliferating granulosa cell lines that retain steroidogenic function. In this chapter, we report the development of a granulosa cell line (Rao-gcl-29) by retrovirus transformation of primary cultures of granulosa cells and its preliminary characterization. Preantral follicle granulosa cells were transformed by a virus produced by the ψ2-SV40-6 cell line that encodes SV40T antigen and the *neo* gene. The transformed clones were selected by G418 resistance and screened for function. One of the clones identified as Rao-gcl-29 can be induced to secrete progesterone after treatment with cAMP analogs (1 mM). cAMP analogs do not induce synthesis of estradiol in this cell line. Treatment with FSH and hCG did not induce progesterone biosynthesis. This cell line would be invaluable in the study of the cAMP regulation of $P450_{scc}$ in ovarian granulosa cells.

Ovarian secretion of estradiol and progesterone during the follicular phase of the ovarian cycle occurs mainly due to synthesis and release of these hormones by the granulosa cells (1). Aromatase is the rate-limiting enzyme in the biosynthesis of estradiol from androstenedione in granulosa cells, and the activity of this enzyme is regulated by *follicle stimulating hormone* (FSH) (2). *Cholesterol side-chain cleavage enzyme* ($P450_{scc}$) converts cholesterol to pregnenolone in the granulosa cells that is then converted to progesterone (3). This enzymatic activity is modulated by FSH, *luteinizing hormone* (LH) and *cyclic adenosine 3'-5'-monophosphate* (cAMP) in vivo and in vitro (2). In human granulosa cells, FSH, hCG, and cAMP have been shown to stimulate the accumulation of $P450_{scc}$ mRNA (4).

Rat granulosa cells have limited growth potential in vitro, but can be induced to express differentiated function by such various agents as FSH, LH, and cAMP (5). It would be advantageous to have continuously proliferating lines of granulosa cells to study the molecular mechanisms of regulation of progesterone biosynthesis. Hence, we have attempted to develop cell lines by retrovirus transformation of preantral rat granulosa cells. One such cell line was isolated and preliminary characterization of its function is reported in this chapter.

Methods

Preparation of Retrovirus-Transformed Granulosa Cells

Diethylstilbestrol (DES) treatment for 2 days of immature rats was shown by our earlier study to yield preantral follicle granulosa cells that secrete progesterone after stimulation by FSH in culture (6). Therefore, 25-day-old immature Sprague-Dawley rats were treated with DES, 2 mg/rat s.c., for 2 days. They were killed 24 h after a second injection of DES, and the ovaries were removed and cleaned of fat. Granulosa cells were isolated as described in Rao et al. (6), and primary cultures were set up in 10-cm dishes in serum-containing medium (10% FBS, 10% HS, 2% UltroSer G in DMEM/F-12 1:1 medium). When grown in culture, ψ2-SV40-6 cells, which were a gift from C. Cepko (7), release a defective retrovirus encoding SV40T antigen and the *neo* gene that confers resistance to the antibiotic G418. The infective medium was incubated with the granulosa cells in the presence of polybrene for 48 h. The cells were then grown in the presence of G418, and the resistant clones were isolated, subcultured, and stored frozen under liquid nitrogen until further characterization.

Cell Culture and Functional Analysis

Clonal lines were plated on fibronectin-coated 35-mm culture dishes in serum-containing medium. To study the response of these clones to cAMP analogs, the cells were incubated in the presence of a mixture of 8-Br-cAMP and N^6-monobutyryl cAMP (1:1; 1 mM) for 48 h in serum-containing medium and also in defined medium (4 F medium) (5). About 1×10^6 primary granulosa cells, obtained from DES-treated rat ovaries, were plated in defined medium in parallel experiments. The dishes were incubated for 48 h at 37°C in 5% CO_2, 5% O_2. Medium was collected at the end of the incubation periods and stored at $-20°C$ for *radioimmunoassays* (RIA). In experiments where the response to FSH and hCG was studied, the cells were incubated in defined medium for 48 h in the presence of FSH (oFSH-17, 100 ng/mL) or cAMP analogs (1 mM). The medium was collected, and the dishes in each group were then incubated with hCG (100 ng/mL; Sigma) or cAMP analogs (1 mM), as

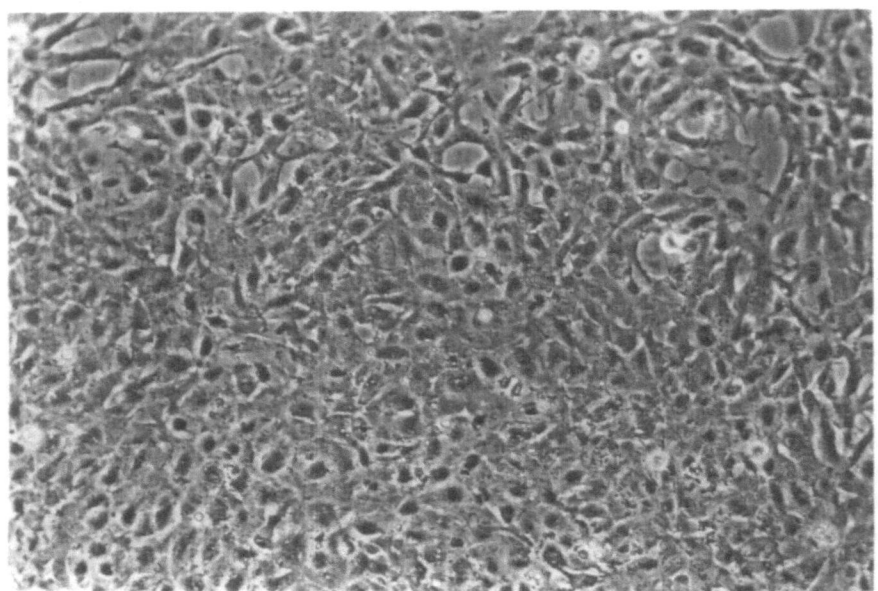

FIGURE 17.1. Phase contrast micrograph of Rao-gcl-29 cells grown in granulosa cell medium.

indicated. At the end, the medium was collected and stored at $-20°C$ until further analysis.

Progesterone in the medium was assayed by a specific RIA as described in Rao and Mahesh (8). The results were expressed as pmol progesterone formed/10^6 cells/48 h.

Results

Light Microscopy

Among the 35 granulosa cell lines developed, after an initial screening, one of the clones identified as Rao-gcl-29 was selected due to progesterone inducibility and studied for function. This cell line shows epithelial appearance when observed under phase contrast microscope (Fig. 17.1). The cells grow densely in serum-containing medium.

Response to cAMP

When grown for 48 h in the presence of 1 mM cAMP analogs in serum-containing medium, as well as in defined medium, Rao-gcl-29 cells responded with an increase in the secretion of progesterone over untreated

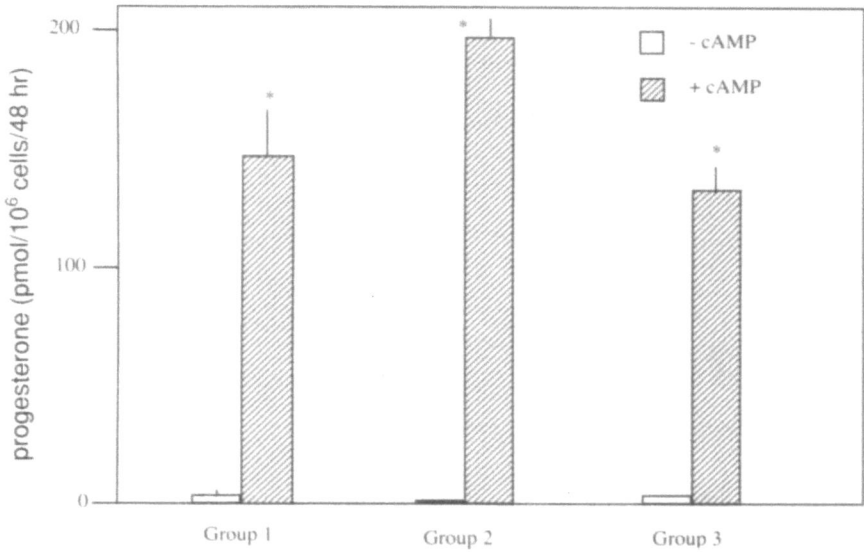

FIGURE 17.2. Induction of progesterone biosynthesis in Rao-gcl-29 and in primary cultures of granulosa cells. Groups 1 and 2 represent Rao-gcl-29 cells grown in serum-containing medium and in defined medium, respectively. Group 3 represents rat granulosa cells cultured in defined medium. Cells were incubated for 48 h in the presence or absence of 1 mM of cAMP analogs (a mixture of 8-Br-cAMP and N^6-monobutyryl cAMP, 1:1), and the progesterone produced was assayed by RIA. (*= P < 0.0001 when compared to corresponding control [n = 3].)

controls (Fig. 17.2, groups 1 and 2). Progesterone production by the cell line was comparable to the response of a primary culture of granulosa cells under similar conditions (Fig. 17.2, group 3). Cyclic AMP analogs did not induce estradiol synthesis in this cell line.

Response to FSH and hCG

Figure 17.3 shows the response of a primary culture of granulosa cells to FSH and hCG. Granulosa cells, when incubated in the presence of FSH (100 ng/mL) for 48 h secrete progesterone. When the same cells are incubated with hCG (100 ng/mL) for a second 48 h, they also respond with progesterone secretion (Fig. 17.3, group 1). Primary cultures of granulosa cells also respond to hCG during the second 48-h incubation after stimulation with cAMP analogs (1 mM) in the first 48-h period (Fig. 17.3, group 2). These cells also respond to cAMP analogs with progesterone secretion at the end of both time periods (Fig. 17.3, group 3). In contrast, Rao-gcl-29 cells do not respond to FSH (100 ng/mL) during the first 48-h incubation. Also, they do not respond to hCG

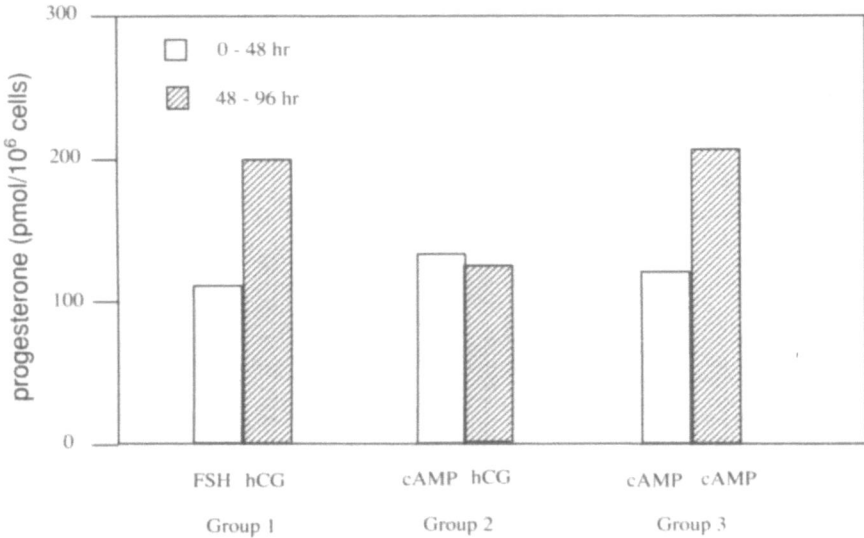

FIGURE 17.3. Effect of FSH, hCG, and cAMP on induction of progesterone biosynthesis in primary cultures of granulosa cells incubated in defined medium. Each of the 3 separate groups of dishes was treated with FSH (100 ng/mL) or cAMP analogs (1 mM) from 0 to 48 h. The medium was collected, and the dishes were then incubated for an additional 48 h with hCG (100 ng/mL) or cAMP analogs (1 mM). At the end of each incubation period, the medium was assayed for progesterone content (n = 2).

(100 ng/mL) during the second 48-h incubation, after treatment in the first 48 h, either with FSH (100 ng/mL) or with 1 mM cAMP analogs (Fig. 17.4, groups 1 and 2). However, they respond to cAMP analogs during both time periods (Fig. 17.4, group 3).

Discussion

Retrovirus transformation of granulosa cells of rat was achieved in this study with the development of a granulosa cell line (Rao-gcl-29) that retained progesterone synthetic ability in response to cAMP analogs. The results showed that in this cell line, cAMP analogs can induce progesterone biosynthesis from a low basal level in the uninduced cells. This would be advantageous, as one can study the molecular events that follow cAMP induction of P450$_{scc}$ gene expression in granulosa cells. The response of this cell line to cAMP compares well with that of primary cultures of granulosa cells from preantral follicles under similar conditions.

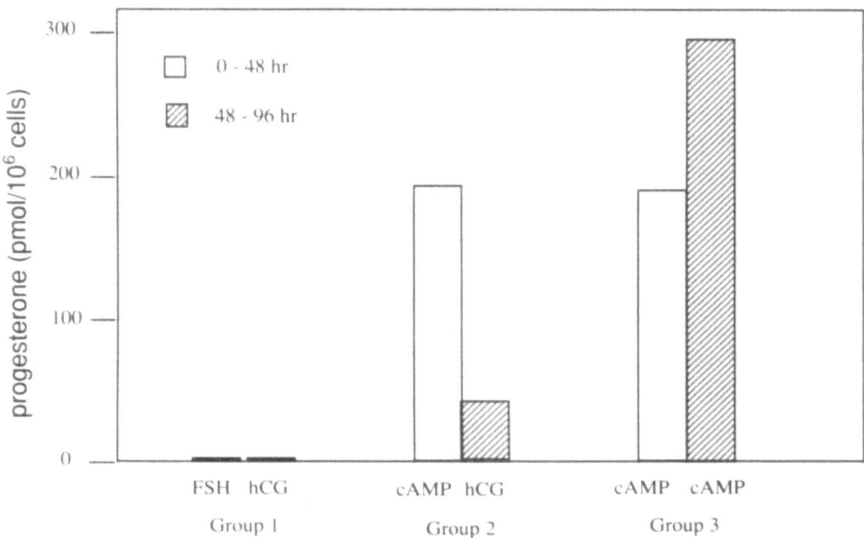

FIGURE 17.4. Effect of FSH, hCG, and cAMP on induction of progesterone biosynthesis in Rao-gcl-29 cells incubated in defined medium. Each of the 3 separate groups of dishes was treated with FSH (100 ng/mL) or cAMP analogs (1 mM) from 0 to 48 h. The medium was collected, and the dishes were then incubated for an additional 48 h with hCG (100 ng/mL) or cAMP analogs (1 mM). At the end of each incubation period, the medium was assayed for progesterone content (n = 2).

This is the first report of development of a granulosa cell line by SV40 transformation of primary cells with retention of steroidogenic function. In previous work, Fitz et al. (9) subcloned a cell line from an SV40-transformed rat granulosa cell line. It produced very low amounts of steroids (6.4 pmol of progesterone/10^6 cells/48 h). Zilberstein et al. (10) have developed granulosa cell lines, by using a temperature-sensitive mutant of SV40, that have no detectable steroidogensis. Amsterdam et al. (11) and Suh and Amsterdam (12) have reported development of granulosa cells lines by transforming preantral and preovulatory follicle granulosa cells of rat. When SV40 alone was used, they did not obtain any steroidogenic clones. However, by cotransfection with Ha-*ras* and SV40, they obtained cell lines that could secrete progesterone and 20α-dihydroprogesterone. Rao-gcl-29 cells compare favorably with the granulosa cell lines developed by Amsterdam et al. (11) and Suh and Amsterdam (12) from preantral follicles in progesterone secretion. Our study suggests that *ras* may not be required for the retention of steroidogenic function.

Initial characterization of Rao-gcl-29 shows that it does not respond to either FSH or hCG in producing progesterone at optimal doses and duration of incubation. Similar observations were made for other granulosa cell lines (11, 12). The lack of response of Rao-gcl-29 cells to hCG in the second 48-h period is not due to cell death, as the cells continue to respond to cAMP in that period (Fig. 17.4, group 3). Also, Rao-gcl-29 cells respond to cAMP analogs for 96 h in defined medium, and such a system may be useful in investigating the role of growth factors in the regulation of the genes for enzymes of the progesterone biosynthetic pathway.

Acknowledgments. This investigation is supported by Research Grant HD-24488 from the National Institute of Child Health and Human Development, NIH.

References

1. Richards JS. Maturation of ovarian follicles: actions and interactions of pituitary and ovarian hormones on follicular cell differentiation. Physiol Rev 1980;60:51–89.
2. Hsueh AJW, Adashi EY, Jones PBC, Welsh, Jr TH. Hormonal regulation of the differentiation of cultured ovarian granulosa cells. Endocr Rev 1984;5: 76–127.
3. Waterman MR, Simpson ER. Regulation of the biosynthesis of cytochromes P-450 involved in steroid hormone synthesis. Mol Cell Endocrinol 1985;39: 81–9.
4. Voutilainen R, Tapaninen J, Chung B-C, Matteson KJ, Miller WL. Hormonal regulation of $P450_{scc}$ (20,22 desmolase) and $P450_{c17}$ (17α-hydroxylase/17-20-lyase) in cultured human granulosa cells. J Clin Endocrinol Metab 1986;63: 202–7.
5. Orly J, Sato G, Erickson GF. Serum suppresses the expression of hormonally induced functions in cultured granulosa cells. Cell 1980;20:817–27.
6. Rao IM, Mills TM, Anderson E, Mahesh VB. Heterogeneity in granulosa cells of developing rat follicles. Anat Rec 1991;229:177–85.
7. Jat PS, Cepko CL, Mulligan RC, Sharp PA. Recombinant retroviruses encoding simian virus 40 large T antigen and polyoma virus large and middle T antigens. Mol Cell Biol 1986;1204–17.
8. Rao IM, Mahesh VB. Role of progesterone in the modulation of the pre-ovulatory surge of gonadotropins and ovulation in the pregnant mare's serum gonadotropin-primed immature rat and the adult rat. Biol Reprod 1986;35: 1154–61.
9. Fitz TA, Wah RM, Schmidt WA, Winkel CA. Physiologic characterization of transformed and cloned rat granulosa cells. Biol Reprod 1989;40:250–8.
10. Zilberstein M, Chou JY, Lowe, Jr WC, et al. Expression of insulin-like growth factor-1 and its receptor by SV40-transformed rat granulosa cells. Mol Endocrinol 1988;3:1488–97.
11. Amsterdam A, Zauberman A, Meir G, Pinhasi-Kimhi O, Suh BS, Oren M. Cotransfection of granulosa cells with simian virus 40 and Ha-*ras* oncogene

generates stable lines capable of induced steroidogenesis. Proc Natl Acad Sci USA 1988;85:7582–6.

12. Suh BS, Amsterdam A. Establishment of highly steroidogenic granulosa cell lines by cotransfection with SV40 and Ha-*ras* oncogene: induction of steroidogenesis by cyclic adenosine 3'-5'-monophosphate and its suppression by phorbol ester. Endocrinology 1991;127:2489–500.

18

Arachidonic Acid as a Second Messenger for Regulation of Aromatase Activity in Porcine Granulosa Cells

F. Ledwitz-Rigby, M.K. Nickerson, S.W. Lin, and Peter C.K. Leung

Arachidonic acid (AA) and its metabolites have many important roles in signal transduction in the reproductive system (1–4). While its ability to stimulate progesterone secretion by rat granulosa cells has been clearly established (3, 4), a role for AA in regulation of aromatase activity in granulosa cells has only been previously reported in preliminary form (5). The present report examines the effects of AA on aromatase activity in porcine granulosa cells.

Granulosa cells were harvested from 1- to 10-mm antral follicles from ovaries obtained at local slaughterhouses as previously described (6). The aromatase assays were performed on freshly collected cells following determination of live cell number by trypan blue dye exclusion and hemocytometer counting. Aromatase activity was assessed by the release of tritiated water from 1β, 2β-^{3}H testosterone in the presence or absence of 5×10^{-6} M unlabeled testosterone during 3-h incubations (7). There were 6 replicates per treatment group per experiment, and each observation was repeated in at least 3 separate experiments. Addition of exogenous AA in the culture media during the 3-h aromatase assay did not consistently influence aromatase activity. In only 3 out of 12 experiments, 10^{-5} M AA significantly (P < 0.05) stimulated aromatase activity 30%–70%. However, manipulations of the cells to increase their endogenous AA contents had consistent significant effects on aromatase activity.

The single most potent stimulator of aromatase activity was indomethacin. *Indomethacin*, a cyclooxygenase inhibitor, blocks the metabolism of arachidonic acid to the prostaglandins and thromboxanes (Fig. 18.1). Indomethacin stimulated aromatase activity by porcine granulosa cells in a dose-dependent manner over a range of 5×10^{-6} to 10^{-4} M and was more effective at stimulating aromatase activity in immature granulosa

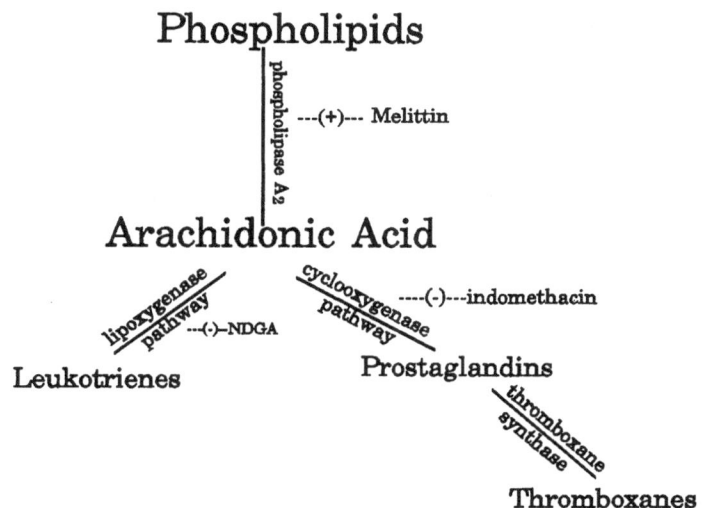

Outline of Arachidonic Acid Cascade

FIGURE 18.1. Arachidonic acid metabolism.

cells from 1- to 3-mm follicles (2.6 ± 0.3-fold) than in cells from more mature follicles (1.5 ± 0.3-fold). Addition of *Melittin* (3 × 10^{-7}M), a stimulator of *phospholipase A_2* (PLA$_2$), or PLA$_2$ itself (0.1–1.0 U) to the culture media had small, but significant, effects in stimulating aromatase activity by porcine granulosa cells. The combined actions of indomethacin and Melittin or indomethacin and PLA$_2$, however, were synergistic and stimulated aromatase activity up to 5 times that of control levels (Fig. 18.2).

The above data suggested that an increase in intracellular AA concentrations achieved by simultaneously increasing AA production and blocking its conversion to prostaglandins and thromboxanes could enhance aromatase activity in porcine granulosa cells. Arachidonic acid could achieve its effects on aromatase through many pathways. Blocking the cyclooxygenase pathway could either remove the production of an inhibitor of aromatase activity (such as a prostaglandin) or make more AA available for conversion to leukotrienes via the lipoxygenase pathway (Fig. 18.1) (8). Alternatively, AA has been shown to influence cell functions by mobilizing calcium (2) and by stimulating the production of *cyclic guanosine monophosphate* (cGMP) (9).

This study examined possible mechanisms by which AA influences aromatase activity. *Nordihydroguaiaretic acid* (NDGA), a lipoxygenase inhibitor, was tested for its ability to reverse the stimulatory effect

Porcine Granulosa Cells

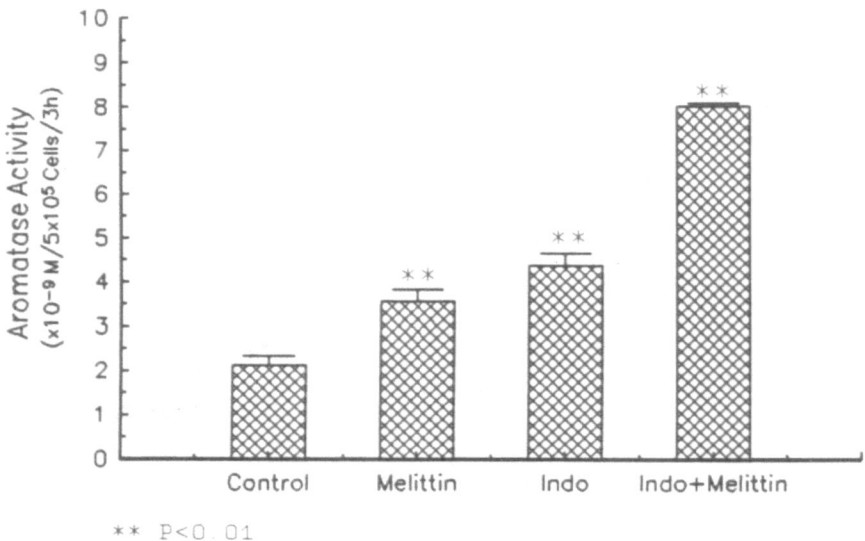

FIGURE 18.2. Stimulation of aromatase activity in porcine granulosa cells by indomethacin and Melittin. (Bars represent $\times \pm$ SE [n = 6]; ** = P < 0.01.)

of indomethacin. The reasoning for these experiments was that indomethacin's ability to enhance aromatase activity might be due to its ability to make more AA available for conversion to a specific leukotriene that would then influence aromatase. Thus, blockage of the production of leukotrienes might interfere with indomethacin stimulation of aromatase activity. NDGA was added to the culture media by itself and in the presence of 10^{-4} M indomethacin. Neither 10^{-5} M nor 10^{-4} M NDGA had any effect on either basal aromatase or on the stimulatory action of indomethacin on aromatase activity, suggesting that a leukotriene was not involved in enhancing aromatase activity.

Involvement of the guanylate cyclase pathway in indomethacin's stimulation of aromatase activity was also found to be unlikely. Addition of 5×10^{-6} to 5×10^{-5} M dibutyryl cGMP had no effect on aromatase activity. Dibutyryl cGMP produced a 50% stimulation of aromatase activity (P < 0.05) at 5×10^{-4} M, but the need to use such a high concentration made us debate the physiological significance of this observation. Furthermore, the addition of an inhibitor of guanylate cyclase, LY83583 (10^{-5} M), had no effect on basal or indomethacin-stimulated aromatase activity.

Whether indomethacin might stimulate aromatase activity by blocking the production of a prostaglandin inhibitor of aromatase activity was subjected to preliminary testing. The addition of prostaglandin $F_{2\alpha}$ (10^{-6} M) or prostaglandin E_2 (10^{-6} to 10^{-4} M) had no effect on aromatase activity. Thus, these well-known inhibitors of progesterone production do not appear to be involved in short-term regulation of aromatase activity. There are numerous other products of the cyclooxygenase pathway that have not been examined that might have an inhibitory role in the regulation of aromatase. Such a mechanism, however, would not explain the synergistic action of indomethacin and a stimulator of AA production (Melittin or PLA_2).

At this point, we do not know the mechanism for AA stimulation of aromatase activity in porcine granulosa cells. Arachidonic acid, however, could play an important role as a second-messenger mediator for the early effects of endocrine signals on aromatase activity.

References

1. Chang JP, Graeter J, Catt KJ. Coordinate actions of arachidonic acid and protein kinase C in gonadotropin-releasing hormone stimulated secretion of luteinizing hormone. Biochem Biophys Res Commun 1986;134:134–9.
2. Kolesnick RN, Musacchio I, Thaw C, Gershengorn MC. Arachidonic acid mobilizes calcium and stimulates prolactin secretion from GH_3 cells. Am J Physiol 1984;246:E458–62.
3. Wang J, Leung PCK. Role of arachidonic acid in luteinizing hormone-releasing hormone action: stimulation of progesterone production in rat granulosa cells. Endocrinology 1988;122:906–11.
4. Wang J, Leung PCK. Arachidonic acid as a stimulatory mediator of luteinizing hormone-releasing hormone action in the rat ovary. Endocrinology 1989; 124:1973–9.
5. Nickerson MK, Ledwitz-Rigby F. Biol Reprod 1990;42(1):113.
6. Channing CP, Ledwitz-Rigby F. Methods for assessing hormone-mediated differentiation of ovarian cells in culture and in short-term incubations. In: Hardman JG, O'Malley BW, eds. Methods in enzymology. New York: Academic Press, 1975;39D:183–250.
7. Ledwitz-Rigby F, Gross TM, Schjeide OA, Rigby BW. The glycosaminoglycan chondroitin-4-sulfate alters progesterone secretion by porcine granulosa cells. Biol Reprod 1987;36:320–7.
8. Yen SC, Jaffe RB. Reproductive endocrinology 2nd ed. Philadelphia: WB Saunders, 1986:154–76.
9. Ignarro LJ, Wood KS. Activation of purified soluble guanylate cyclase by arachidonic acid required absence of enzyme bound heme. Biochem Biophys Acta (Copenh) 1987;928:160–70.

19

Inhibition of Progesterone Production in Cultured Porcine Granulosa Cells by Transforming Growth Factor β_1

WILLIAM Y. CHANG, HIROSHI OHMURA, SERDAR COSKUN, AND YOUNG C. LIN

Transforming growth factor-β (TGFβ) affects the functions of a variety of steroidogenic cells. Porcine thecal cells have been shown to secrete TGFβ in vitro, and it has been proposed that TGFβ may function as an autocrine-paracrine regulator in follicular development (1). The purpose of this experiment is to evaluate the regulatory effect of $TGF\beta_1$ on *progesterone* (P_4) production in cultured porcine granulosa cells pre-exposed to *follicle stimulating hormone* (FSH). Porcine ovaries were obtained from a local slaughterhouse. The porcine granulosa cells used for the current studies were harvested from medium-sized ovarian follicles (3–8 mm in diameter). Approximately 2.0×10^5 viable cells/well were cultured in 1-mL serum-free Dulbecco's Modified Eagle Medium/Ham's F-12 (DME/F-12) supplemented with 1-μg insulin, 1-IU thrombin, and 10-ng low-density lipoprotein (DME/F-12/S). The granulosa cells were incubated under 5% CO_2, 95% air at 37°C. After 2 days of culture, the wells were washed 3 times with unsupplemented DME/F-12, and fresh DME/F-12/S containing 100-ng/mL FSH was added. Then, after 2 days of FSH treatment, new DME/F-12/S containing 0-, 0.001-, 0.01-, 1-, or 10-ng/mL $TGF\beta_1$ was added to each well and allowed to incubate for 42 h. Progesterone (P_4) secreted into the culture media by the porcine granulosa cells was measured by radioimmunoassay, and protein content of the granulosa cells was measured by dye binding method. The P_4 contents (pg/μg cell protein, mean ± SEM) for the porcine granulosa cells treated with 0-, 0.001-, 0.01-, 0.1-, 1-, and 10-ng/mL $TGF\beta_1$ were 28.6 ± 3.0, 23.4 ± 4.4, 24.6 ± 1.0, 16.3 ± 1.8, 18.9 ± 2.4, and 19.1 ± 0.7, respectively. $TGF\beta_1$ at 0.001 and 0.01 ng/mL decreased P_4 production by the granulosa cells, but insignificantly. Treatment with 0.1-ng/mL

TGFβ_1 maximally and significantly (P < 0.05) decreased P$_4$ production from control. Higher levels of TGFβ_1 (1 and 10 ng/mL) did not further attenuate P$_4$ production as compared to 0.1-ng/mL TGFβ_1. The steroidogenic effects of TGFβ are known to be cell specific, but this finding suggests that its effects within granulosa cells may also be species specific. Data from our laboratory by Ohmura et al. (2) shows that TGFβ_1 stimulates P$_4$ production in cultured granulosa cells from super-stimulated adult rats. Additional research is needed to elicit the precise regulatory mechanism of TGFβ on steroidogenesis in granulosa cells before this species specificity can be understood.

In 1978, Joseph DeLarco and George Todaro discovered that the retroviral transformation of murine 3T3 fibroblast cells by the Moloney murine sarcoma virus was associated with a factor they called *sarcoma growth factor* (SGF) (3). Today, SGF is known to have 2 active components and has been renamed transforming growth factor (TGF) for its ability to transform normal fibroblast cells in culture (4, 5). The 2 active components, TGFα and TGFβ, have been extracted from normal mammalian tissues and have been shown to exhibit numerous regulatory actions. TGFβ, which is present in nearly all cells of every organ (6), is produced in rat ovaries by theca-interstitial cells (7, 8), as well as by granulosa cells (9, 10) in vitro, and seems to stimulate rat granulosa cell proliferation and differentiation (11–15). In the bovine ovaries, TGFβ is secreted by thecal cells (7) and enhances FSH-induced aromatase activity and P$_4$ synthesis (16). However, in porcine ovaries, TGFβ is known to be secreted by thecal cells (1), but has profound effects on proliferation and differentiation of granulosa cells (17, 18).

In the mammalian estrous cycle, only a small number of follicles are growing at one time, and the majority of them undergo atresia. The selection of only a fraction of the developing follicles to ovulate in any one cycle suggests a complex regulatory mechanism must exist to locally control further development or atresia of follicles within an ovary. The ovarian follicles consist of an outer layer of theca cells and inner layers of granulosa cells. Inevitably, it is the control of these two cell types that determines follicular function. It has already been shown that many factors, such as FSH, *luteinizing hormone* (LH), *epidermal growth factor* (EGF), estrogens, and androgens, act to regulate granulosa cell function (19). TGFβ, a factor produced and secreted by porcine theca cells (1), is another likely candidate to play a role in the control of porcine granulosa cell function. TGFβ is a polypeptide in a family of structurally related regulatory peptides that includes inhibins (20). TGFβs are homodimeric peptides consisting of two 112-amino acid monomers and possessing a molecular weight of 25,000 (21). Within the TGFβ superfamily, there exist several forms of TGFβ. The first form to be discovered, TGFβ_1, is used in this study. The current study was designed to determine the role of TGFβ_1 on steroidogenesis in porcine granulosa cells.

Materials and Methods

Chemically defined medium was used for the current study. DME/F-12, insulin, thrombin, and *low-density lipoprotein* (LDL) were obtained from Sigma Chemical Company (St. Louis, MO). *Fetal calf serum* (FCS) was obtained from Hyclone (Logan, UT). TGFβ₁ from porcine platelets was obtained from R&D Systems (Minneapolis, MN). FSH was obtained from the National Institutes of Health. Antibiotic-antimycotic (100×) containing penicillin, streptomycin, and amphotericin B was purchased from Gibco Laboratories (Grand Island, NY). Corning 24-well plates #25820 were obtained from Corning Glass Works (Corning, NY). The Bio-Rad protein assay dye reagent concentrate was obtained from Bio-Rad Laboratories (Richmond, CA).

Granulosa Cell Culture

All media used contained 250-µg/L amphotericin B, 100-IU/mL penicillin, and 100-µg/mL streptomycin. Porcine ovaries obtained from a local slaughterhouse were soaked in 70% ethanol for approximately 1 min and then washed 10 times with double-distilled water. Granulosa cells were obtained by aspirating medium-sized follicles (3–8 mm in diameter) with a 20-gauge needle. The fluid was centrifuged for 5 min at approximately 200 × g. The supernatant was removed, and cells were resuspended by gentle mixing in DME/F-12 and recentrifuged. Removal of supernatant, resuspension, and centrifugation were repeated again. After removing the remaining supernatant, the cells were suspended in DME/F-12 supplemented with 1-µg/mL insulin, 1-IU/mL thrombin, and 10-ng/mL LDL (DME/F-12/S). Cells were then seeded into 24-well Corning culture plates at a density of 2.0×10^5 live cells/well in 10% FCS-coated wells. The wells were coated by adding 1.0-mL DME/F-12 plus 10% FCS to each well 18 h prior to seeding. The 10% FCS medium was completely removed 1 h before granulosa cell seeding. FCS coating was used to enhance cell attachment. Each well contained a total volume of 1.0-mL DME/F-12/S. The cells were allowed to culture for 48 h in 5% CO_2, 95% air at 37°C. The wells were then washed 3 times with DME/F-12, and 1.0-mL DME/F-12/S containing 100-ng/mL FSH was added to each well. The cells were allowed to incubate for another 48 h. Media were then removed, and the cells were treated with 0-, 0.001-, 0.01-, 1-, or 10-ng/mL TGFβ₁ in DME/F-12/S for 42 h. Media and cells were collected for P_4 *radioimmunoassay* (RIA) and protein assay, respectively.

Progesterone RIA and Protein Assay

Concentration of P_4 was determined by RIA as described by our previous publication (22). Protein content of the cells was determined by Bio-Rad microassay techniques. Briefly, 0.8 mL of 0.1N NaOH in double-distilled water was added to each well after removal of media. The NaOH solution was then removed and mixed with 0.2-mL Bio-Rad protein assay dye reagent concentrate. The optical densities were measured with a Beckman DU-70 Spectrophotometer using visible light at a wavelength of 595 nm.

Statistical Analysis

Statistical analyses were performed by one-way *analysis of variance* (ANOVA) followed by Bonferroni posttest. The data are presented as the mean ± SE. Values of P less than 0.05 are considered as statistically significant.

Results

The P_4 content of the media was chosen as an end parameter for indication of granulosa cell function. Protein content of the cells was used to approximate the number of granulosa cells that were still attached to the

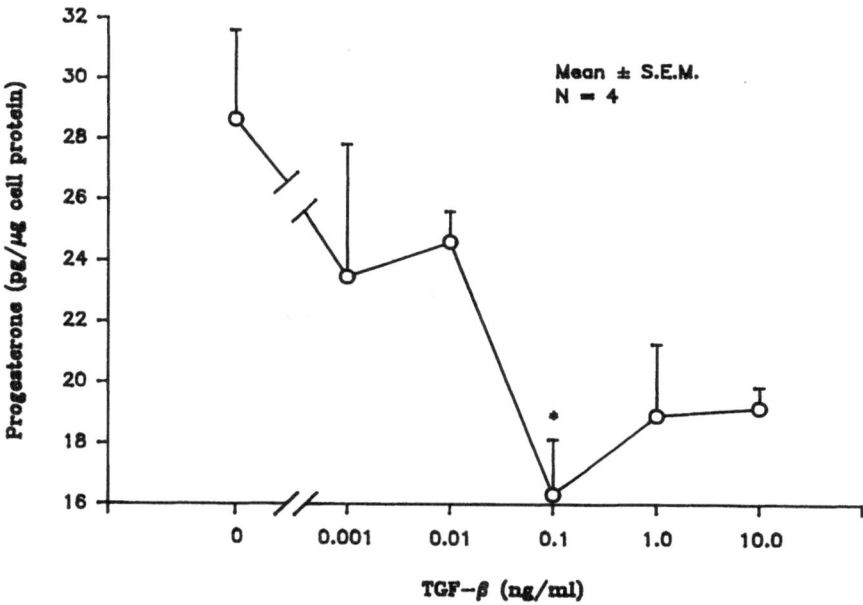

FIGURE 19.1. Progesterone production (pg/μg cell protein) by porcine granulosa cells collected from medium-sized follicles (3–8 mm diameter) after 48-h pretreatment with 100-ng/mL FSH followed by 42-h treatment with 0-, 0.001-, 0.01-, 0.1-, 1-, or 10-ng/mL TGFβ₁ (mean ± SEM). Treatment with 0.1-ng/mL TGFβ₁ significantly (*) decreased progesterone production as compared to control.

wells when the media were collected. Variability in cell content will exist due to variabilities in seeding samples, attachment rates, and proliferation rates. As shown in Figure 19.1, $TGF\beta_1$ significantly ($P < 0.05$) attenuated P_4 production per unit cell content. Productions of P_4 (pg/µg protein, mean $\pm$ SEM) for cells treated with 0-, 0.001-, 0.01-, 0.1-, 1-, and 10-ng/mL $TGF\beta_1$ are 28.615 $\pm$ 2.98, 23.440 $\pm$ 4.36, 24.587 $\pm$ 1.00, 16.302 $\pm$ 1.82, 18.878 $\pm$ 2.36, and 19.125 $\pm$ 0.70, respectively. $TGF\beta_1$ at 0.001 and 0.01 ng/mL decreased P_4 production by the porcine granulosa cells, but insignificantly. Treatment with 0.1-ng/mL $TGF\beta_1$ maximally and significantly ($P < 0.05$) decreased P_4 production from control. Higher levels of $TGF\beta_1$ (1 and 10 ng/mL) did not further attenuate P_4 production and were insignificant from control and 0.1-ng/mL $TGF\beta_1$ treatment.

Discussion

A number of research articles have addressed the effect of $TGF\beta$ on rat granulosa cell proliferation and differentiation (8, 11–15). However, few articles have been published about porcine granulosa cells (1, 17, 18). Generally, it has been shown that $TGF\beta$ acts on the rat granulosa cells to promote growth and differentiation (11–14). On the other hand, one study on porcine granulosa cells obtained from small follicles (<3 mm diameter) showed that $TGF\beta$ inhibited FSH-stimulated P_4 production (18). Our finding determined that in pigs, $TGF\beta_1$ inhibits P_4 production from FSH-pretreated granulosa cells obtained from medium-sized follicles (3–8 mm diameter).

The porcine ovaries used in this experiment were obtained from a local slaughterhouse, and medium-sized immature ovarian follicles (3–8 mm in diameter) were aspirated. With this source of ovaries, the conditions of the donor are not experimentally controlled. Therefore, correlation of the estrous cycle stage with the granulosa cells cannot be defined. The collection of granulosa cells used in this study must be considered a general representation of granulosa cells from all stages of the estrous cycle. Previous work from our laboratories has shown the granulosa cells obtained from the slaughterhouse to be functional and appropriate for experimentation (23). This conglomerate of granulosa cells, however, may explain the high variability in P_4 production seen in the control (0-ng/mL $TGF\beta_1$) and at low $TGF\beta_1$ concentrations (0.001 ng/mL). The in vitro porcine granulosa cell model used in this study is convenient since no animal sacrifice is necessary, but standardization of follicle selection and collection methods is essential.

Our study using cultured porcine granulosa cells grown in a chemically defined medium shows that after 42-h treatment with $TGF\beta_1$, P_4 production was maximally suppressed at 0.1-ng/mL $TGF\beta_1$. Different treatment times may yield different levels of suppression. Adashi et al. (24) showed that TGFβ augmentation of FSH-supported aromatase activity in murine granulosa cells required a minimum of 48 h. An even greater response was seen after 72 h. Also, Mondschein et al. (18) reported significant inhibition of FSH-stimulated P_4 production in porcine granulosa cells treated for 72 h with 0.1 ng/mL and higher concentrations of $TGFβ_1$. The latter study supports our finding that P_4 production is maximally suppressed with 0.1-ng/mL $TGFβ_1$. We can only speculate that the significance of this concentration may lie within the TGFβ receptor function. Perhaps the receptors are saturated or are down-regulated as TGFβ concentration increases. However, a complex ligand-receptor association exists with different forms of TGFβ and the individual forms of TGFβ receptors (25), and it is beyond the scope of this chapter.

A significant finding in the study is the contradictory effects of $TGFβ_1$ on porcine granulosa cells as compared to adult rat granulosa cells. In another study performed in our laboratory, $TGFβ_1$ was found to stimulate P_4 production in granulosa cells collected from ovaries of mature rats treated with pregnant mare serum gonadotropin (2). These findings suggest that the effect of $TGFβ_1$ may be species specific.

The mechanism of action of TGFβ in ovarian cells is poorly understood. In general, it is known that TGFβ inhibits LH-induced androsterone accumulation and stimulates P_4 production in thecal cells (26). In the granulosa cell, it stimulates FSH-induced estrogen and P_4 production in the rat. More specifically, TGFβ alone does not alter cAMP generation in granulosa cells (8, 24) and probably acts at a site distal to cAMP generation. In the rat, TGFβ has been speculated to decrease the 17α-hydroxylase/17–20-lyase enzyme (24, 27).

The mechanism of action in porcine granulosa cells is still unknown; thus, its profound effects cannot be explained. Perhaps TGFβ in porcine granulosa cells acts through a different mechanism involving an intermediate factor. Mondschein et al. (28) have reported that $TGFβ_1$ inhibits *insulin-like growth factor binding protein 3* (IGFBP-3) secretion by porcine granulosa cells. IGFBPs can either enhance or inhibit IGF-dependent cell function in other in vitro cell systems. Their effects in porcine granulosa cells are still unknown.

Whatever its mechanism of action, the production and secretion by porcine thecal cells and the lack of detectable amounts in follicular fluid suggest that TGFβ acts in a autocrine-paracrine fashion (1). Undoubtedly, TGFβ plays an important role in regulating follicular development. Understanding of its function and of regulation of its production are essential for understanding the ovarian cycle. The findings in this study raise the possibility that TGFβ possesses species-specific effects, which implies a complex mechanism of action.

The difference in effects of this relatively new polypeptide on porcine and rat granulosa cells is profound and, as of yet, not understood. Perhaps the effect of TGFβ is truly species dependent, which only raises the question about the significance of this difference. There are even conflicting findings within the porcine species about its effects on granulosa cell proliferation (17, 18). These uncertainties are not surprising considering the many interactions of gonadotropins and other local factors upon ovarian function. Further investigation is needed before we can fully understand all the factors and their interactions involved in the regulation of granulosa cell function.

Acknowledgments. This work was supported by grants from the March of Dimes Birth Defects Foundation and the Food and Drug Administration. Porcine ovaries were provided by the Ohio Packing Company, Columbus, OH.

References

1. Gangrade BK, May JV. The production of transforming growth factor-β in the porcine ovary and its secretion in vitro. Endocrinology 1990;127:2372–80.
2. Ohmura H, Uzumcu M, Coskun S, Akira S, Chang W, Lin YC. Transforming growth factor-β (TGFβ) stimulates progesterone production in cultured granulosa cells aspirated from pregnant mare serum gonadotropin (PMSG)-stimulated adult rats. In: Leung CK, Hsueh AJW, Friesen HG, eds. Molecular basis of reproductive endocrinology. New York: Springer-Verlag, 1992. (*See* Chapter 16, this volume.)
3. DeLarco JE, Todaro GJ. Growth factors from murine sarcoma virus-transformed cells. Proc Natl Acad Sci USA 1978;78:4001–5.
4. Roberts AB, Anzano MA, Lamb LC, Smith JM, Sporn MB. New class of transforming growth factors potentiated by epidermal growth factor: isolation from non-neoplastic tissues. Proc Natl Acad Sci USA 1981;78:5339–43.
5. Roberts AB, Lamb LC, Newton DL, et al. Transforming growth factors: isolation of peptides from virally and chemically transformed cells by acid/ethanol extraction. Proc Natl Acad Sci USA 1980;77:3494–8.
6. Sporn MB, Roberts AB, Wakefield LM, De Crombrugghe B. Some recent advances in the chemistry and biology of transforming growth factor-beta. J Cell Biol 1987;105:1039–45.
7. Skinner MK, Keski-Oja J, Osteen KG, Moses HL. Ovarian thecal cells produce transforming growth factor-β which can regulate granulosa cell growth. Endocrinology 1987;121:786–92.
8. Hernandez ER, Hurwitz A, Payne DW, Dharmarajan AM, Purchio AF, Adashi EY. Transforming growth factor-β₁ inhibits ovarian androgen production: gene expression, cellular localization, mechanism(s), and site(s) of action. Endocrinology 1990;127:2804–11.
9. Kim I-C, Schomberg DW. The production of transforming growth factor-β activity by rat granulosa cell cultures. Endocrinology 1989;124:1345–51.
10. Mulheron GW, Schomberg DW. Rat granulosa cells express transforming growth factor-β type 2 messenger ribonucleic acid which is regulatable in follicle-stimulating hormone in vitro. Endocrinology 1990;126:1777–9.

11. Bendell JJ, Dorrington J. Rat theca/interstitial cells secrete a transforming growth factor-β-like factor that promotes growth and differentiation in rat granulosa cells. Endocrinology 1988;123:941–8.

12. Knecht M, Feng P, Catt K. Bifunctional role of transforming growth factor-β during granulosa cell development. Endocrinology 1987;120:1243–9.

13. Dorrington J, Chuma AV, Bendell JJ. Transforming growth factor β and follicle-stimulating hormone promote rat granulosa cell proliferation. Endocrinology 1988;123:353–9.

14. Dodson WC, Schomberg DW. The effect of transforming growth factor-β on follicle-stimulating hormone-induced differentiation of cultured rat granulosa cells. Endocrinology 1987;120:512–6.

15. Feng P, Catt KJ, Knecht M. Transforming growth factor β regulates the inhibitory actions of epidermal growth factor during granulosa cell differentiation. J Biol Chem 1986;261:14167–70.

16. Hutchinson LA, Findlay JK, de Vos FL, Robertson DM. Effects of bovine inhibin, transforming growth factor-β and bovine activin-A on granulosa cell differentiation. Biochem Biophys Res Commun 1987;146;1405–12.

17. May JW, Frost JP, Schomberg DW. Differential effects of epidermal growth factor, somatomedin-C/insulin-like growth factor I, and transforming growth factor-β on porcine granulosa cell deoxyribonucleic acid synthesis and cell proliferation. Endocrinology 1988;123:168–79.

18. Mondschein JS, Canning SF, Hammond JM. Effects of transforming growth factor-β on the production of immunoreactive insulin-like growth factor I and progesterone and on [^{3}H]thymidine incorporation in porcine granulosa cell cultures. Endocrinology 1988;123:1970–6.

19. Hsueh AJ, Adashi EY, Jones PBC, Welsh TH. Hormonal regulation of the differentiation of cultured ovarian granulosa cells. Endocr Rev 1984;5:76–127.

20. Pfeilschifter J. Transforming growth factor-β. In: Habenicht A, ed. Growth factors, differentiation factors, and cytokines. New York: Springer-Verlag, 1990:56–64.

21. Derynck R, Jarrett JA, Chen EY, et al. Human transforming growth factor-beta cDNA sequence and expression in tumor cell lines. Nature 1985;316:701–5.

22. Gu Y, Lin YC, Rikihisa Y. Inhibitory effect of gossypol on steroidogenic pathways in cultured bovine luteal cells. Biochem Biophys Res Commun 1990;169:455–61.

23. Akira S, Ohmura H, Araki T, Lin YC. Aromatase activity (AA) in immature porcine granulosa cells (IPGC) under different culture conditions. FASEB J 1991;5:1804A.

24. Adashi EY, Resnick CE, Hernandez ER, May JV, Purchio AF, Twardzik DR. Ovarian transforming growth factor-β (TGFβ): cellular site(s), and mechanism(s) of action. Mol Cell Endocrinol 1989;61:247–56.

25. Massague J, Cheifetz S, Ignotz RA, Boyd FT. Multiple type-β transforming growth factors and their receptors. J Cell Physiol [Suppl] 1987;5:43–7.

26. Flanders KC, Marascalco BA, Roberts AB, Sporn MB. Transforming growth factor β: a multifunctional regulatory peptide with actions in the reproductive system. In: Schomberg DW, ed. Growth factors in reproduction. Proc

Symposium on Growth Factors in Reproduction, Savannah, GA, April 1–4, 1990. New York: Springer-Verlag, 1991:23–37.

27. Magoffin DA, Gancedo B, Erickson GF. Transforming growth factor-β promotes differentiation of ovarian thecal-interstitial cells but inhibits androgen production. Endocrinology 1989;125:1951–8.

28. Mondschein JS, Smith SA, Hammond JM. Production of insulin-like growth factor binding proteins (IGFBPs) by porcine granulosa cells: identification of IGFBP-2 and -3 and regulation by hormones and growth factors. Endocrinology 1990;127:2298–306.

20

Biological Relevance and Analysis of c-*myc* Gene Expression in Normal Human Ovary

Hironori Tashiro, Manabu Fukumoto, Kohji Miyazaki, and Hitoshi Okamura

Among oncogenes, the c-*myc* gene is most frequently activated in a wide spectrum of tumors (1). Recently, we demonstrated that the c-*myc* gene is overexpressed in primary human ovarian cancer tissues, especially in serous tumors (unpublished observation). The c-*myc* gene is also known to be involved in cell proliferation and differentiation other than carcinogenic process. It has been reported that granulosa cells in rat ovary express the c-*myc* transiently after *pregnant mare serum gonadotropin* (PMSG) administration, suggesting that the c-*myc* gene is involved in steroidogenesis (2). To elucidate the function of the c-*myc* gene in normal human ovary, c-*myc* expression was investigated by Northern hybridization and immunohistochemistry in 11 independent, normal ovarian tissues.

Materials and Methods

Materials

Normal ovarian tissues from 10 independent cases that were dissected by medically warranted oophorectomy and 1 autopsy case were snap frozen in liquid nitrogen. A summary of cases is shown in Table 20.1. Mature follicles and a corpus luteum were dissected from other components before frozen in cases 7 and 8, respectively. The samples were stored at −80°C for subsequent analysis.

Oncogene Expression

Frozen tissues were homogenized under nitrogen using a magnetic vibrator (Mikro-Dismembrator II, B. Braun Instruments) and in 4 M

TABLE 20.1. Age, menstrual status or day of cycle, and c-*myc* expression in ovarian tissues.

Case	Age (years)	Menstrual status or day of cycle	c-*myc* expression (fold)
1	53	Menopause at age 50	7.6
2	47	Menopause at age 46	1.5
3	45	13th day	0.7
4	33	27th day	1.6
5[a]	47	Treated with the pill	1.0 (basal level)
6	39	22nd day	1.1
7[b]	47	13th day	0.3
8[c]	34	28th day	0.8
9	54	Menopause at age 51	0.8
10	59	Menopause at age 55	1.8
11	13	Premenarche	2.0

[a] Combined progesterone-estrogen therapy (5-mg norgestrel and 0.05-mg ethynylestradiol) was used in the form of the pill for 1 month. This c-*myc* expression was defined as basal level.
[b] Mature follicles were dissected.
[c] A corpus luteum was dissected.

guanidine isothiocyanate. The lysate was layered on a CsCl gradient and spun in an ultracentrifuge (SW41 rotor, Beckman), and total RNA was isolated (3). Twenty micrograms of total RNA was subjected to electrophoresis in an 0.8% agarose gel containing 2.2 M formaldehyde. After Northern blotting onto a nylon membrane (Biodyne B, Pall Biosupport), RNA was hybridized to a c-*myc* probe (exon 3 of HSR-1) (4) labeled with ^{32}P-dCTP by oligolabeling. The value of densitometric determination (Ultroscan XL laser densitometer, LKB) on autoradiogram was carried out at the control level of expression. The amount of RNA electrophoresed was normalized by ethidium bromide profile after electrophoresis.

Immunohistochemical Detection of c-*myc* Protein

Frozen tissues were sectioned at 6 μm thick and fixed in 6% neutral formalin. Monoclonal mouse antibody against c-*myc* protein produced in *Escherichia coli*, *myc*-1 (IgG$_2$) (5) was kindly provided by Dr. Shiku (Nagasaki University, Japan). After application of biotinylated horse antimouse immunoglobulin, colorization was carried out by Vectorstain ABC kit (Vector). Nonimmunized mouse IgG$_2$ was used as negative control for staining.

Results

The profile of Northern hybridization to the c-*myc* probe is shown in Figure 20.1. The expression of case 5 was defined as the basal level

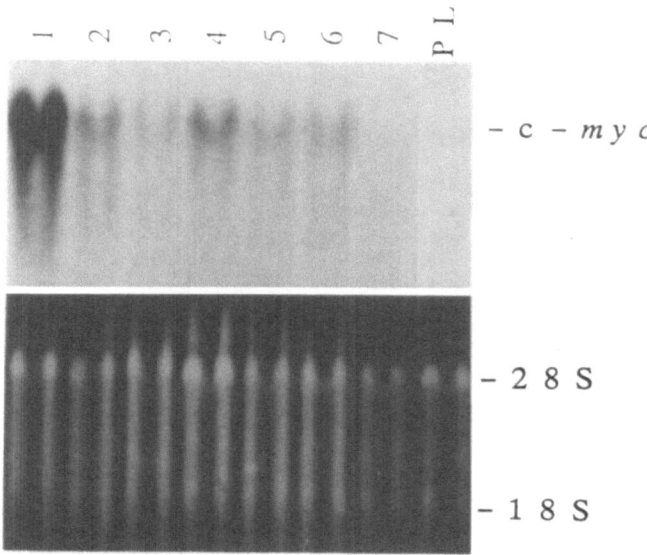

FIGURE 20.1. Northern blot analysis of c-*myc* expression in ovarian tissues. Each lane number corresponds respectively to the case numbers in Table 20.1. (PL = placental tissue at full term.)

because hormonal state was thought to be most stable. The expression normalized against the basal level is shown in Table 20.1.

As measured by tissue mass, c-*myc* expression in ovarian tissues did not show significant fluctuation during the menstrual cycle (cases 3, 4, and 6). However, in dissected tissues of mature follicles (case 7) and corpus luteum (case 8), the c-*myc* was expressed less than the basal level. The expression of premenarchal ovarian tissue from the 13-year-old autopsy case (case 11) was twice as high as the basal level.

Immunohistochemically, c-*myc* protein was localized in granulosa cells of preantral and early antral follicles (less than 1 mm in diameter) (Fig. 20.2). Follicles larger than 1 mm were stained weakly, and mature follicles and corpus luteum showed negative staining (data not shown).

In case 1, a postmenopausal case that showed the highest expression of the c-*myc* gene by Northern hybridization, many inclusion cysts were characteristically found on histology of serial sections. Immunohistochemical detection of c-*myc* protein in case 1 is shown in Figure 20.3. Nuclei of lining cells of the inclusion cysts were positively stained for c-*mcy* protein. The other postmenopausal ovarian tissues (cases 2, 9, and 10) did not show high expression of the c-*myc* gene by Northern hybridization.

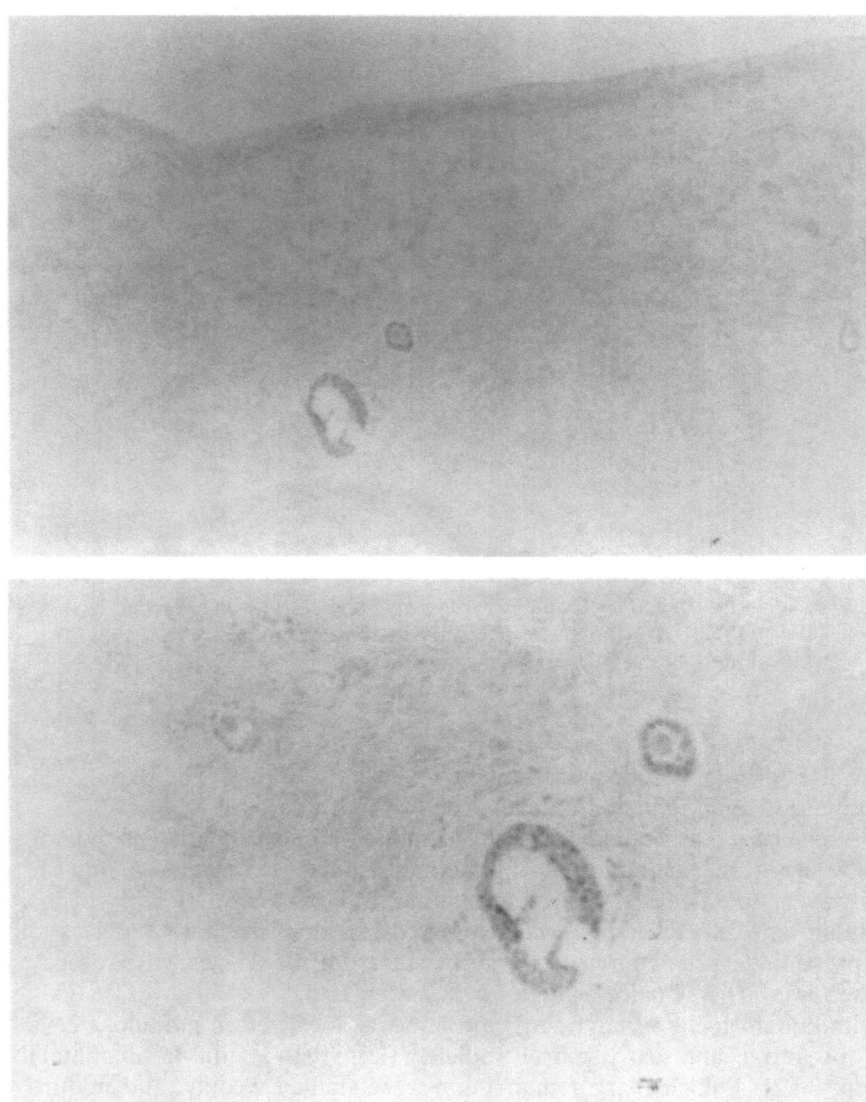

FIGURE 20.2. Immunohistochemistry of c-*myc* protein in case 11. Granulosa cells of preantral follicles and an early antral follicle are positively stained. Top: 100×; bottom: 200×.

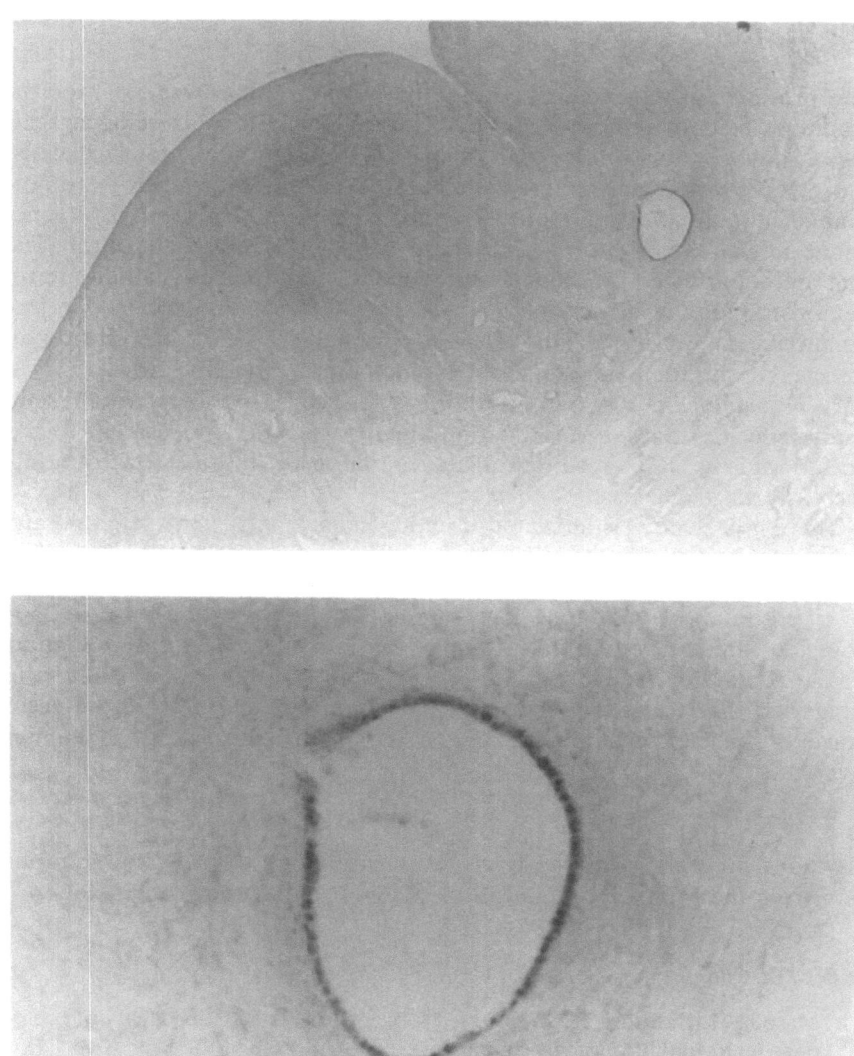

FIGURE 20.3. Immunohistochemistry of c-*myc* protein in case 1. Lining cells of an inclusion cyst are positively stained. Top: 40×; bottom: 200×.

Discussion

The present study has demonstrated that c-*myc* protein was localized in granulosa cells of preantral and early antral follicles in a premenarchal case. However, granulosa cells of mature follicles and luteinized cells expressed c-*myc* gene less than the basal level, and these cells were not stained by anti-c-*myc* antibody. These data suggest that the c-*myc* gene might play an important role in the early stage of development of follicles that are not affected by endogenous gonadotropin. Recently, studies have demonstrated that the *nerve growth factor* (NGF) gene is expressed in the immature rat ovary (6) and that NGF is necessary for the growth of follicles during the early stage of development (7, 8). Moreover, NGF is known to induce c-*myc* transcription in PC12 cells (9). Thus, the present study indicates that c-*myc* activation in early antral granulosa cells might be induced by NGF, resulting in the initiation of follicle growth during the early stage.

The c-*myc* gene was activated in lining cells of inclusion cysts that are known to be the first stage in the neoplastic process. We have found a case of low malignant potential of serous tumor with the remarkable overexpression of the c-*myc* gene (unpublished observation). These data coincide with the notion that c-*mcy* expression may reflect differentiation or the neoplastic process in tumor lineages (10). In the present study, it is suggested that c-*myc* expression has an important role both in the early stage of follicle development and in the neoplastic process in human ovary.

Acknowledgments. This work was partly supported to M.F. by Grant-in-Aid from the Ministry of Education, Science, and Culture in Japan.

References

1. Masuda H, Battifora H, Yokota J, Meltzer S, Cline MJ. Specificity of proto-oncogene amplification in human malignant disease. Mol Biol Med 1987; 4:213–27.
2. Delidow BC, White BA, Peluso JJ. Gonadotropin induction of c-*fos* and c-*myc* expression and deoxyribonucleic acid synthesis in rat granulosa cells. Endocrinology 1990;126:2302–6.
3. Davis LG, Dibner MD, Battery JF. Basic methods in moleculular biology. New York: Elsevier Science, 1986.
4. Schwab M, Alitalo K, Klempnauer KH, et al. Amplified DNA with limited homology to *myc* cellular oncogene is shared by human neuroblastoma cell lines and a neuroblastoma tumour. Nature 1983;305:245–8.
5. Naoe T, Nozaki N, Yamada K, et al. Diversity of cellular molecules in human cells detected by monoclonal antibodies reactive with c-*myc* proteins produced in *Escherichia coli*. Jpn J Cancer Res 1989;80:747–53.

6. Lara HE, Hill DF, Katz KH, Ojeda SR. The gene encoding nerve growth factor is expressed in the immature rat ovary: effect of denervation and hormonal treatment. Endocrinology 1990;126:357–63.
7. Lora HE, McDonald JK, Ojeda SR. Involvement of nerve growth factor in female sexual development. Endocrinology 1990;126:364–75.
8. Lora HE, McDonald JK, Ojeda SR. Guanethidine-mediated destruction of ovarian sympathetic nerves disrupts ovarian development and function in rats. Endocrinology 1990;127:2199–209.
9. Greenberg ME, Greene LA, Ziff EB. Nerve growth factor and epidermal growh factor induce rapid transient changes in proto-oncogene transcription in PC12 cells. J Biol Chem 1985;260:14101–10.
10. Semsei I, Ma S, Cutler RG. Tissue and age specific expression of the *myc* proto-oncogene family throughout the life span of the C57BL/6J mouse strain. Oncogene 1989;4:465–70.

21

Signal Transduction of $PGF_{2\alpha}$ in Rat Corpora Lutea

GILLIAN L. STEELE AND PETER C.K. LEUNG

The mechanisms involved in *prostaglandin $F_{2\alpha}$* ($PGF_{2\alpha}$) action in the corpus luteum remain to be clearly defined. Certainly, this eicosanoid is implicated in the process of *luteolysis*—that is, inhibiting the adenylate cyclase-cyclic AMP system—thereby blocking gonadotropin-induced steroidogenesis in corpora lutea (1, 2). An induction of inositol lipid metabolism has been observed following binding of $PGF_{2\alpha}$ to its membrane receptor, as well as subsequent mobilization of intracellular free calcium (3, 4). Furthermore, intracellular unesterified *arachidonic acid* (AA) levels are reported to become elevated in response to $PGF_{2\alpha}$ receptor binding (5, 6). This introduces the potential for a positive feedback loop whereby AA would provide the substrate for de novo generation of eicosanoids, including $PGF_{2\alpha}$. Such a model is attractive in view of the rate at which luteolysis occurs. In this chapter, recent studies are reviewed in order to evaluate the possibility that these intracellular mechanisms are responsible for the observed physiological effects of $PGF_{2\alpha}$.

Signal Transduction via Inositol Lipid Metabolism

Our understanding of the mechanisms by which hormones induce a cellular response has increased dramatically in the past decade. It is now evident that in addition to the well-characterized adenylate cyclase-cAMP pathway, another prominent signaling pathway relies on the metabolism of membrane phosphoinositides: the stimulation of inositol lipid breakdown. Hormone-specific binding initiates a conformational change in the receptor that leads to the activation of *phospholipase C* (PLC). *Phosphatidylinositol 4,5-bisphosphate* (PIP_2) is hydrolyzed by PLC to form *inositol 1,4,5-trisphosphate* (IP_3) and *diacylglycerol* (DAG). The former is released into the cytosol, while the latter remains membrane bound.

IP$_3$ stimulates the release of calcium from intracellular stores—specifically, the endoplasmic reticulum—and has also been shown to facilitate calcium transport across the plasma membrane (7). DAG also functions as a second messenger, activating protein kinase C (8). This activation appears to require intracellular free calcium as a cofactor, which may be involved in the translocation of this protein kinase from the cytosol to the membrane. Additionally, DAG may be converted to either phosphatidic acid or monoacylglycerol and AA. The former is resynthesized into phosphatidylinositol, while AA provides the substrate for the cyclooxygenase and lipoxygenase pathways. In this respect, it stimulates the generation of additional second messengers, such as prostaglandins, leukotrienes, and thromboxane.

PGF$_{2\alpha}$-Induced Hydrolysis of Membrane Polyphosphoinositides

In larger domestic animals like the sheep, the source of PGF$_{2\alpha}$ involved in luteolysis has been shown to be the uterus (9). In the rat ovary however, PGF$_{2\alpha}$ synthesis in several cell types is maximal close to the time of luteolysis, suggesting this tissue to be the source (10). Two binding sites for PGF$_{2\alpha}$ have been identified in rat corpora lutea, one with high and one with lower affinity for the ligand (11). With regard to the affinity and binding capacity of this prostaglandin, its binding to the receptor appears to parallel the luteolytic action that it induces. In rat luteal and granulosa cells, radiolabeling studies have shown that PGF$_{2\alpha}$ binding decreases phosphatidylinositol 4-phosphate and PIP$_2$ concomitant with rapid and significant increases in *phosphatidylinositol* (PI), *phosphatidic acid* (PA), and *inositol phosphates* (IP) (3, 12). These results are consistent with the notion that hydrolysis of polyphosphoinositides plays a role in the early signal transduction of PGF$_{2\alpha}$ action. Furthermore, treatment of these cells with a PLC preparation had a similar effect on IP accumulation (3).

In bovine luteal cells, PGF$_{2\alpha}$ treatment also induced inositol phospholipid turnover as evidenced by increased incorporation of radiolabeled phosphate into PA and PI (13). The prostaglandin provoked rapid and sustained increases in IP, IP$_2$, and IP$_3$. Furthermore, the isomers were shown to be formed by the action of a specific calcium/calmodulin-regulated kinase (IP$_3$-3-kinase) that phosphorylates IP$_3$ to form inositol 1,3,4,5-tetrakisphosphate (14). The latter may be implicated in the mobilization of free calcium due to its apparent ability to aid calcium influx (15), activate Ca^{++}-dependent K^{+}-currents (16), and regulate the flow of calcium between intracellular calcium pools (7).

In a temporal study, the stimulation of inositol phospholipid metabolism by PGF$_{2\alpha}$ was investigated in rat corpora lutea. Unlike the significant stimulation of PLC observed in young (day 2) corpora lutea, the prostaglandin was ineffective in mature (day 7) tissue (17). Considering that the luteolytic effects of $PGF_{2\alpha}$ increase greatly with luteal age, it seems likely that enhanced inositol phosphate metabolism in itself is not sufficient to explain the signal transduction mechanism. However, second messengers generated as a result of this PLC activity may also be involved.

$PGF_{2\alpha}$-Induced Increases in Intracellular Free Calcium

The calcium-mobilizing properties of IP_3 are well recognized, and given the evidence for $PGF_{2\alpha}$-stimulated accumulation of this phospholipid, it is not surprising that intracellular free calcium levels are also enhanced by this hormone. Using the calcium-sensitive fluorescent dye Fura-2, rat luteal cells have been shown to respond to $PGF_{2\alpha}$ with a significant increase in intracellular free calcium (4). This was a single, transient response that could be mimicked by perifusion with the hormone, suggesting either that the source of calcium was a limited store or that receptor desensitization was rapid. In bovine luteal cells, $PGF_{2\alpha}$ appears to enhance intracellular calcium in a similar rapid fashion (13). However, the response in bovine cells is reported to be sustained for as long as 10 min (13) and may occur only in the large luteal cell population (18).

Despite the attractive hypothesis of an IP_3-stimulated release of calcium from intracellular stores, there remains considerable controversy regarding the actual source of calcium. Some find the observed increase to be independent of extracellular calcium (13), while others attribute it to an influx across the plasma membrane (18). These conclusions are drawn from experiments utilizing specific channel blockers, chelators, and calcium-free media. Furthermore, the relevance of changes in intracellular free calcium to the luteolytic effect of the hormone is also a point of contention. Support for such a paradigm comes from studies in which calcium alone inhibits LH-sensitive adenylate cyclase activity in isolated luteal membranes (19). Addition of the calcium ionophore A23187 in calcium-replete media is similarly effective in mimicking $PGF_{2\alpha}$ action. In the ewe, the luteolytic effects of $PGF_{2\alpha}$ appear to be mediated in the large cell by the ability of this hormone to produce elevations in cystolic free calcium (20). However, others claim that the luteolytic action of $PGF_{2\alpha}$ is not mediated by calcium based on several observations. The intracellular calcium chelator BAPTA is not effective in inhibiting the effects of the hormone (21); the calcium ionophore ionomycin increases intracellular calcium to a similar extent to that of $PGF_{2\alpha}$, but does not mimic its action; and a putative inhibitor of intracellular calcium release or action (TMB-8) also did not abolish $PGF_{2\alpha}$ activity (22).

PGF$_{2\alpha}$-Induced Activation of DAG and PKC

In addition to the generation of inositol phosphates, PLC activity results in the accumulation of DAG. In order to investigate the possible role of DAG and PKC in the luteolytic action of PGF$_{2\alpha}$, phorbol esters have been used to mimic DAG activation of PKC. In one study, treatment of isolated rat luteal cells with *phorbol 12-myristate 13-acetate* (PMA) resulted in a marked inhibition of LH-induced cAMP and progesterone production (23). This supports the theory of PKC-mediated PGF$_{2\alpha}$ action in these cells. Furthermore, a PGF$_{2\alpha}$-induced calcium-dependent phosphorylation of endogenous proteins has also been identified in luteal tissue that was presumed to be the result of activated PKC (24). In contrast, cumulative effects on the inhibition of LH-dependent cAMP accumulation were observed using PMA in conjunction with PGF$_{2\alpha}$ (25). In the same study, staurosporine (a PKC inhibitor) was effective in reversing this inhibition by PMA, but had no effect on PGF$_{2\alpha}$-induced inhibition. These results raise questions as to the relevance of PGF$_{2\alpha}$-induced DAG and PKC activity with respect to the luteolytic effects of the hormone. Certainly, our understanding of the intracellular mechanism of action of PGF$_{2\alpha}$ is far from complete.

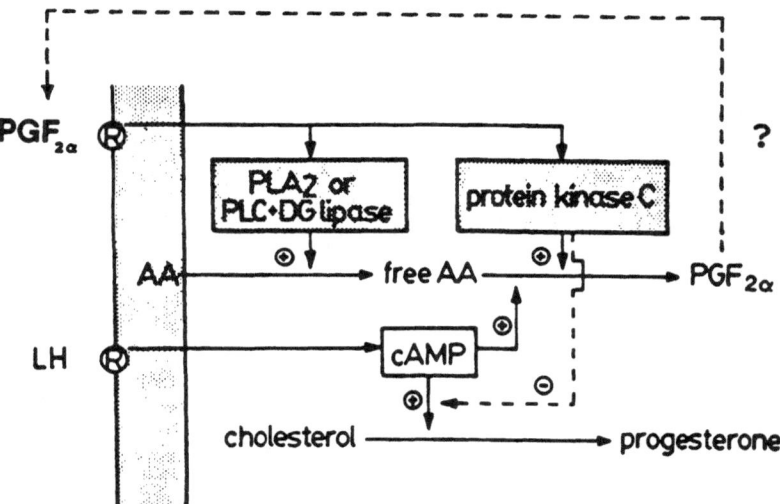

FIGURE 21.1. A model of a positive feedback loop for PGF$_{2\alpha}$ synthesis in the corpus luteum. Binding of PGF$_{2\alpha}$ to its membrane receptor (R) stimulates liberation of arachidonic acid (AA) from the membrane phospholipids by the combined action of phospholipase C (PLC) and diacylglycerol lipase (DG lipase) or by phospholipase A$_2$ (PLA$_2$). As substrate for eicosanoids, AA aids in further synthesis of PGF$_{2\alpha}$. The resultant cascade may contribute in part to the subsequent demise of the corpus luteum.

PGF$_{2\alpha}$-Induced AA Release

Consistent with the generation of DAG, PGF$_{2\alpha}$ has been reported to induce a stimulation of AA release in rat corpora lutea cells (5, 6). The stimulation of the release of both IP and AA was observed in cells of young (day 1) as well as mature (day 5) luteal tissue. Arachidonic acid may be generated by the successive actions of PLC and DAG-lipase or, alternatively, by receptor-stimulated *phospholipase A$_2$* (PLA$_2$) activity, an enzyme that liberates AA from membrane phospholipids (26). In the event of the latter, PGF$_{2\alpha}$ may be acting through a single class of receptor that is coupled to both PLC and PLA$_2$. As hypothesized in Figure 21.1, the effect of PGF$_{2\alpha}$ on AA mobilization is an exciting aspect to pursue considering the potential positive feedback loop that would be created by the generation of additional eicosanoid substrate.

Summary

As the primary luteolytic agent in the ovary, PGF$_{2\alpha}$ plays a crucial role in the regulation of normal reproductive cycles. Although the effects of its action are clear, our understanding of the signal transduction is incomplete. There is strong evidence for the stimulation of inositol lipid metabolism by receptor binding, which results from an activation of PLC. The subsequent generation of IP$_3$ and mobilization of intracellular free calcium have been implicated in the luteolytic actions of the hormone. DAG, another product of PLC activity, may also elicit PGF$_{2\alpha}$ action by the activation of PKC. Enhanced activity of PLC and/or PLA$_2$ by PGF$_{2\alpha}$ may lead to further generation of AA metabolites. Therefore, there may be several second messengers that are integrated in the signal transduction of this important ovarian regulator.

References

1. Lahav M, Freud A, Lindner HR. Abrogation by prostaglandin F$_{2\alpha}$ of LH-stimulated cyclic AMP accumulation in isolated rat corpora lutea of pregnancy. Biochem Biophys Res Commun 1976;68:1294–1300.
2. Thomas JP, Dorflinger LJ, Behrman HR. Mechanism of the rapid antigonadotropic action of prostaglandins in cultured luteal cells. Proc Natl Acad Sci USA 1978;75:1344–8.
3. Leung PCK, Minegishi T, Ma F, Zhou F, Ho-Yuen B. Induction of polyphosphoinositide breakdown in rat corpus luteum by prostaglandin F$_{2\alpha}$. Endocrinology 1986;119(1):12–8.
4. Rodway MR, Baimbridge KG, Ho-Yuen B, Leung PCK. Effect of prostaglandin F$_{2\alpha}$ on cytosolic free calcium ion concentrations in rat luteal cells. Endocrinology 1991;129:889–95.

5. Watanabe H, Tanaka S, Akino T, Hasegawa-Sasake H. Evidence for coupling of different receptors for gonadotropin-releasing hormone to phospholipases C and A$_2$ in cultured rat luteal cells. Biochem Biophys Res Commun 1990;168(1):328–34.

6. Wang J, Minegishi T, Leung PCK. Mechanism of prostaglandin F$_{2\alpha}$ action in rat corpus luteum [Abstract 169.06]. Proc 30th cong International Union of Physiological Sciences; Vancouver, BC, July 13–19, 1986:112.

7. Berridge MJ, Irvine RF. Inositol phosphates and cell signalling. Nature 1989;341:197–205.

8. Farago A, Nishizuka Y. Protein kinase C in transmembrane signalling. FEBS Lett 1990;268(2):350–4.

9. Niswender GD, Nett TM. The corpus luteum and its control. In: Knobil E, Neill J, eds. The physiology of reproduction. New York: Raven Press, 1988: 489–525.

10. Olofsson J, Norjavaara E, Selstam G. In vivo levels of prostaglandin F$_{2\alpha}$, E$_2$ and prostacyclin in the corpus luteum of pregnant and pseudopregnant rats. Biol Reprod 1990;42:792–800.

11. Wright K, Luborsky-Moore JL, Behrman HR. Specific binding of prostaglandin F$_{2\alpha}$ to membranes of rat corpora lutea. Mol Cell Endocrinol 1979; 13:25–34.

12. Minegishi T, Leung PCK. Effects of prostaglandins and luteinizing hormone on phosphatidic acid-phosphatidylinositol labeling in rat granulosa cells. Can J Physiol Pharmacol 1985;63(4):320–4.

13. Davis JS, Weakland LL, Weiland DA, Farese RV, West LA. Prostaglandin F$_2$ alpha stimulates phosphatidylinositol 4,5-bisphosphate hydrolysis and mobilizes intracellular Ca^{++} in bovine luteal cells. Proc Natl Acad Sci USA 1987;84(11):3728–32.

14. Duncan RA, Davis JS. Prostaglandin F$_{2\alpha}$ stimulates inositol 1,4,5-trisphosphate and inositol 1,3,4,5–tetrakisphosphate formation in bovine luteal cells. Endocrinology 1991;128(3):1519–26.

15. Irvine RF, Moor RM. Micro-injection of inositol 1,3,4,5-tetrakisphosphate activates sea urchin eggs by a mechanism dependent on external Ca^{++}. Biochem J 1986;240:917–20.

16. Morris AP, Gallacher DV, Irvine RF, Peterson OH. Synergism of inositol trisphosphate and tetrakisphosphate in activating Ca^{++}-dependent K^{+} channels. Nature 1987;330:653–5.

17. Lahav M, West LA, Davis JS. Effects of prostaglandin F$_{2\alpha}$ and a gonadotropin-releasing hormone agonist on inositol phospholipid metabolism in isolated rat corpora lutea of various ages. Endocrinology 1988;123(2): 1044–52.

18. Wiltbank MC, Guthrie PB, Mattson MP, Kater SB, Niswender GD. Hormonal regulation of free intracellular calcium concentrations in small and large ovine luteal cells. Biol Reprod 1989;41:771–8.

19. Dorflinger LJ, Albert PJ, Williams AT, Behrman HR. Calcium is an inhibitor of luteinizing hormone-sensitive adenylate cyclase in the luteal cell. Endocrinology 1984;114(4):1208–15.

20. Wegner JA, Martinez-Zaguilan R, Gillies RJ, Hoyer PB. Prostaglandin F$_{2\alpha}$-induced calcium transient in ovine large luteal cells, II. Modulation of the

transient and resting cytosolic free calcium alters progesterone secretion. Endocrinology 1991;128(2):929–36.

21. Pepperell JR, Preston SL, Behrman HR. The antigonadotropic action of prostaglandin $F_{2\alpha}$ is not mediated by elevated cytosolic calcium levels in rat luteal cells. Endocrinology 1989;125(1):144–51.

22. Lahav M, Rennert H, Sabag K, Barzilai D. Calmodulin inhibitors and 8-(N,N-diethylamino)-octyl-3,4,5-trimethoxybenzoate do not prevent the inhibitory effect of prostaglandin F2 alpha on cyclic AMP production in isolated rat corpora lutea. J Endocrinol 1987;113(2):205–12.

23. Baum MS, Rosberg S. A phorbol ester, phorbol 12-myristate 13-acetate, and a calcium ionophore, A23187, can mimic the luteolytic effect of prostaglandin $F_{2\alpha}$ in isolated rat luteal cells. Endocrinology 1987;120(3):1019–26.

24. Baum MS. Prostaglandin $F_{2\alpha}$ administered in vivo induces Ca^{++}-dependent protein phosphorylation in rat luteal tissue. Endocrinology 1989;124(1): 555–7.

25. Musicki B, Aten RF, Behrman HR. The antigonadotropic actions of prostaglandin F2 alpha and phorbol ester are mediated by separate processes in rat luteal cells. Endocrinology 1990;126(3):1388–95.

26. Lapetina EG. The inositide and arachidonic acid signal system. Adv Exp Med Biol 1989;261:285–93.

22

Protein Changes in the Rat Corpus Luteum During Luteolysis

MASAAKI SAWADA, ULRIKE SESTER, XIU MEI WU, AND
JOHN C. CARLSON

Ovarian changes in protein synthesis, nucleic acid content, and Na^+/K^+ ATPase activity were examined during luteolysis in the rat. Rats were treated with $PGF_{2\alpha}$ 7 days after injection of hCG to induce super-ovulation, and ovaries were removed at various intervals. $PGF_{2\alpha}$ caused a significant decrease in plasma progesterone concentration within 1 h ($P < 0.05$), but there was no change in total *messenger RNA* (mRNA) or total DNA content in ovarian extracts for 48 h. However, at 48 h a significant decline in ouabain-sensitive Na^+/K^+ ATPase activity in plasma membrane samples occurred. Alterations in protein synthesis in ovarian extracts were examined using an EF-1α activity assay. Protein synthesis activity decreased in stages. An initial decline occurred within 60 min of $PGF_{2\alpha}$ treatment ($P < 0.05$), and a second reduction, which was around 60% of pretreatment activity, appeared by 48 h ($P < 0.05$). In addition, injection of actinomycin D, an inhibitor of protein synthesis, into the ovarian bursa caused a significant reduction in plasma progesterone and protein synthesis within 1 h. Membrane samples from luteinized ovaries were also examined by SDS-PAGE for protein changes. At 96 h after $PGF_{2\alpha}$ treatment, there was a clear decrease in a 49-kd protein band in plasma membrane samples. In contrast, in ovarian microsome samples, there was a prominent increase in a 35-kd protein band. These results indicate that one of the early luteolytic alterations is a decrease in protein synthesis activity. This occurs before membrane protein changes that are detectable by SDS-PAGE.

Corpus luteum (CL) regression is characterized by a number of cellular changes during the loss of steroidogenesis. These include reduction in gonadotropin binding and generation of cAMP, activation of phospholipases, membrane perturbation, loss in Na^+/K^+ ATPase activity, and increases in free-radical formation and lipid peroxidation. These alterations are well documented, but it is not known whether changes in proteins do occur. Posttranslational modification represents one of the means by which proteins may be altered during luteolysis. This is indicated by the recent finding that administration of $PGF_{2\alpha}$ induced a Ca^{++}-dependent phosphorylation of a 45-kd protein in luteal cells (1). However, it is uncertain whether the loss of steroidogenesis is also accompanied by changes in protein synthesis or degradation. The purpose of the present study was to determine if $PGF_{2\alpha}$ could disrupt protein synthesis and to examine whether this could result in changes in the protein profile.

Methods

Sample Preparation

Immature 3.5-week-old rats were superovulated with 50-IU pregnant mare serum gonadotropin and 50-IU hCG. $PGF_{2\alpha}$ (500 µg) was injected 7 days later to induce regression (2, 3). Ovaries were removed at various time points after $PGF_{2\alpha}$ treatment, and homogenates of whole ovaries were prepared or plasma membrane samples were isolated using a modified dextran-polyethylene glycol system (2, 3). In addition, microsomes were isolated from ovarian homogenates (4). In order to block protein synthesis, 2-µg actinomycin D was injected into the ovarian bursa (3).

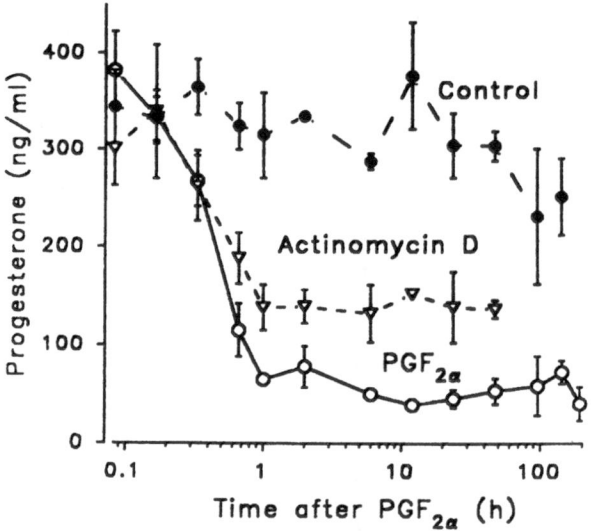

FIGURE 22.1. Plasma progesterone concentration in control, $PGF_{2\alpha}$-, and actinomycin D-treated rats. A decrease was seen starting 40 min after treatment with $PGF_{2\alpha}$ (P < 0.02) or administration of actinomycin D (P < 0.05). Each point represents the mean ± SE of 2–3 measurements.

Protein levels were determined by the method outlined by Bradford (5). All samples were maintained at 4°C under nitrogen to minimize oxidation during preparation.

Assays

For mRNA isolation, total RNA was isolated by guanidinium thiocyanate-phenol-chloroform extraction (6) and passed through an oligo(dt)-cellulose column to select for mRNA (7). The activity of *elongation factor 1α* (EF-1α) was measured to determine protein synthesis (8). DNA levels were determined as indicated by Kirby (9). The activity of Na^+/K^+ ATPase in plasma membrane samples was determined by the method of Cole and Waddell (10). To examine changes in the protein profile, SDS-PAGE was performed using a 9% resolving gel with a 4% stacking gel (11).

Results

Subcutaneous injection of $PGF_{2\alpha}$ caused a significant decrease in plasma progesterone concentration (Fig. 22.1) within 1 h ($P < 0.05$). Thereafter, circulating levels remained at low concentrations. In plasma membrane samples, ouabain-sensitive Na^+/K^+ ATPase activity declined slowly, and this decline did not appear significant until 48 h after $PGF_{2\alpha}$ treatment (Table 22.1). During this time course, there were no changes in total mRNA or total DNA content in ovarian extracts (Table 22.1). However, protein synthesis decreased in stages (Fig. 22.2). An initial decline occurred within 1 h ($P < 0.05$), and a second reduction, to approximately

TABLE 22.1. Alterations in Na^+/K^+ ATPase activity (μg phosphate produced/mg protein/h), and concentrations of mRNA and DNA during luteolysis.

Time after $PGF_{2\alpha}$	Na^+/K^+ ATPase activity	mRNA (μg/g)	DNA (mg/g)
0	9.0 ± 0.6	16 ± 3	1.7 ± 0.1
5 min	8.4 ± 0.5	ND	1.7 ± 0.4
10 min	8.2 ± 1.4	21 ± 5	1.9 ± 0.2
20 min	8.1 ± 0.6	18 ± 2	1.7 ± 0.1
40 min	7.4 ± 0.3	16 ± 1	1.5 ± 0.1
1 h	8.0 ± 0.7	20 ± 2	1.8 ± 0.1
2 h	7.1 ± 0.3	21 ± 4	ND
6 h	ND	16 ± 3	1.5 ± 0.2
12 h	6.9 ± 0.3	18 ± 1	1.4 ± 0.1
24 h	6.2 ± 1.4	14 ± 1	1.1 ± 0.2
48 h	5.7 ± 0.8	13 ± 2	1.2 ± 0.3

A significant decrease in Na^+/K^+ ATPase activity occurred at 48 h. No significant alterations were found for mRNA or DNA levels with regression. Each point represents the mean ± SE of 2–3 measurements. Time points marked ND were not determined.

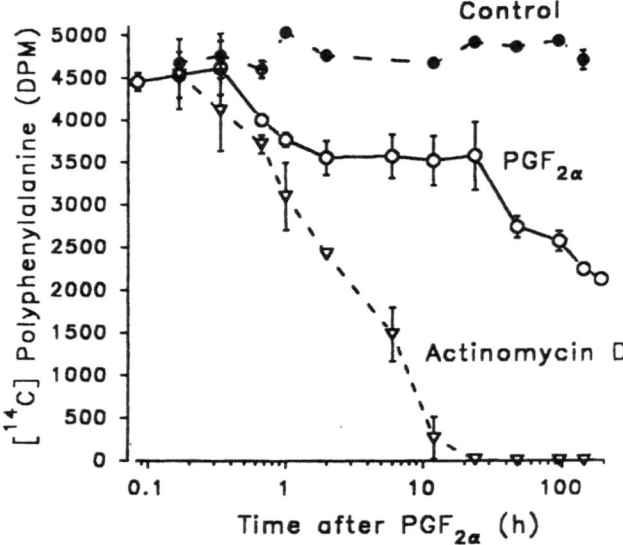

FIGURE 22.2. Alterations in protein synthesis activity in CL of control, $PGF_{2\alpha}$-, and actinomycin D-treated rats. The $PGF_{2\alpha}$ treatment resulted in a biphasic decline starting after 1 h ($P < 0.05$) and after 48 h ($P < 0.05$). Actinomycin D decreased protein synthesis activity 20 min after injection. Each point represents the mean ± SE of 2–3 measurements.

60% of pretreatment activity, appeared by 48 h post-$PGF_{2\alpha}$ ($P < 0.05$). Within 40 min, administration of actinomycin D, an inhibitor of protein synthesis, into the ovarian bursa resulted in both a decline in EF-1α activity ($P < 0.05$) (Fig. 22.2) and a significant reduction in plasma progesterone concentration ($P < 0.05$) (Fig. 22.1). Examination of the protein profile following SDS-PAGE indicated a clear decrease in a protein band with an apparent molecular weight of 49,000 in plasma membrane samples removed 96 h after $PGF_{2\alpha}$ treatment. In contrast, in microsome samples, there was an increase in the amount of a protein with an apparent molecular weight of 35,000 (Fig. 22.3).

Discussion

The activity of EF-1α, a protein that is involved in polypeptide chain elongation (8), was used as a measure of protein formation. Treatment of rats with $PGF_{2\alpha}$ caused a rapid decline in plasma progesterone and a decrease in EF-1α activity within 1 h of administration. This indicates that a decline in protein synthesis activity is an early event associated with the luteolytic process. It is not known whether this decline is due to

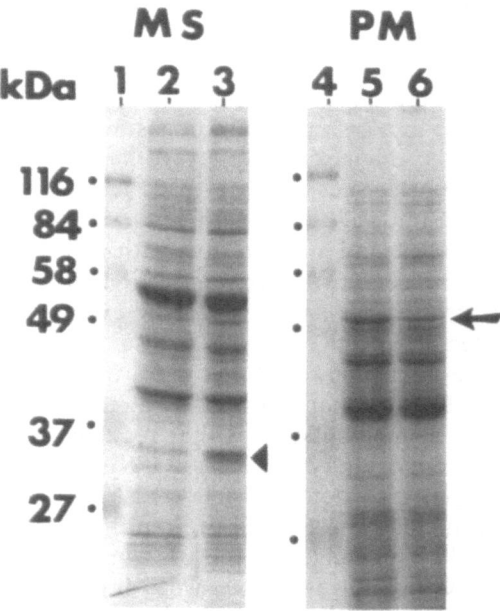

FIGURE 22.3. SDS-PAGE analysis of proteins from microsomes (MS) and plasma membrane (PM) from control and regressing rat CL 4 days after $PGF_{2\alpha}$ treatment. There was an apparent increase in a 35-kd protein band (arrowhead) in microsomes after $PGF_{2\alpha}$ treatment (lane 3) as compared to the nonregressing CL (lane 2). In plasma membrane samples, there was an apparent decrease in a 49-kd protein band (arrow) after $PGF_{2\alpha}$ treatment (lane 6) as compared to the nonregressing control (lane 5). Prestained molecular weight marker proteins appear in lanes 1 and 4, and their corresponding sizes in kilodaltons are shown on the left.

cytoplasmic or nuclear changes. However, no alterations were observed in total mRNA or DNA concentration for 48 h. Thus, $PGF_{2\alpha}$ may be interfering with protein synthesis at a posttranscriptional level.

Changes in ouabain-sensitive Na^+/K^+ ATPase activity were also monitored. This enzyme transports Na^+ and K^+ across the plasma membrane and is required for maintenance of ionic balance. It consumes a considerable proportion of total energy produced by the cell (12), and it could represent a control site for luteolysis. Changes in Na^+/K^+ ATPase activity, however, were not observed until 48 h, which diminishes the possibility that this enzyme is directly involved in CL regression.

Studies of the luteolytic mechanism indicate that $PGF_{2\alpha}$ works at 2 sites. One involves stimulation of cAMP by luteinizing hormone, and the second appears distal to the second messenger (13). Details of the second site(s) are unknown. The present study indicates the possibility of disruption of protein synthesis on the ribosome. Interestingly, inhibition of EF-1α activity by actinomycin D also reduced progesterone secretion similar to $PGF_{2\alpha}$ treatment. Thus, inhibition of protein synthesis could be closely coupled to a disruption in progesterone secretion.

The cause for the changes in protein synthesis activity during luteolysis is unknown. Recent studies indicate that the luteolytic process involves production of free radicals, lipid peroxidation, and H_2O_2 (2, 3, 14, 15). These agents are toxic and may interfere with various intracellular events. Free radicals increase shortly after induction of luteolysis (2) and can inhibit cAMP production and post-cAMP events (16). Interruption of progesterone secretion may occur as a result of production of SOR or H_2O_2, which are damaging, or of the more toxic OH^-, which is formed from SOR and H_2O_2 spontaneously or by an iron-catalyzed reaction. The presence of iron at metabolic steps that regulate progesterone synthesis may provide the opportunity for site-specific attack (17).

Interestingly, we also detected a decrease in the amount of a 49-kd protein and an increase in a 32-kd protein in membrane samples of the regressing CL. These changes in the protein pattern became very clear after 96 h post-$PGF_{2\alpha}$; however, they were already apparent after 24 h. It is not known whether these proteins are directly involved in the luteolytic process. However, since it is known that specific cytosolic proteins are required for free-radical generation in human neutrophils (18, 19), it could well be that the protein changes observed in our study after $PGF_{2\alpha}$ administration are involved in free-radical formation in luteal cells. In summary, this study demonstrates that CL regression is associated with a significant decrease in protein synthesis activity within 1 h of $PGF_{2\alpha}$ treatment and later changes in membrane proteins.

References

1. Baum MS. Prostaglandin $F_{2\alpha}$ administered in vivo induces Ca^{2+}-dependent protein phosphorylation in rat luteal tissue. Endocrinology 1989;124:555–7.
2. Sawada M, Carlson JC. Rapid plasma membrane changes in superoxide radical formation, fluidity and phospholipase A_2 activity in the corpus luteum of the rat during induction of luteolysis. Endocrinology 1991;128:2992–8.
3. Sawada M, Carlson JC. Superoxide radical production in plasma membrane samples from regressing rat corpora lutea. Can J Physiol Pharmacol 1989; 67:465–71.
4. Wu XM, Carlson JC. Alterations in phospholipase A_2 activity during luteal regression in pseudopregnant and pregnant rats. Endocrinology 1990; 127:2464–8.
5. Bradford MM. A rapid and sensitive method for the quantification of microgram quantities of protein utilizing the principle of protein-dye binding. Anal Biochem 1976;72:248–54.

6. Chomczynski P, Sacchi N. Single-step method of RNA isolation by acid guanidinium thiocyanate-phenol-chloroform extraction. Anal Biochem 1987; 162:156–9.

7. Aviv H, Leder P. Purification of biologically active globin messenger RNA by chromatography on oligothymidylic acid-cellulose. Proc Natl Acad Sci USA 1972;69:1408–12.

8. Crechet JB, Canceill D, Bocchini V, Parmeggiani A. Characterization of the elongation factors from calf brain, 1. Purification, molecular and immunological properties. Eur J Biochem 1986;161:635–45.

9. Kirby KS. A new method for the isolation of deoxyribonucleic acids: evidence on the nature of bonds between deoxyribonucleic acid and protein. Biochem J 1957;66:495–504.

10. Cole CH, Waddell RW. Alteration in intracellular sodium concentration and ouabain-sensitive ATPase in erythrocytes from hyperthyroid patients. J Clin Endocrinol Metab 1976;42:1056–63.

11. Laemmli UK. Cleavage of structural proteins during the assembly of the head of bacteriophage T4. Nature 1970;227:680–5.

12. Hulbert A, Else PL. Comparison of the "mammal machine" and the "reptile machine": energy use and thyroid activity. Am J Physiol 1981;241:R350–5.

13. Jordan AW. Effects of prostaglandin $F_{2\alpha}$ treatment on LH and dibuyrul cyclic AMP-stimulated progesterone secretion by isolated rat luteal cells. Biol Reprod 1981;25:327–33.

14. Sawada M, Carlson JC. Association of lipid peroxidation during luteal regression in the rat and natural aging in the rotifer. Exp Gerontol 1985; 20:179–86.

15. Riley J, Behrman HR. In vivo generation of hydrogen peroxide during luteolysis. Endocrinology 1991;128:1749–53.

16. Gatzuli E, Aten RF, Behrman HR. Inhibition of gonadotropin action and progesterone sythesis by xanthine oxidase in rat luteal cells. Endocrinology 1991;128:2253–8.

17. Fridovich I. Biological effects of the superoxide radical. Arch Biochem Biophys 1986;247:1–11.

18. Volpp BD, Nauseef WM, Clark RA. Two cytosolic neutrophil oxidase components absent in autosomal chronic granulomatous disease. Science 1988;242:1295–7.

19. Nunoi H, Rotrosen D, Gallin JI, Malech HL. Two forms of autosomal chronic granulomatous disease lack distinct neutrophil cytosol factors. Science 1988;242:1298–1301.

23

Detection of Relaxin Secretion and Gene Expression in Individual Porcine Luteal Cells

Carol A. Bagnell, Kathleen Ohleth, Cheryl L. Clark, and Michael J. Taylor

Although large luteal cells of pregnancy contain immunoreactive *relaxin* (RLX), only about 50% of the cells release RLX in vitro in the *reverse hemolytic plaque assay* (RHPA) (1). Whether the heterogeneity in RLX secretion by subpopulations of luteal cells can be explained by differences in RLX gene expression is unknown. In situ evidence indicates the RLX gene is expressed by the corpus luteum of pregnancy, but whether a specific subset of luteal cells contains RLX transcript was not determined (2). In this study, RHPA, followed by *in situ hybridization histochemistry* (ISHH), was used to study the relationship between RLX secretion and gene expression, respectively, by the same luteal cell. These techniques have been used to measure peptide secretion and gene expression in individual pituitary lactotrophs (3). The objective here was to determine whether differences exist between RLX secreting and nonsecreting luteal cells in terms of expression of RLX message.

Methods

Reverse Hemolytic Plaque Assay

Ovaries from pregnant pigs (days 30–35 of gestation; term is day 114 ± 2) were collected within 30 min of death at the Meat Laboratory, Iowa State University. Luteal cell dispersion and RHPA methodology were as described by Taylor et al. (1). In this procedure, monodispersed luteal cells (0.25×10^6 large cells/mL) were mixed 1:1 with a 12% suspension of ovine red blood cells to which protein A had been chemically attached. This mixture was infused into Cunningham chambers by capillary action and rinsed 45 min later to remove unattached cells. After overnight incubation, the monolayers were washed again and chambers filled with DMEM-0.1% BSA containing 1:80 porcine RLX antiserum and *prostaglandin E_2* (PGE$_2$) (0.01 µM). The monolayers were incubated with PGE$_2$ (a stimulator of RLX secretion) to ensure that maximal plaque formation was complete in 3 h of incubation (4). Chambers were then filled with guinea pig serum as a source of complement (1:40) and incubated for 50 min to allow plaque formation. Slides were fixed in 4% paraformaldehyde in *phosphate-buffered saline* (PBS) at 4°C overnight, air-dried, rinsed in 0.1 M phosphate buffer, and dried and stored at −70°C. In this study, a *plaque* was defined as a minimum of a single, concentric ring of lysed red blood cells around a luteal cell and served as an indicator of RLX secretion by the cell. The specificity of the porcine RLX antibody has been characterized (5). Plaque formation was abolished by preabsorption of RLX antibody with purified porcine RLX (10-µg RLX/mL combined with 1:80 RLX antibody) and by omission of RLX antibody or complement.

In Situ Hybridization

Relaxin gene expression in luteal cells was detected using 48-mer oligonucleotide probes complementary to 3 different regions of the porcine prorelaxin molecule and previously used to detect RLX mRNA in sections of porcine luteal tissue (2). The probes were labeled with [α-^{35}S]dATP (>1300 Ci/mmol) at the 3′ end using terminal deoxynucleotidyl transferase. To check the specificity of hybridization, a 48-mer oligomer to a region of the human *prolactin* (PRL) molecule was used as a negative control. Hybridization was detected as described by Bagnell et al. (2) with modifications (3). Slides from RHPA were thawed, washed twice in PBS, acetylated to reduce nonspecific hybridization, and dehydrated in graded alcohols. Slides were incubated with [^{35}S]-labeled probe (3–5 × 10^5 dpm/50 µL/section) in hybridization buffer overnight at 37°C. Following formation of the oligomer-mRNA hybrid, slides were washed twice (15 min each) in a solution containing 0.15 M NaCl and 0.015 M sodium citrate (1× SSC) at 40°C followed by two 15–min washes in 0.1× SSC at 40°C. This was followed by two 30-min washes in 0.1× SSC and one 10 min wash in 1× SSC at room temperature. Slides were dehydrated in graded alcohols, dipped in NTB-2 emulsion, and exposed for 1–2 weeks. After photographic development, slides were stained with 4% methyl green. Monolayers were viewed microscopically to determine the percentage of (1) plaque- (i.e., RLX-releasing) and nonplaque-forming *large luteal cells* (LLC) and (2) LLC with autoradiographic grains indicating hybridization and presence of mRNA. The diameter of the luteal cells counted was greater than 25 µm, corresponding to the size of LLC (6). Small luteal cells were excluded from cell counts since plaque formation is restricted to LLC (1). Results are expressed as mean ± SEM. Statistical differences between 2 means were determined by Student's t-test.

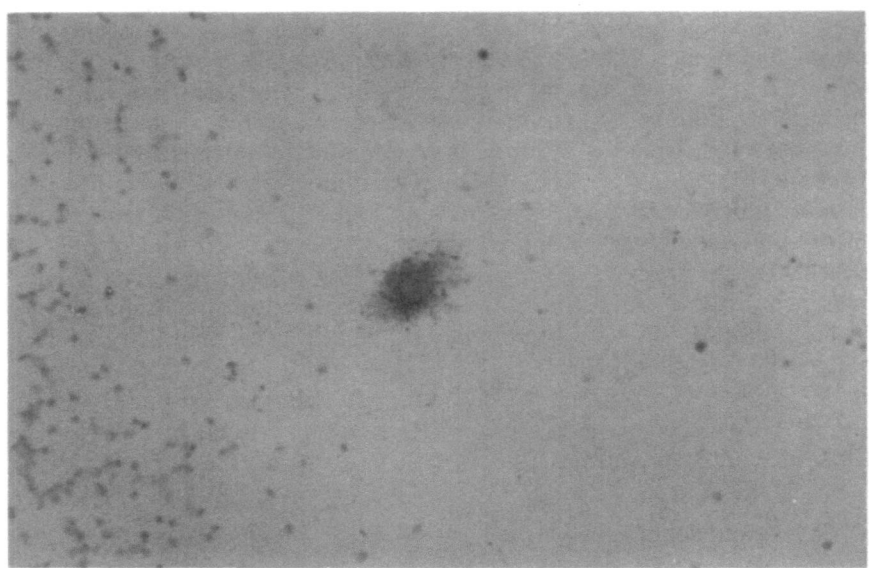

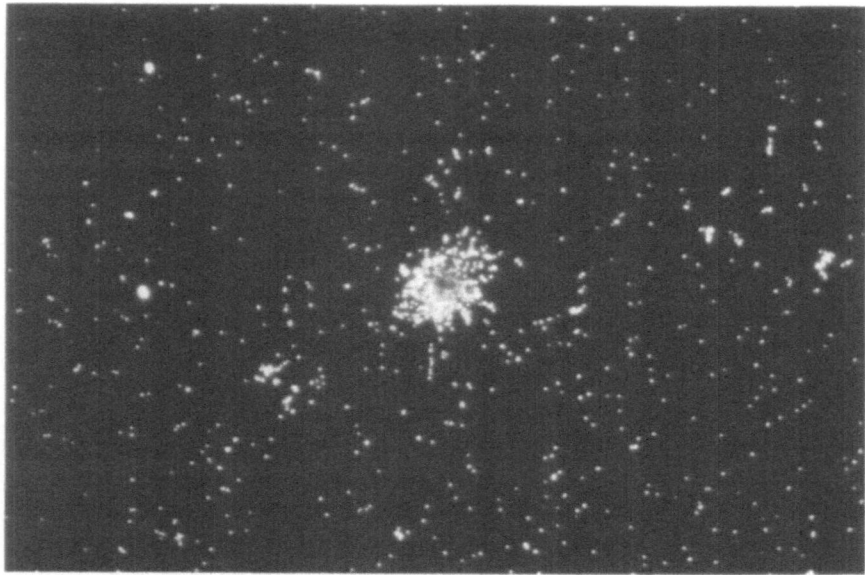

FIGURE 23.1. In situ localization of RLX mRNA in porcine luteal cell monolayers using a [³⁵S]-labeled pRLX probe following RHPA. Top: Bright-field photomicrograph of a plaque-forming luteal cell indicating RLX secretion. Note black grains concentrated over the cell, demonstrating the presence of RLX mRNA. Bottom: Dark-field view of the same cell shown at top. Hybridization to detect RLX mRNA is indicated by silver grains that appear as clusters of white grains over the luteal cell. Note intense labeling of the luteal cell in comparison to background (450×).

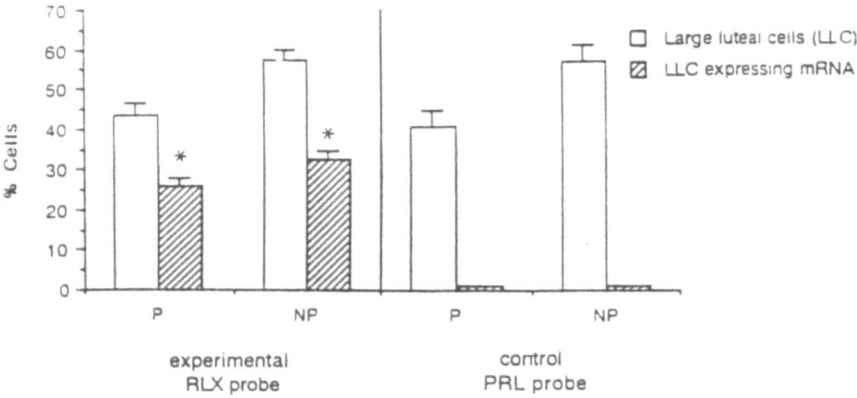

FIGURE 23.2. Percentages of porcine luteal cell subtypes as identified by the combined use of the RHPA and ISHH. Following RHPA, monolayers were incubated with either a [³⁵S]-pRLX (experimental) or hPRL (control) probe. After a 2-week exposure to emulsion, LLC were counted and scored in terms of the percentage of LLC (1) that were plaque (P) or nonplaque (NP) formers and (2) with autoradiographic grains indicating the presence of RLX mRNA. Values are the mean ± SEM of 6 slides/probe tested from a representative experiment. The percentage of RLX mRNA containing cells in both P and NP cell groups was significantly lower than total P and NP cells (P < 0.001).

Results

Figure 23.1 illustrates bright- and dark-field views of an RLX-secreting cell (i.e., plaque-forming cell) following combined RHPA/ISHH. Auto-radiographic grains that appear black in the bright-field view (Fig. 23.1, top) and white in the dark-field view (Fig. 23.1, bottom) are an indicator of hybridization to RLX mRNA. Grains were found in close association with both plaque- (shown here) and nonplaque-forming LLC. However, RLX mRNA was not detectable in all LLC. Figure 23.2 shows that although 43.9% of LLC formed plaques, of the monolayers incubated with a labeled pRLX probe, only 30.5% ± 5.0% of the total LLC both formed plaques and expressed RLX message. Similarly, whereas 56.1% of the LLC were non-RLX secreting cells (i.e., nonplaque-forming cells), only 32.6% ± 4.5% of total LLC were nonplaque-forming cells that also expressed RLX message. Control monolayers incubated with a labeled heterologous probe to hPRL showed the expected percentage of plaque-forming cells (41.7%); however, none of the cells hybridized to the hPRL probe (Fig. 23.2). When RLX gene expression in the 2 luteal cell subtypes was compared, the proportion of plaque-forming cells containing mRNA (62.4 ± 3.9) was not significantly different from the proportion of nonplaque cells containing mRNA (58.3 ± 4.0). Thus, no correlation between RLX secretion by individual cells and the presence or absence of RLX mRNA was found.

Discussion

Recently, Scarbrough (3) reported use of the RHPA in combination with in situ hybridization to detect PRL secretion and gene expression in individual pituitary lactotrophs. In this study, we have applied these techniques to the luteal cell system to demonstrate RLX secretion and gene expression by individual cells. Both plaque- and nonplaque-forming cells expressed RLX mRNA, indicating that in luteal cells RLX gene expression and hormone secretion are independent events. A similar phenomenon was reported in pituitary lactotrophs in that no correlation was found between the amount of PRL secretory activity and level of PRL gene expression (3).

The basis for this lack of correlation between gene expression and secretion is unknown. However, Wise and colleagues (3) suggest that a temporal dissociation between hormone secretion and transcription may exist if, for example, changes in the rate of secretion occur more rapidly than levels of mature mRNA. In addition, we have found that although RLX has been localized in LLC of the pregnant sow by immuno-histochemistry (7), only 58%–62% of LLC express RLX message. The absence of a transcript in the presence of protein product may be due to a variety of factors. For example, if a hormone is synthesized then stored in secretory granules prior to release, as is the case for RLX, the presence or absence of mRNA in the same cell would depend on the stability of the transcript. In addition, it is possible that our inability to detect RLX transcript in all LLC may be due to an insufficiently sensitive detection system. Since the concentration of RLX transcript may differ markedly between individual cells, we are planning to address this issue by using RNA probes that are reported to improve the sensitivity of detection. In conclusion, we have employed the plaque assay in combination with ISHH to study RLX secretion and gene expression in individual luteal cells. These studies indicate that posttranscriptional events are involved in determining the RLX secretory response of the luteal cell.

Acknowledgments. The authors would like to thank Dr. O.D. Sherwood for providing the porcine relaxin antiserum and Drs. W.S. Young III and M.J. Brownstein, NIMH, for providing the oligonucleotide probes. This work was supported by the New Jersey Agriculture Experiment Station (publication F-06115-1-91) and NIH HD-20624 (C.A.B.) and NIH HD-22786 (M.J.T.).

References

1. Taylor MJ, Clark CL, Frawley LS. Analysis of relaxin release from cultured porcine luteal cells by reverse hemolytic plaque assay: influence of gestational age and prostaglandin $F_{2\alpha}$. Endocrinology 1987;120:2085–91.
2. Bagnell CA, Tashima L, Tsark W, Ali S, McMurtry JP. Relaxin gene expression in the sow corpus luteum during the cycle, pregnancy and lactation. Endocrinology 1990;126:2514–40.
3. Scarbrough K, Weiland NG, Larson GH, et al. Measurement of peptide secretion and gene expression in the same cell. Mol Endocrinol 1991;5:134–42.
4. Taylor MJ, Clark CL. Detection of relaxin release by porcine luteal cells using a reverse hemolytic plaque assay: effect of prostaglandin E_2 and $F_{2\alpha}$, human chorionic gonadotrophin and oxytocin. Biol Reprod 1987;37:377–84.
5. Sherwood OD, Rosentreter KR, Birkhimer ML. Development of a radio-immunoassay for porcine relaxin using ^{125}I-labeled poly-tyrosyl-relaxin. Endocrinology 1975;96:1110–4.
6. Lemon M, Loir M. Steroid release in vitro by two luteal cell types in the corpus luteum of the pregnant sow. J Endocrinol 1977;72:351–9.
7. Fields PA, Fields MJ. Ultrastructural localization of relaxin in the corpus luteum of the nonpregnant, pseudopregnant and pregnant pig. Biol Reprod 1985;32:1169–79.

24

Oxytocin Receptors in Myometrium: Preliminary Evidence for a Receptor Switch Mechanism

Vladimir Pliška, Hildegard Kohlhauf Albertin, and Denis J. Crankshaw

The phasic contractile activity of mammalian uterus is controlled by several hormones, among them oxytocin. *Oxytocin* (OT) increases contractile activity in a rapid, concentration-dependent manner. This effect is of importance under certain physiological circumstances, most obviously during delivery, and a precise control of the onset and cessation of OT action is a necessary feature of this regulatory system. How can this control be achieved?

Two regulatory pathways are feasible. The primary pathway operates via pituitary secretion, tissue compartmentalization, and inactivation of the hormone and regulates the OT concentration in the vicinity of its target tissue receptors. Out of these processes, only pituitary secretion responds instantaneously to various exteroceptive and interoceptive stimuli. Therefore, the onset of the response, rather than its cessation, is efficiently controlled. The secondary pathway exercises control over the OT response at the level of the interacting cells. Direct or circumstantial evidence speaks in favor of up- and down-regulation processes in OT action. The up-regulation may proceed via an increasing number (and/or affinity) of receptors, most likely controlled by ovarian steroids, notably during estrous cycles (1) or close to the term (2). On the other hand, receptors may be readily down-regulated by internalization or by an as yet unclarified switch-on-and-off mechanism in case of an exaggerated tissue response; for example, as a consequence of an excess of the hormone at the receptors. Thus, myometrial contractility via a system involving OT may be regulated in the long term by control of the sensitivity of the uterus to the hormone (receptor regulation), in the short term by control of the release of the hormone, and in the very short term by a receptor switch—or a similar—mechanism. Over the past few years, we have collected some evidence that a receptor switch operating between 2 thermodynamic states of a single receptor may indeed explain some of the observed phenomena (3). The problem is discussed in this chapter.

Oxytocin Binding Sites on Myometrial Cells

Steady State Conditions

A high-affinity site with a dissociation constant K_d of about 2×10^{-9} mol/L was identified in the myometrium of various species by early binding studies (2, 4, 5). More recently, utilization of [^{3}H]oxytocin with variable specific radioactivity (isotopic dilution) and introduction of considerably improved computational techniques in the place of the inadequate—and misused—Scatchard plot, led to the detection of another saturable, medium-activity site with K_d between 2 and 5×10^{-7} mol/L. The 2 sites were observed on both cultured myometrium cells (6) and on cell membranes (7) of sheep and rat. The high-affinity site displays the Hill coefficient of about 2, indicating probably a positively cooperative binding mechanism. A detailed analysis has shown that magnesium, which was claimed to enhance OT-receptor binding considerably (4), has a predominant influence upon the cooperativeness, with less effect on the K_d and receptor density (both in nonmonotonic manner) of the high-affinity site (8). The effects of magnesium on the medium-affinity, non-cooperative site seem to be much less pronounced. In terms of receptor density, the ratio of high- to medium-activity sites is approximately 1:20. None of these features in the cultured cells were changed when 1–3 μg/mL of estradiol acetate was added into the culture media for 3 days preceding the binding experiment (experimental conditions were described earlier [6]; Kohlhauf Albertin, Pliška, unpublished results).

Under conditions of steady state binding experiments, the 2 binding sites appear as separate entities. However, their possible interdependence cannot be detected in this way.

Time Course of OT Binding

Assuming that the binding of the hormone H on 2 binding sites R_1 and R_2 (see above) follows the canonic scheme

$$H + R_1 \rightleftharpoons X_1$$
$$H + R_2 \rightleftharpoons X_2$$

(X_1 and X_2 are corresponding hormone-receptor complexes), the rise of bound ligand concentration should follow a simple exponential. However, this is not so in the case of OT binding to myometrial cells in culture. Figure 24.1 shows that the time-binding curves are triphasic. The binding is initially rather rapid and reaches a peak within 2–5 min; then follows a

c_b(rel.)

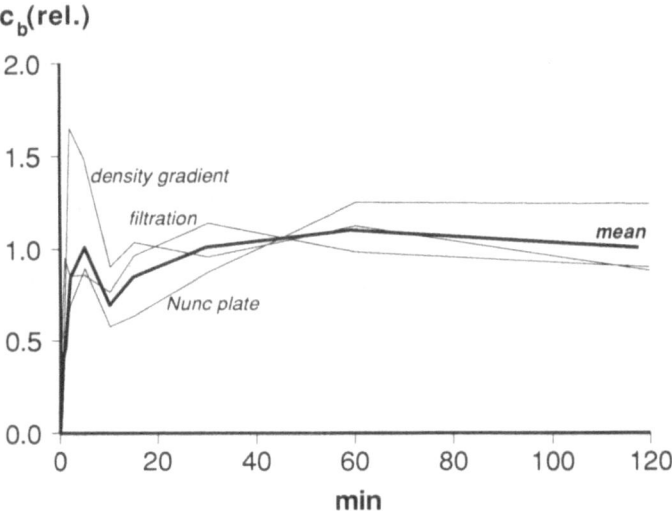

FIGURE 24.1. Time course of OT binding to cultured sheep myometrial cells. Three techniques were utilized. The first two are (i) filtration of the cell suspension after incubation with tritium-labeled OT through the Millititer filtration plates (Millipore), pore size 5 μm; and (ii) separation of the cells in the same suspension in a density gradient (1-vol dibutyl phthalate +1.5-vol bis[2-ethylhexyl] phthalate), which causes a rapid separation of the liquid phase from the cells. After centrifugation, the cells were collected as a pellet, and their radioactivity was measured; (iii) incubation was carried out with cell cultures grown on cover slips placed in Nunc plates without suspending the cells. Radioactivity of each cover slip was measured after denaturation in 0.1% sodium dodecylsulphate. All incubations were at 22°C; OT concentration was 10^{-7}M. Each measurement was repeated 4 times.

rebound phase during which the binding drops significantly; and, finally, there is a second increase that reaches a plateau after 60 min. Similar time profiles were obtained in our laboratory with myometrial cell membranes. In nonoscillating systems, such a concentration overshoot can appear if and only if a transition from one steady state to another occurs in the course of the process (9). It is likely that the system follows an alternative scheme:

$$H + R_1 \underset{K_1}{\rightleftharpoons} X_1 \overset{k_{12}}{\underset{k_{21}}{\rightleftharpoons}} X_2 \underset{K_2}{\rightleftharpoons} R_2 + H$$

where K_1 and K_2 are equilibrium dissociation constants and k_{12} and k_{21} are rate constants of the "receptor switch." The triphasic time course suggests that at least 2 rate parameters are mutually independent functions of time, with rather different time constants. One can hypothesize that the time changes of these coefficients are due to a feedback regulation and that they are controlled either by the steps R_1, R_2, X_1, and X_2 or, equally plausibly, by a futher step of the cellular response. (Conceivable steps would be, for instance, concentrations of second messengers; modulators of intracellular response, such as GTP; and the like.) To consider a particular feedback mechanism would be, at the present state of knowledge, farfetched. We can only infer from these profiles that the 2 receptor sites identified in equilibrium experiments are interdependent and that the basis of this interdependence is probably a switch from one steady state to another.

Inositol Phosphates Generation in Myometrial Cells: Effect of OT

It has been recently demonstrated that OT stimulates inositol phospholipid hydrolysis in the myometrial cells (10–12); products of this process—*inositol mono-, di-, and triphosphates* (InsP1, InspP2, and

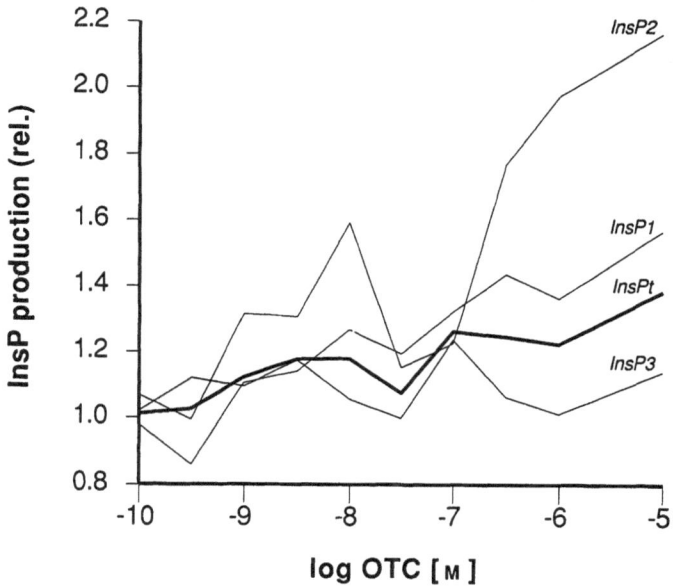

FIGURE 24.2. Generation of inositol phosphates (abbreviations cf. text) in cultured sheep myometrial cells (14 days) by OT (22°C). The ordinate is the concentration of inositol phosphates relative to preincubation values for each moiety. For assays of individual InsPs, the protocol by Berridge et al. (16) was followed. Individual measurements were repeated 2–10 times. The abscissa is the logarithm of OT concentration (OTC).

InsP3, respectively)—act as second messengers in the uterotonic response. The OT concentration that elicits the half-maximal production of the *total inositol phosphates* (InsPt) in rat uterine tissue was estimated from the published data (12) as 5×10^{-9} M. We have carried out similar experiments with sheep myometrial cells grown on a Nunc plate for 2 weeks. The concentration range of OT was extended from 10^{-10} to 10^{-5} M. We could verify the results mentioned above, even though the production of inositol phosphates after 1-h incubation was considerably lower in the cultured cells. The highest percentage increase was registered for InsP2. This rise, as well as the one of InsPt, was extended over a broad concentration range, with a "surge" at about 10^{-8} M OT and a second branch with a half-maximal concentration of about 3×10^{-7} M (Fig. 24.2). At present, we have no definitive interpretation of this profile. It is, however, noteworthy that the profile resembles those of the binding isotherms (see above). The range of greatest production of inositol phosphates is identical to that of the medium-affinity site.

Response of Uterus to OT

Finally, a role of the 2 receptors was demonstrated pharmacologically. When an isolated rat uterus is treated stepwise by an irreversible inhibitor of OT, the uterotonic response displays the Nickerson shift (3, 13), a phenomenon ascribed, not quite unambiguously, to the so-called spare receptors: Up to a certain degree, the receptor inactivation causes a shift of the half-maximal OT concentration from about 10^{-9} mol/L in the noninhibited uterus to higher values, but not—or not visibly—the expected decrease of maximal uterotonic response. Applying the Furchgott-Bursztyn method (14) based on comparison of equipotent doses on noninhibited and partially inhibited uteri, K_d values and degrees of inactivation (not measurable a priori) can be computed. The result is quite astonishing: In a narrow range of "mild" irreversible receptor blocking, the K_d values are close to those of the high-affinity binding site, but below this narrow range, they are close to the medium-affinity site. This latter site is the one to which structural analogs of OT acting as competitive inhibitors of uterotonic response to OT are bound (15).

Conclusion

The 2 binding sites for OT found on myometrial cells of various species seem to be interdependent and parts of a "receptor switch." The function of such a switch is most likely a down-regulation of uterotonic response, required in physiological states in which OT action may take an undesired dominance or in which a rapid cessation of the response is necessary.

Acknowledgments. The support by the Swiss National Science Foundation, by the FFVFF Foundation, Zürich, and by the Medical Research Council of Canada is gratefully acknowledged. Our thanks are due to Dr. Urs. T. Rüegg and Mr. J. Peters for their assistance with the assays of inositol phosphates.

References

1. Soloff MS, Fields MJ. Changes in uterine oxytocin receptor concentrations throughout the estrous cycle of the cow. Biol Reprod 1989;40:283–7.
2. Fuchs A-R, Fuchs F, Husslein P, Soloff MS, Fernström MJ. Oxytocin receptors and human parturition: a dual role for oxytocin in the initiation of labor. Science 1982;215:1396–8.
3. Pliška V. Pharmacological versus binding analysis of receptor systems: how do they interplay? Myometrial cell receptors for oxytocin as a paradigm. Experientia 1991;47:216–21.
4. Soloff MS, Swartz TL. Characterization of a proposed oxytocin receptor in the uterus of the rat and sow. J Biol Chem 1974;249:1376–81.
5. Crankshaw DJ, Branda LA, Matlib MA, Daniel EE. Localization of the oxytocin receptor in the plasma membrane of rat myometrium. Eur J Biochem 1978;86:481–6.
6. Pliška V, Heiniger J, Müller-Lhotsky A, Pliska P, Ekberg B. Binding of oxytocin to uterine cells in vitro. Occurence of several binding site populations and reidentification of oxytocin receptors. J Biol Chem 1986;261: 16984–9.
7. Crankshaw DJ, Gaspar V, Pliška V. Multiple [^{3}H]-oxytocin binding sites in rat myometrial plasma membranes. J Recept Res 1990;10:269–85.
8. Pliška V, Kohlhauf Albertin H. Effect of magnesium on the binding of oxytocin to sheep myometrial cells. Biochem J 1991;277:97–101.
9. Burton AC. The properties of steady state compared to those of equilibrium as shown in characteristic biological behaviour. J Cell Comp Physiol 1939; 14:327–49.
10. Marc S, Leiber D, Harbon S. Carbachol and oxytocin stimulate the generation of inositol phosphates in the guinea pig myometrium. FEBS Lett 1986; 201:9–14.
11. Schrey MP, Read AM, Steer PJ. Oxytocin and vasopressin stimulate inositol phosphate production in human gestational myometrium and decidua cells. Biosci Rep 1986;6:613–9.
12. Ruzycky AL, Crankshaw DJ. Role of inositol phospholipid hydrolysis in the initiation of agonist-induced contractions of rat uterus: effect of domination by 17β-estradiol and progesterone. Can J Physiol Pharmacol 1988;66:10–7.
13. Pliška V. Pharmacological approaches to the identification and classification of myometrial oxytocin receptors. J Receptor Res 1988;8:245–59.
14. Furchgott RF, Bursztyn P. Comparison of dissociation constants and of relative efficacies of selected agonists acting on parasympathetic receptors. Ann NY Acad Sci 1967;144:882–99.

15. Pliška V, Heiniger J. Structural requirements of the oxytocin receptor in rat uterus. Free-Wilson analysis in a series of competitive oxytocin inhibitors. Int J Pept Protein Res 1988;31:520–36.
16. Berridge MJ, Downes CP, Hanley MR. Lithium amplifies agonist-dependent phosphatidylinositol responses in brain and salivary glands. Biochem J 1982; 206:587–95.

25

GnRH- and cAMP-Stimulated Human Chorionic Gonadotropin Secretion from Perifused Placental Cells

W. David Currie, Gillian L. Steele, Basil Ho-Yuen, and Peter C.K. Leung

Gonadotropin releasing hormone (GnRH) stimulates *human chorionic gonadotropin* (hCG) secretion in a specific, dose-dependent manner in vitro (1–8). GnRH in maternal circulation and stimulating hCG secretion may be primarily of placental origin; GnRH synthesized and secreted by cytotrophoblast stimulates hCG synthesis and secretion by syncytiotrophoblast (1, 3, 7–13). A component of basal hCG secretion may be GnRH independent (9). Cultured placental tissue obtained throughout gestation responds to GnRH, but may be most responsive at about 10 weeks (13–15). Prior to obtaining data presented, repeated attempts were made to demonstrate GnRH-stimulated hCG secretion from perifused placental cells. Initially, trophoblast dissected from placenta in buffer or culture media on ice were dissociated with collagenase (1 h, 37°C). Hematocytes were removed and cells plated as detailed below. Cell viability was limited (<65%). Within 2–5 days, basal hCG secretion was evident (>10-mIU/mL media/h), and cells responded to 24-h GnRH (10^{-9} M) exposure in static culture, but not to 5-min GnRH pulses (10^{-9} M) in perifusion. Subsequently, enzyme-free physical dissociation at 21°C has continuously provided cultures of greater viability (>90%) and allowed hCG secretion in response to GnRH pulses in perifusion. This study examined basal, cAMP-, and GnRH-stimulated hCG secretion from term and first-trimester placental cells cultured for 4–5 days and basal and GnRH-stimulated hCG secretion from first-trimester placental cells cultured for 12–14 days.

Materials and Methods

Cell Preparation

Sterile conditions were maintained throughout. Term placentae were
from elective cesarean section, first-trimester placentae from therapeutic
abortion at 8–12 weeks. Membranes were removed and trophoblast

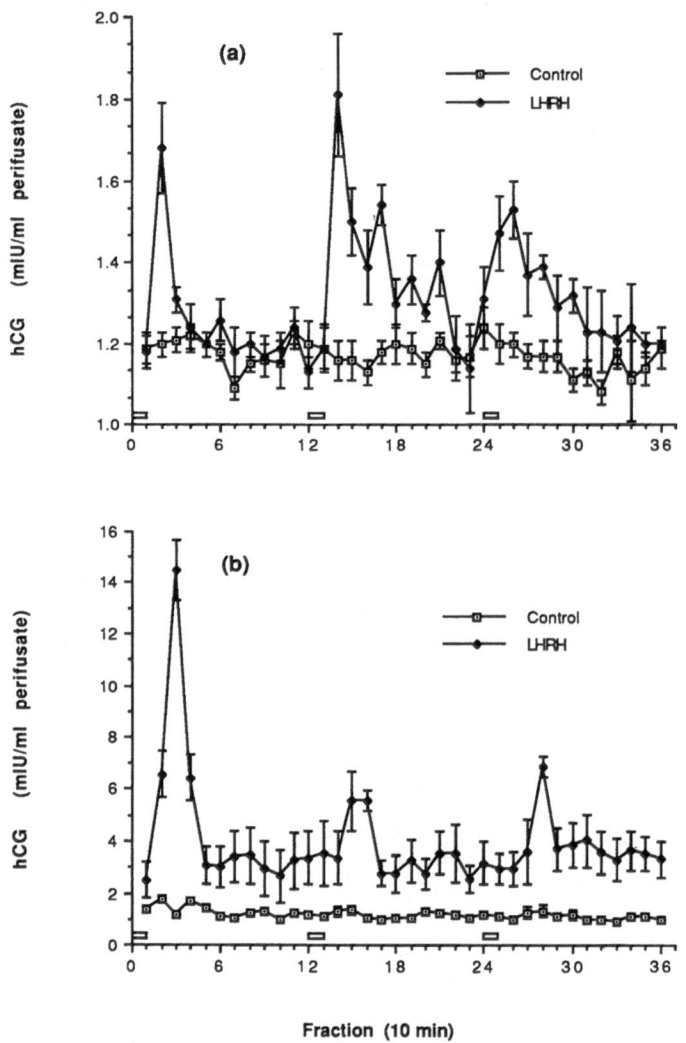

FIGURE 25.1. hCG concentrations (mean ± SE) in perifusate from (a) term
trophoblast (n = 4 chambers) and (b) first-trimester placental cells (n = 4
chambers) treated with GnRH (10^{-9} M) at 2-h intervals. Bars over horizontal
axes indicate GnRH treatment periods.

dissected from term decidua. Term trophoblast or first-trimester decidua were dissociated in Media 199 (1% v/v fetal calf serum, 100-U penicillin G sodium and 100-µg streptomycin/mL, 25 mM NaHCO$_3$, 15 mM Hepes; supplemented M199, Gibco, Burlington, Ontario) through a 150 µ-screen (Sigma, St. Louis, MO). Suspensions were filtered (48 µ-mesh, B&SH

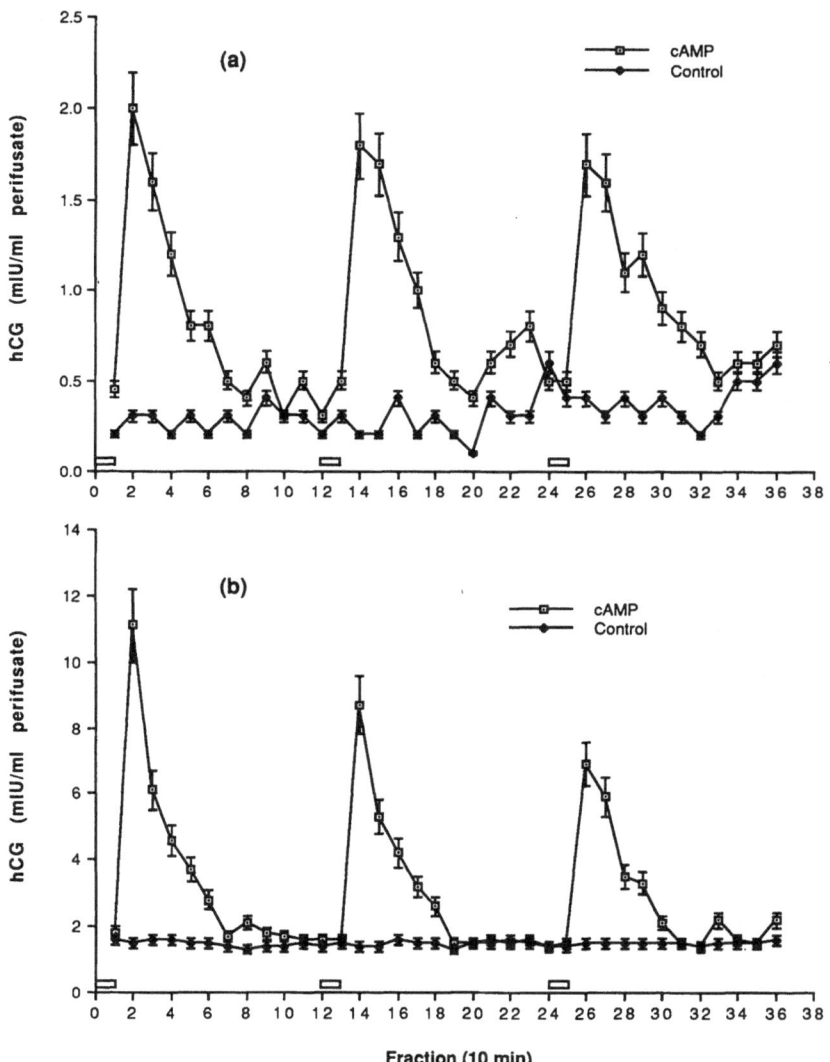

FIGURE 25.2. hCG concentrations (mean ± SE) in perifusate from (a) term trophoblast and (b) first-trimester placental cells treated with cAMP (10^{-9}M; n = 4 chambers each). Bars over horizontal axes indicate GnRH treatment periods.

Thompson, Montreal, Quebec) and hematocytes removed by centrifuging (1700 × g for 20 min) the filtrate on 40% Percoll in Hanks' balanced salt solution (Gibco). Cell viability exceeded 90%. Cells from 2–4 term or first-trimester placentae were mixed and plated on carrier beads (~1 × 10^6 cells/19-mg beads, Cytodex-3, Sigma) at 37°C in humidified air with 5% CO_2.

Perifusion and Sampling

Beads were loaded into chambers (1.5 × 10^6 cells/1.5-mL chamber, Endotronics, Minneapolis, MN) in a 36°C water bath 24 h pretrial. Perifusion was with supplemented M199. Beginning 1 h before and during trials, chambers were perifused at 15 mL/h. Between trials, perifusion was at 5 mL/h for 6 h or more.

Trial 1: 4- to 5-Day Cultures

Control and treatment perifusions were over consecutive 6-h periods. Four chambers with cells from 4 different term preparations and 4 chambers with cells from 4 different first-trimester preparations were perifused with cAMP or GnRH (10^{-9} M, Sigma) for the first 5 min of each 2-h period during the treatment perifusion. Effluent was collected in 10-min fractions during GnRH (Fig. 25.1) and cAMP treatment (Fig. 25.2).

Trial 2: 4- to 5-Day Cultures

Control and treatment perifusions were over consecutive 3-h periods. Four chambers with cells from 4 different first-trimester preparations were perifused with GnRH (10^{-9} M) for the first 5 min of each 1 h-period during treatment perifusion. Effluent was collected in 5-min fractions (Fig. 25.3).

Trial 3: 4- to 5-Day and 12- to 14-Day Cultures

Control and treatment perifusions were over consecutive 1-h periods. Fifteen chambers with cells from 5 different first-trimester preparations cultured for 4–5 days or 12–14 days were perifused with GnRH (10^{-9} M) for the first 5 min of the treatment perifusion. Effluent was collected in 5-min fractions (Fig. 25.4).

hCG Assay

Standards were assayed in quadruplicate and samples were assayed in duplicate using diagnostic reagents (Serono, Allentown, PA). Mixed mouse monoclonal β-hCG antibodies labeled with ^{125}I and fluorescein were added (100 μL) to 50-μL samples and incubated (24 h, 4°C).

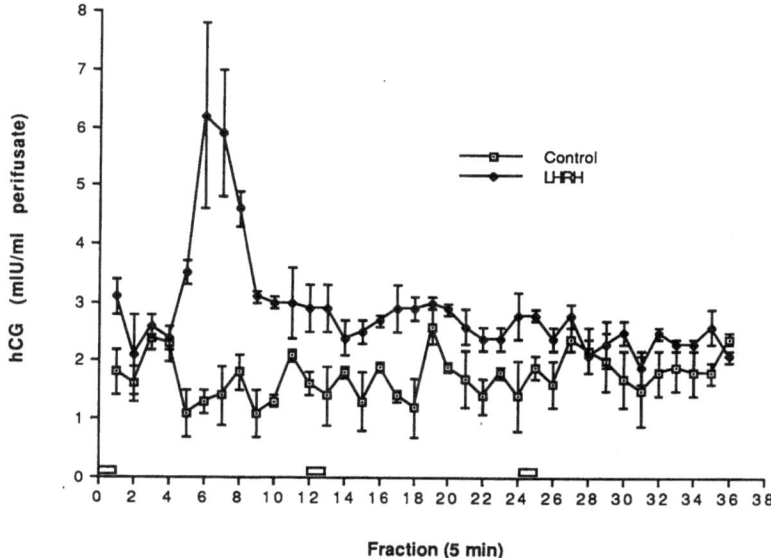

FIGURE 25.3. hCG concentrations (mean ± SE) in perifusate from first-trimester placental cells treated with GnRH (10^{-9}M) at 1-h intervals (n = 4 chambers). Bars over horizontal axes indicate GnRH treatment periods.

Sheep anti-fluorescein covalently bound to magnetic particles was added (75 μL), and the assay incubated (24 h, 4°C). Supernatant was decanted with magnetic racks (Serono). Pellets were washed with Tris buffer (10 mM, 500 μL) and counted for 1 min (LKB γ-counter, Turku, Finland). Concentrations of hCG determined by spline function are expressed in terms of 1st IRP/3rd IS 75/537. Assay sensitivity was 0.5-mIU/mL media (by Student's t-test, P < 0.05). Intra- and interassay coefficients of variation for pregnancy serum diluted in M199 were (1) 7.1% and 9.9%, 7.75 mIU/mL; (2) 7.9% and 9.6%, 5.26 mIU/mL; and (3) 12.7% and 13.3%, 3.43 mIU/mL, respectively.

Results

Trial 1: (4- to 5-Day Cultures)

Increased hCG concentrations were detected following each of 3 GnRH or cAMP treatment pulses at 2-h intervals. The peaks of GnRH- and cAMP-stimulated hCG concentrations were less from term trophoblast than from first-trimester placental cells (peak minus basal, 0.49 ± 0.05 and 0.53 ± 0.08 mIU/mL [Figs. 25.1a and 25.2a] vs 6.99 ± 1.47 and 8.82

± 2.05 [Figs. 25.1b and 25.2b], respectively, P < 0.01). Peaks in hCG concentration from term trophoblast cells in response to consecutive GnRH treatment pulses were consistent. Peaks in hCG concentration from first-trimester placental cells were greater following initial, rather than subsequent, GnRH stimuli (9.5 ± 1.3 vs 3.8 ± 1.1 mIU/mL, P < 0.05) (Fig. 25.1b). Peaks in hCG concentration from term trophoblast and first-trimester placental cells in response to consecutive cAMP treatment pulses were consistent (Figs. 25.2a and 25.2b). Basal hCG concentrations from term trophoblast cells returned to baseline between GnRH and cAMP pulses (Figs. 25.1a and 25.2a). Basal hCG concentrations from first-trimester cells returned to baseline between cAMP treatment pulses (Fig. 25.2b) but not between GnRH treatment pulses (Fig. 25.1b).

Trial 2: (4- to 5-Day First-Trimester Cultures)

Increased hCG concentrations were detected only following the first of 3 GnRH treatment pulses at 1-h intervals (Fig. 25.3). The increase in hCG concentration following the first GnRH treatment pulse was less than that by first-trimester cells in response to the first GnRH in trial 1, but similar to that following the second and third GnRH pulses in trial 1 (3.7 ± 1.6 vs 3.8 ± 1.1 mIU/mL, P < 0.05).

Trial 3: (4- to 5-Day vs 12- to 14-Day First-Trimester Cultures)

Baseline hCG concentrations from first-trimester placental cells cultured for 4–5 days were less than from cells cultured for 12–14 days (5.1 ± 1.3 vs 150 ± 18 mIU/mL, P < 0.01) (Figs. 25.4a and 25.4b). Peak hCG concentrations in response to GnRH pulses were less from 4- to 5-day than from 12- to 14-day cultures (10.4 ± 1.9 vs 220 ± 90 mIU/mL, P < 0.01).

Discussion

In this study, cell filtration prior to culturing may have excluded syncytiotrophoblast. In keeping with the 2-day period required for cytotrophoblast transformation to syncytiotrophoblast in culture (16), basal and GnRH-stimulated hCG secretion were detectable within 2–5 days of culture. In agreement with a study using explants (15), first-trimester placental cells were more responsive to GnRH than term trophoblast cells. Diminished or lost hCG response to GnRH stimulation of first-trimester placental cells at 2-h or 1-h intervals, respectively, suggested reduced GnRH receptor numbers following initial GnRH stimulation. Consistent hCG secretory responses to cAMP stimulation of first-trimester placental cells supported this conclusion and suggested that the reduced response to

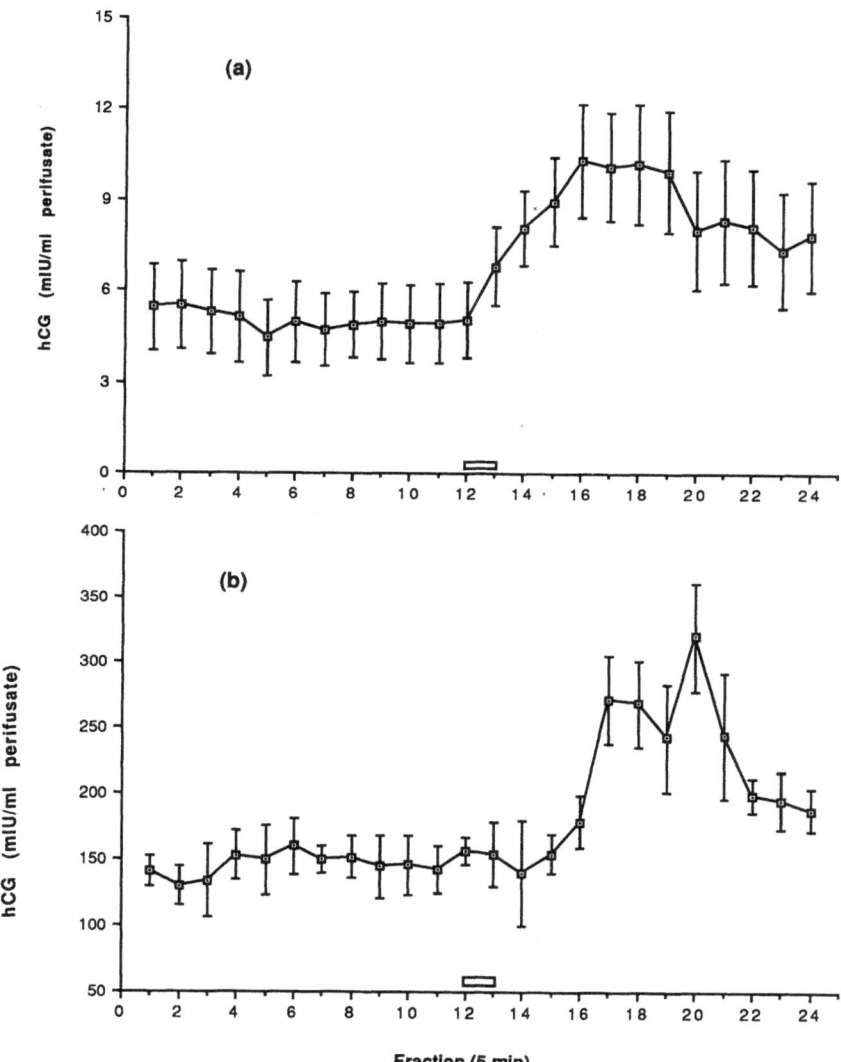

FIGURE 25.4. hCG concentrations (mean ± SE) in perifusate from first-trimester placental cells treated with GnRH (10^{-9} M) following (a) 4–5 days or (b) 12–14 days in culture (n = 15 chambers each). Bars over horizontal axes indicate GnRH treatment periods.

consecutive GnRH stimuli was not due to reduced hCG stores. Placental GnRH receptors may not be upregulated as quickly as predicted by studies using pituitary gonadotropes (17). The consistent, but lesser, response of term trophoblast cells to cAMP suggested that term trophoblast cells contained smaller hCG secretory stores than first-trimester placental cells. The lesser response of term trophoblast to GnRH likely reflected the lesser hCG secretory stores and may have reflected lessened efficiency of syncytiotrophoblast development. The consistent response of term trophoblast to repeated GnRH stimulation may have reflected lower initial GnRH receptor numbers and relatively undertaxed receptor recycling mechanisms.

In this study, basal hCG secretion did not depend upon exogenous GnRH stimulation. Cytotrophoblast cells synthesize and secrete GnRH (9), which may be responsible for hCG secretion in the absence of GnRH treatment. Increased basal and GnRH-stimulated hCG secretion from first-trimester cells cultured for 12–14 days suggested increased hCG synthesis. The results may have reflected continued syncytiotrophoblast development, increased GnRH synthesis, and increased GnRH receptor numbers with time in culture.

In summary, perifusion confirmed cAMP- and GnRH-stimulated hCG secretion from term trophoblast and first-trimester placental cells. Term trophoblast cells may have fewer GnRH receptors and lesser hCG stores than 1st trimester placental cells. Where fresh placental tissues are obtained, physical dissociation at 21°C may be more optimal than enzymatic dissociation at temperatures approaching 0°C. Cell plating in moderate dilution on carrier beads provided cultures that despite active secretory changes, required no supplementation other than noted for up to 2 weeks in culture.

Acknowledgments. Funding was by the Medical Research Council of Canada and British Columbia Health Care Research Foundation. W.D.C. is a recipient of a Natural Science and Engineering Research Council of Canada postdoctoral fellowship. P.C.K.L. is a Career Investigator of the British Columbia Children's Hospital and recipient of a Medical Research Council of Canada Scientist Award. G.L.S. is a recipient of a Ph.D. scholarship from the British Columbia Children's Hospital Foundation.

References

1. Butzow R. Luteinizing hormone releasing factor increases release of human chorionic gonadotropin in isolated cell columns of normal and malignant trophoblasts. Int J Cancer 1982;29:9–11.
2. Haning RV, Choi L, Kiggens AJ, Kuzma DL, Summerville JW. Effects of dibutyryl adenosine 3'-5'-monophosphate, luteinizing hormone-releasing hormone, and aromatase inhibitor on simultaneous outputs of progesterone, 17β-estradiol, and human chorionic gonadotropin by term placental explants. J Clin Endocrinol Metab 1982;55:213–8.

3. Khodr GS, Siler-Khodr TM. The effect of luteinizing hormone-releasing factor on human chorionic gonadotropin secretion. Fertil Steril 1978;30: 301–4.

4. Khodr GS, Siler-Khodr TM. Placental LRF and its synthesis. Science 1980; 207:315–7.

5. Kim SJ, Namkoong SE, Lee JW, Jung JK, Kang BC, Park JS. Response of human chorionic gonadotropin to luteinizing hormone-releasing hormone stimulation in the culture media of normal human placenta, choriocarcinoma cell lines and in the serum of patients with gestational trophoblastic disease. Placenta 1987;8:257–64.

6. Petraglia F, Lim ATW, Vale W. Adenosine 3′,5′-monophosphate, prostaglandins, and epinephrine stimulate the secretion of immunoreactive gonadotropin-releasing hormone from cultured human placental cells. J Clin Endocrinol Metab 1987;65:1020–5.

7. Siler-Khodr TM, Khodr GS. Dose response analysis of Gn-RH stimulation of hCG release from human term placenta. Biol Reprod 1981;25:353–8.

8. Siler-Khodr TM, Khodr GS, Vickery BH, Nestor JJ. Inhibition of hCG, α-hCG and progesterone release from human placental tissue in vitro by a GnRH antagonist. Life Sci 1984;32:2741–5.

9. Petraglia F, Volpe A, Genazzani AR, Rivier J, Sawchenko PE, Vale W. Neuroendocrinology of the human placenta. Front Neuroendocrinol 1990; 11:6–37.

10. Miyake A, Sakumoto T, Aono T, Kawamura Y, Maeda T, Kurachi K. Changes in luteinizing hormone-releasing hormone in human placenta throughout pregnancy. Obstet Gynecol 1984;60:444–9.

11. Siler-Khodr TM, Khodr GS, Valenzuela G. Immunoreactive gonadotropin releasing hormone level in maternal circulation throughout pregnancy. Am J Obstet Gynecol 1984;150:376–9.

12. Khodr GS, Siler-Khodr TM. Localization of luteinizing hormone releasing factor (LRF) in the human placenta. Fertil Steril 1978;29:523–6.

13. Belisle S, Guevin JF, Bellabarba D, Lehoux JG. Luteinizing hormone-releasing hormone binds to enriched human placental membranes and stimulates in vitro the synthesis of bioactive human chorionic gonadotropin. J Clin Endocrinol Metab 1984;59:119–26.

14. Currie AJ, Fraser HM, Sharpe RM. Human placental receptors for luteinizing hormone releasing hormone. Biochem Biophys Res Commun 1981;99: 332–8.

15. Siler-Khodr TM, Khodr GS, Valenzuela G, Rhode J. GnRH effects on placental hormones during gestation; 1. αhCG, hCG and hCS. Biol Reprod 1986;34:245–54.

16. Kliman HJ, Nestler JE, Sermasi E, Sanger JM, Strauss JF. Purification, characterization, and in vitro differentiation of cytotrophoblasts from human term placentae. Endocrinology 1986;118:1567–82.

17. Conn PM, Staley D, Jinnah H, Bates M. Molecular mechanism of gonadotrophin releasing hormone action. J Steroid Biochem 1985;23:703–10.

26

3β-Hydroxysteroid Dehydrogenase Activity of Human Placental Cytotrophoblast Following Differentiation In Vitro

K. Rajkumar and R. Zhai

The placenta originates from the trophectoderm of the implanting blastocyst. The villous cytotrophoblast and syncytiotrophoblast cells of human placenta are the main sites of hormone production. The cytotrophoblast cells mainly elaborate the hypothalamic releasing factors synthesized by placenta including gonadotropin releasing hormone, corticotropin releasing factor, somatostatin, and inhibin-like peptides (1–5). The syncytiotrophoblast cells are the main site for *human chorionic gonadotrophin* (hCG), *human placental lactogen* (hPL), *progesterone* (P_4), and estrogen synthesis (6–8). Given the different pattern of endocrine activities of the mononuclear cytotrophoblast and syncytium, it is evident that the differentiation of cytotrophoblast to syncytiotrophoblast entails major changes in expression of genes encoding hormones and in the machinery needed to synthesize them.

The biosynthesis of steroid hormones from cholesterol requires the participation of several steroidogenic enzymes. In human placental trophoblast, 3β-hydroxy-5-ene-steroid dehydrogenase (EC 1.1.1.145) and steroid 5-4-ene-isomerase (EC 5.3.3.1), hereafter referred to as 3βHSD, is an enzyme complex found in both microsomes and mitochondria (9–11). The enzyme 3βHSD is responsible for the conversion of all the Δ^5-3β-hydroxysteroids into Δ^4-3-ketosteroids. This enzymatic step is thus essential for the biosynthesis of all biologically active steroid hormones; namely, progesterone, glucocorticoids, mineralocorticoids, estrogens, and androgens (12). In view of the crucial role of 3βHSD in placental steroidogenesis, it is important to evaluate the changes in the activity of this enzyme as well as the gene expression in human placental cytotrophoblast before and following differentiation into syncytiotrophoblast.

Recent studies of Kliman et al. (13) demonstrate that when mononuclear cytotrophoblast culture in serum-supplemented medium, they aggregate and ultimately fuse to form large multinucleated syncytial structures. However, when cells were cultured in serum-free medium, the cells failed to aggregate and fuse (14). In the present study, we have used the cell culture model developed by Kliman et al. (13) to investigate the changes in 3βHSD in cytotrophoblast before and following differentiation into syncytiotrophoblast.

Materials and Methods

Preparation and Culture of Cytotrophoblasts

Normal term placenta were obtained immediately after uncomplicated cesarean section. The cytotrophoblast cells were isolated as described by Kliman et al. (13). The cytotrophoblast cells were cultured on Falcon 24-well culture dishes at a density of 2 million cells/mL either in the presence or absence of 10% *fetal bovine serum* (FBS) for 4 or 6 days. The medium and serum were replaced every 48 h. Some cultures were terminated at 96 h, and the media between 48 and 96 h were collected for the estimation of hCG (Amerlex-M βhCG RIA kit, Amersham), hPL (Amerlex hPL IRMA kit, Amersham), and progesterone (15). The remaining cultures were terminated at 144 h, and the media were collected for hormone estimation. At the termination of the experiment, the cells were collected for protein estimation (16).

To examine and compare the 3βHSD activity, the mononuclear cyto-trophoblast cells were cultured in Falcon 24-well culture dishes for 4 days in DMEM either in the presence or absence of serum. The cells were then provided with graded doses of exogenous pregnenolone (0.3–10 μg/mL) as substrate. Aminoglutethimide (50 μM), was added to cultures 24 h prior to the addition of exogenous substrate to inhibit endogenous pregnenolone production. The media were collected at 24 h following the provision of substrate for progesterone estimation. Progesterone levels in the media were expressed as μg/mg protein. In these experiments, media containing different doses of pregnenolone were also incubated with no cells to determine the blank values that were subtracted for estimates of the progesterone levels. To determine the time course effect of serum on changes in 3βHSD activity in cyto-trophoblast, the cells were cultured in the presence or absence of 10% FBS for 1, 2, 4, and 6 days. At designated time intervals, the 3βHSD activity was assessed as described above. To determine whether the continued presence of serum factors is necessary for increased 3βHSD activity in syncytiotrophoblast, the following experiment was performed. Cytotrophoblast cells were cultured in the presence of 10% FBS for 4 days to induce differentiation to syncytiotrophoblast. The cultures were then divided into 2 subgroups. In one subgroup of cultures, serum was withdrawn, while the other subgroup was cultured with serum. At 48, 96, and 144 h of serum withdrawal, the 3βHSD activity was assessed as described above.

Quantification of 3βHSD by Immunoblotting

Placental cytotrophoblast cells were cultured in Falcon 24-well culture dishes for 4 days in DMEM either in the presence or absence of 10% FBS. The cells were collected from the culture dishes and sonicated in 1-mL *phosphate buffered saline* (PBS) containing 1% cholate and 0.1% SDS. The samples (5 and 10 μg of protein) were applied to nitrocellulose filters using dot blot apparatus. The blots were then treated with a wash solution (5% fat-free milk [Carnation] and 0.1% Nonidet P-40 in PBS) and incubated for 2 h at room temperature with a 1:2000 dilution of rabbit antiserum raised against purified human placental 3βHSD (17). The blots were then extensively washed and further incubated an additional 2 h with ^{125}I-labeled antirabbit immunoglobulin (0.4 μCi/mL). At the end of incubation, the membrane was washed again and processed by autoradiography. The radioactivity adhering to the protein applied on the membrane was quantitated with a gamma counter.

RNA Isolation and Blot Analysis

Total cellular RNA from cytotrophoblast cells cultured in the presence or absence of serum for 4 days and from human placental tissue was obtained by ultracentrifugation through 5.7 M CsCl (18). The RNA was then extracted with phenol and chloroform, precipitated with ethanol, resuspended in DEPC-treated water, and quantified by absorbance at 260 nM. For RNA Northern blot analysis, placental tissue RNA was denatured in 50% formamide, 6% formaldehyde, and 1 × MOPS buffer (0.02 M MOPS, 5 mM sodium acetate, 1 mM EDTA [pH 7.0]) at 65°C and loaded onto 1% horizontal agarose gel and transferred to Hybond nylon membranes (Amersham, Arlington Heights, IL) by standard procedures. Membranes were crosslinked in a UV Stratalinker 2400 (Stratagene, La Jolla, CA). For RNA slot blot hybridization, RNA samples from cultures were denatured as described above, and 0.5 and 5 μg of sample was blotted onto nylon membranes using Bio-Rad slot blot apparatus. The membranes were then crosslinked as described above. The membranes were hybridized with a probe isolated from the plasmid hp3β-HSD63 containing the 1.5-kb cDNA clone for the human placental 3βHSD (17). A human γ-actin cDNA probe was used as a control. All cDNA were labeled by random primer synthesis (19) with [α-^{32}P]dCTP (3000 Ci/mmol) to a specific activity of 1–4 × 10^9 dpm/μg/DNA. The membranes were prehybridized at 45°C for 8 h in a buffer containing 50% formamide, 1 M NaCl, 10% dextran sulphate, 0.05 M Tris (pH 7.5), 1% SDS, and 100-μg salmon sperm DNA. The

[32]P-labeled cDNA probe was heat denatured before hybridization. After 12 h of hybridization at 45°C, the blots were washed once with 5× SSC, 3 times with 2× SSC + 0.1% SDS at room temperature, and 3 times with 2× SSC + 0.1% SDS at 65°C. The blots were then autoradiographed.

Statistical Analysis

The results of this study were subjected to nested analysis of variance (20). When a significant F-value was present, Fisher's least-significant difference test was used for individual comparison of means.

Results

When placental cells were separated on Percoll gradient, 3 distinct bands were visible. Red blood cells were found at the base of the gradient (density 1.138 g/mL). Connective tissue and small vessels were present at the top of the gradient (density 1.018 g/mL). In the middle of the gradient (density 1.05–1.07 g/mL), mononuclear cells were located. Light microscopic examination of the pellets from Percoll gradient-purified cells revealed that most of the cells were mononuclear, with few larger

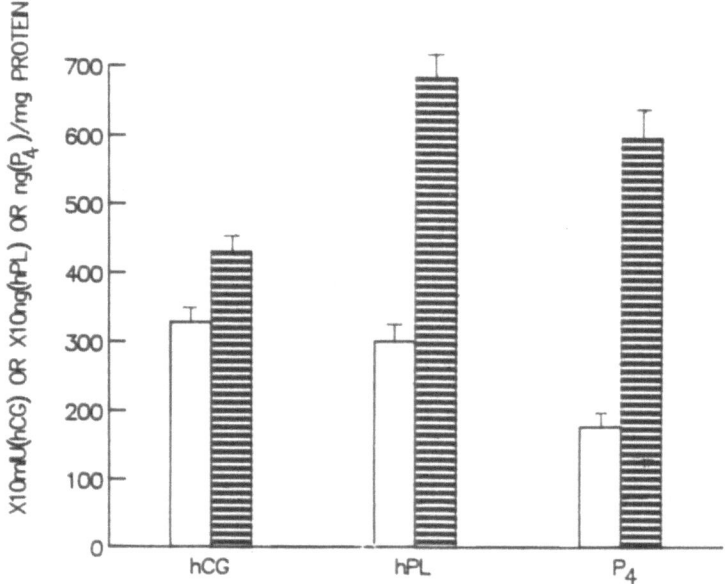

FIGURE 26.1. Effect of serum on hCG, hPL, and progesterone production in vitro. Cytotrophoblast cells were cultured in DMEM either in the presence (striped column) or absence (open column) of 10% fetal bovine serum. Media were collected between 48 and 96 h of culture for hormone estimation. Each bar represents (the mean ± SEM) of 5 replicates.

binucleated cells. Previous studies have characterized these cells as purified mononuclear cytotrophoblast cells (13). Cytotrophoblast cells cultured in the absence of serum secreted significant amounts of hCG, hPL, and progesterone between 48 and 96 h of culture (Fig. 26.1). When cultures continued for 2 additional days, the secretion of hCG, hPL, and progesterone was decreased (P < 0.01) (Fig. 26.2). In contrast, when cytotrophoblast cells were cultured in the presence of serum, hCG and hPL secretion between 96 and 144 h of culture was greater than the secretion levels between 48 and 96 h of culture (P < 0.01) (Figs. 26.1 and 26.2). The secretion of progesterone did not change significantly with time when cells were cultured in the presence of serum (P > 0.05) (Figs. 26.1 and 26.2). Analysis of variance revealed that the addition of serum to cytotrophoblast cells in culture significantly increased the secretion of hCG, hPL, and progesterone (P < 0.01) (Figs. 26.1 and 26.2). Cytotrophoblast cells following 4 days of culture with serum secreted hCG, hPL, and progesterone 5.8-, 13.7-, and 5.9-fold more, respectively, than did serum-free cultures (Fig. 26.2).

Figure 26.3 depicts the ability of cytotrophoblast cells cultured in the presence or absence of serum for 4 days to convert exogenous

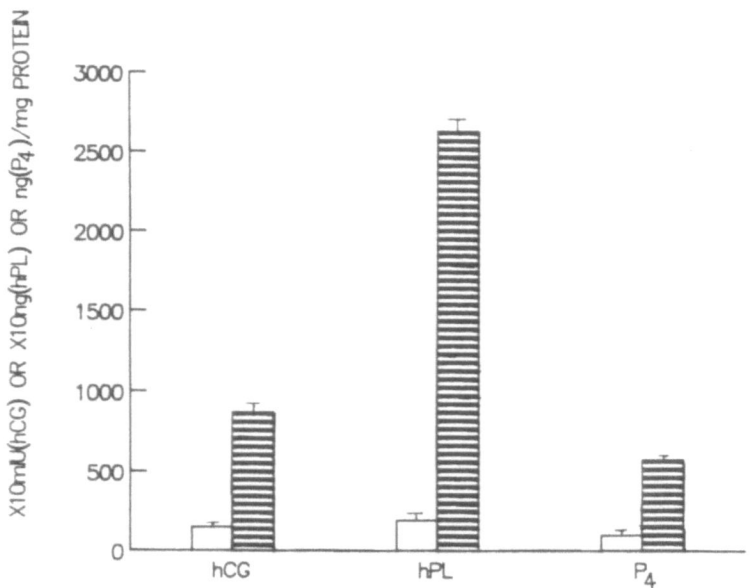

FIGURE 26.2. Effect of serum on hCG, hPL, and progesterone production in vitro. Cytotrophoblast cells were cultured in DMEM either in the presence (striped column) or absence (open column) of 10% fetal bovine serum. Media were collected between 96 and 144 h of culture for hormone estimation. Each bar represents the mean ± SEM of 5 replicates.

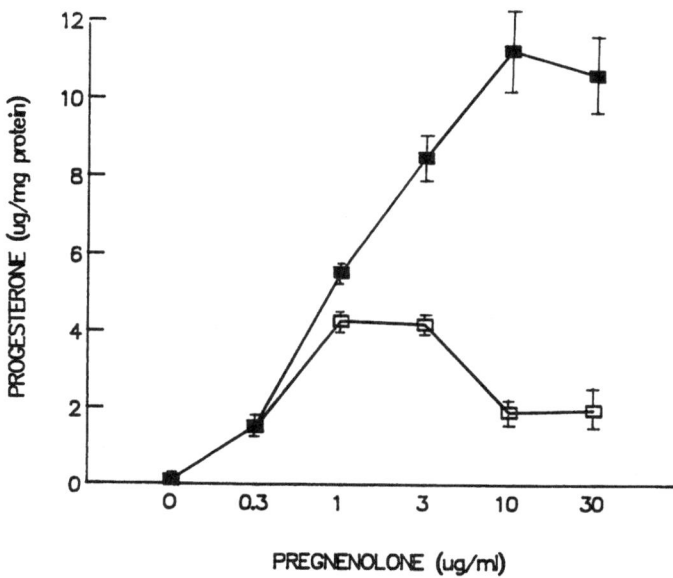

FIGURE 26.3. Effect of serum on 3βHSD activity in placental cells. Cytotrophoblast cells were cultured in DMEM either in the presence (solid squares) or absence (open squares) of 10% fetal bovine serum for 4 days. The ability of cells to convert graded doses of exogenous pregnenolone to progesterone was determined as described in "Materials and Methods." The mean ± SEM of 6 replicates is presented.

pregnenolone to progesterone. When cells were cultured in the absence of serum, the provision of 0.3-μg/mL pregnenolone as substrate significantly enhanced progesterone secretion ($P < 0.01$). The secretion of progesterone further increased ($P < 0.01$) with an increase in the amount of substrate to 1 μg/mL. However, further increase in the substrate availability (3 to 30 μg/mL) did not further enhance progesterone secretion. Progesterone secretion by cytotrophoblast cells was decreased ($P < 0.01$) with 10 and 30 μg/mL of pregnenolone as substrate. When cytotrophoblast cells were cultured in the presence of serum, the provision of pregnenolone as exogenous substrate had a dose-dependent, stimulatory effect on progesterone secretion. The maximal stimulatory effect on progesterone secretion was observed with 10-μg/mL substrate (Fig. 26.3). When 0.3-μg/mL pregnenolone was provided as substrate, no significant difference was observed in the ability of cytotrophoblast cells to convert it to progesterone, whether the cells were cultured with or without serum. However, when the amount of substrate provided was increased (1–30 μg/mL), cells cultured in the presence of serum had an increased ability to secrete progesterone ($P < 0.01$). The maximal secre-

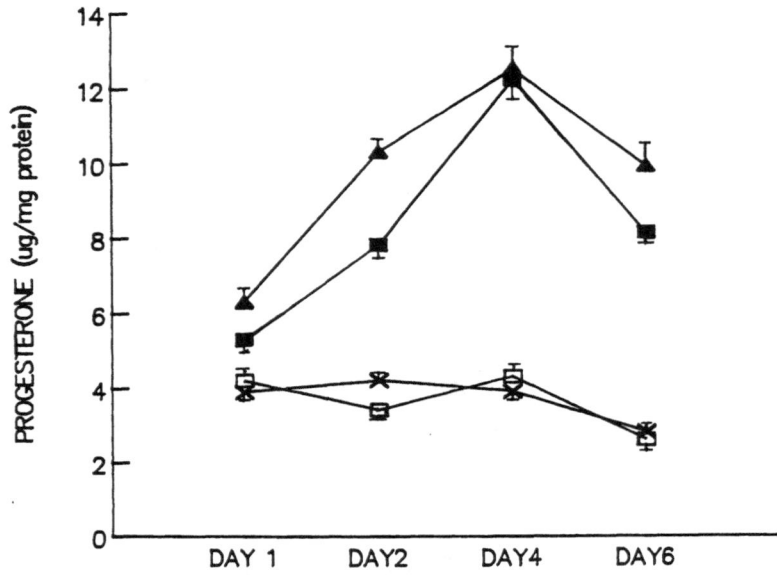

FIGURE 26.4. Time course effect of serum on 3βHSD activity. Cytotrophoblast cells were cultured in DMEM either in the presence (solid squares and triangles) or absence (open squares and X's) of 10% fetal bovine serum. 3βHSD activity was determined at different time periods of exposure to serum by providing either 3 μg/mL (solid and open squares) or 10 μg/mL (solid triangles and X's) of pregnenolone as substrate. The mean ± SEM of 6 replicates is presented.

tion of progesterone was 2.7-fold higher in cells cultured with serum compared to serum-free cultures.

The time course effect of serum on changes in 3βHSD activity in cytotrophoblast is depicted in Figure 26.4. When cytotrophoblast cells were cultured with serum for 1 day, a slight but significant ($P < 0.01$) increase in the ability of cells to convert 3 and 10 μg of pregnenolone to progesterone was observed as compared to serum-free culture. With an increase in the time of culture with serum, a further increase in the ability of cells to convert substrate to progesterone was observed ($P < 0.01$). Cells cultured for 4 days with serum had the maximal ability to convert pregnenolone to progesterone. When cells were cultured for 6 days in serum, a slight decrease ($P < 0.01$) in the ability of cells to convert pregnenolone to progesterone was observed. In the experiments done to examine the effect of serum withdrawal on the ability of differentiated cells to convert pregnenolone to progesterone, a 30% decrease was observed within 48 h of serum withdrawal ($P < 0.01$) as compared to the ability of cells that continued to be cultured with serum. At 96 and 144 h of serum withdrawal, the ability of cells to convert 10-μg pregnenolone to

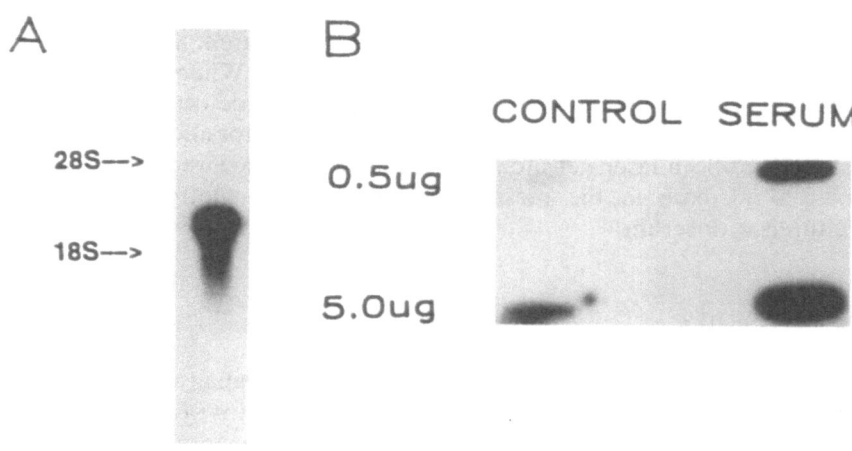

FIGURE 26.5. Northern blot hybridization of RNA from human placenta (column A). Ten-micrograms total RNA from placenta was separated in agarose-formaldehyde gels and transferred to nylon membranes. In column B, total RNA from cytotrophoblast cells cultured in the presence or absence of serum for 4 days was denatured and applied to nylon membranes at the indicated concentration. Hybridization to a 3βHSD cDNA probe and washes were performed under highly stringent conditions, as described in "Materials and Methods." Exposure time was 8 h.

progesterone was decreased by 40% ($P < 0.01$). However, even after 6 days of serum withdrawal, the cells had a 2-fold increased ability to convert pregnenolone to progesterone as compared to cells that were maintained in serum-free media from the day of plating ($P < 0.01$).

The autoradiograph from 3βHSD protein immunoblot studies indicated an increase in the quantity of 3βHSD protein in cytotrophoblast cells cultured in the presence of serum as compared to serum-free cultures. When 5 and 10 μg of total cellular protein was applied to the nitrocellulose membrane, the radioactivity associated with the protein following immunoblot studies was 881 ± 57 and 1308 ± 58 cpm, respectively, for cytotrophoblast cells cultured with serum. However, when the same amount of cellular protein was applied from serum-free cultures, the radioactivity associated with the protein was 440 ± 17 and 646 ± 27 cpm, respectively, which was significantly less ($P < 0.01$) than that of cells cultured with serum.

Figure 26.5, column A, illustrates that in Northern analysis, an α^{32}P-cDNA probe against human placental 3βHSD hybridized to a single band in placental RNA preparation. The size of the message for which the cDNA probe was specific was 1.8 kb, as determined by DNA size markers. Figure 26.5, column B, depicts the autoradiograph from slot blot hybridization. When cytotrophoblast cells were cultured in the presence of serum for 4 days, there was an apparent increase in the 3βHSD mRNA as compared to serum-free cultures. When the membrane was rehybridized with γ-actin probe, there was no detectable difference between the two preparations. The scanning of autoradiograms with an LKB Ultro-Scan laser densitometer (LKB, Bromma, Sweden) revealed a 5.8-fold increase in the message for 3βHSD in cytotrophoblast cells cultured with serum.

Discussion

The salient finding of this study is that when human placental cytotrophoblasts undergo functional differentiation in vitro, an increase in 3βHSD mRNA as well as 3βHSD enzyme quantity ensues in the cells. The physiological consequence of such an effect is the enhanced capability of cells to convert the Δ^5-3β-hydroxysteroids into Δ^4-3-ketosteroids. Therefore, an increase in the quantity of 3βHSD in cells catalyzes the synthesis of progesterone from maternal pregnenolone and the production of androstenedione from fetal dehydro-epiandrosterone sulfate (21), which is eventually aromatized to estrogen.

Cytotrophoblast cells, when cultured in the presence of serum, aggregate and fuse. At 96 h after plating, light microscopic examination demonstrated the presence of multinucleated cells. These changes were not evident in cells cultured without serum. This corroborates the findings of Kliman et al. (13). Cytotrophoblast cells cultured in the presence of serum secreted increased quantities of hCG, hPL, and progesterone compared to serum-free culture. This implies that the cells cultured in the presence of serum undergo not only morphological but also functional differentiation. Previously, Kliman et al. (13) demonstrated with immunoperoxidase staining technique that cytotrophoblast cells cultured in the presence of serum display a progressive increase in hCG and hPL with increasing numbers of aggregates and syncytia.

Cytotrophoblast cells cultured in the absence of serum secreted significant amounts of hCG and hPL. This indicates that secretion of hCG and hPL is not specific to syncytial cells. In situ hybridization studies have shown that hCG is present in both cytotrophoblast and syncytial cells. Indeed, hCGα mRNA is maximal in the syncytial cells, but is also initially expressed in the cytotrophoblast (22, 23). Although the synthesis of hPL has been shown only in the syncytial cells, there is evidence suggesting that the transcription of hormonal genes is not typical of a single stage of trophoblast differentiation, but may occur in all cell types, with a more pronounced expression in a specific cell population (24).

When cytotrophoblast cells undergo functional differentiation, they acquire an increased capacity to convert pregnenolone to progesterone compared to serum-free cultures. When graded amounts of substrate were provided to cultures, differentiated cells could convert 10 times the quantity of substrate compared to serum-free cultures. However, when low amounts of substrate were provided (0.3 µg/mL), the ability of cells to convert the substrate to progesterone was not different in the two groups. This indirectly suggests that following functional differentiation, the 3βHSD enzyme quantity increases, but the rate at which the enzyme can convert pregnenolone to progesterone may not be altered. The results of 3βHSD immunoblot studies support the hypothesis that functional differentiation leads to an increased quantity of 3βHSD. Further quantification of 3βHSD mRNA indicates an increase in cells following functional differentiation. The close correlation observed in the 3βHSD mRNA, protein content, and activity levels suggests that changes of 3βHSD activity in cytotrophoblast following differentiation into syncytiotrophoblast cells are controlled at the level of 3βHSD gene expression and/or 3βHSD mRNA stability.

Time course studies indicate that there was a slight increase in the activity of 3βHSD within 24 h of culture. The activity progressively increased with time, reaching maximum by 96 h. When cells were cultured for an additional 2 days, a decrease in 3βHSD activity was observed. We believe this decrease is an experimental artifact, as we have not observed this phenomenon in other experiments when the cells were cultured up to 10 days. The morphological studies of Kliman et al. (13) demonstrate the presence of few syncytia by 24 h of culture, which progressively increased at 48- and 72-h serum. A good correlation between the morphological findings of Kliman et al. (13) and the present time course studies on 3βHSD activity supports the view that the changes in enzyme quantity are a consequence of cell differentiation. It is well known that serum contains various growth factors. Insulin and insulin-like growth factor 1 have been demonstrated to stimulate 3βHSD activity in placental cytotrophoblasts when the cells were exposed for 1 day (24). The observed increase in 3βHSD activity in functionally differentiated cells may not be entirely due to the presence of insulin and insulin-like growth factor in the serum, as in time course studies, a 1-day exposure to serum had a minimal increase in 3βHSD activity as compared to 4 days of culture. Further, the experiments indicate that even after 6 days of serum withdrawal, the cells had increased 3βHSD activity compared to serum-free cultures. However, as serum-withdrawal experiments indicate a partial decrease (40%) in 3βHSD activity in cells where serum is withdrawn as compared to cells that continue to be cultured with serum, we suggest that serum factors (insulin and insulin-like growth factors?) may be necessary for an optimal increase in 3βHSD activity.

In summary, we conclude that placental cytotrophoblast cells, when cultured with serum, undergo functional differentiation with an increased ability to secrete hCG, hPL, and progesterone. An increase quantity of

3βHSD correlates well with functional differentiation. However, serum factors are necessary for the optimal induction of the enzyme in functionally differentiated cells.

Acknowledgments. We thank D. Caird and S. Lakshman for excellent technical assistance and Dr. F. Labrie for the provision of the human 3βHSD probe. This study was supported by a grant from the Canadian Diabetes Association to K. Rajkumar.

References

1. Petraglia F, Vaughn J, Vale W. Inhibin and activin modulate the release of gonadotropin-releasing hormone, human chorionic gonadotropin and progesterone from cultured placental cells. Proc Natl Acad Sci USA 1989; 86:5114–7.
2. Petraglia F, Lim ATW, Vale W. Adenosine 3′,5′-monophosphate, prostaglandins and epinephrine stimulate the secretion of immunoreactive gonadotropin releasing hormone from cultured human placental cells. J Clin Endocrinol Metab 1987;65:1020–5.
3. Robinson BG, Emanuel RL, Frim DM, Majzoub JA. Glucocorticoid stimulates expression of corticotropin releasing hormone gene in human placenta. Proc Natl Acad Sci USA 1988;85:5244–8.
4. Petraglia F, Sawchenko PE, Rivier J, Vale W. Evidence for local stimulation of ACTH secretion by corticotropin-releasing factor in human placenta. Nature 1987;328:717–9.
5. Watkins WB, Yen SSC. Somatostatin in cytotrophoblasts of immature human placenta: localization by immunoperoxidase cytochemistry. J Clin Endocrinol Metab 1980;50:969–71.
6. Hoshina M, Boothby M, Boime I. Cytological localization of chorionic gonadotropin and placental lactogen mRNA during development of the human placenta. J Cell Biol 1982;93:190–8.
7. Beck TG, Schweikhart G, Stolz E. Immunohistochemical location of hPL, spl, β-hCG in normal placentas of varying gestational age. Arch Gynecol 1986;239:63–74.
8. Fournet-Dulguerov N, Maclusky NJ, Leranth CZ, et al. Immunohisto-chemical localization of aromatase, cytochrome P-450 and estradiol dehydrogenase in the syncytiotrophoblast of the human placenta. J Clin Endocrinol Metab 1987;65:757–64.
9. Koide SS, Torres MT. Distribution of 3β-hydroxysteroid dehydrogenase and Δ^5-3-oxosteroid isomerase in homogenate fractions of human term placenta. Biochim Biophys Acta 1965;105:115–20.
10. Ferre F, Breuiller M, Cedard L, et al. Human placental Δ^5-3β hydroxysteroid dehydrogenase activity (Δ^5-3β HSDH): intracellular distribution, kinetic properties, retroinhibition and influence of membrane delipidation. Steroids 1975;26:551–70.
11. Rabe T, Brandstetter K, Kellerman J, Runnebaum B. Partial characterization of placental 3β-hydroxysteroid dehydrogenase (EC 1.1.1.145) Δ^{4-5} isomerase

(EC5.3.3.1) in human term placental mitochondria. J Steroid Biochem 1982;17:427–33.

12. Lipsett MB. In: Yen SSC, Jaffe RB, eds. Reproductive endocrinology. Philadelphia: Saunders, 1986:140–53.

13. Kliman HJ, Nestler JE, Sermasi E, Sanger JM, Strauss III JF. Purification, characterization and in vitro differentiation of cytotrophoblasts from human term placentae. Endocrinology 1986;118:1567–82.

14. Feinman MA, Kliman HJ, Caltabiano S, Strauss III JF. 8-bromo-3',5'-adenosine monophosphate stimulates the endocrine activity of human cytotrophoblasts in culture. J Clin Endocrinol Metab 1986;63:1211–7.

15. Rajkumar K, Malinek J, Murphy BD. Effect of lipoproteins and luteotrophins on progesterone accumulation by luteal cells from the pregnant pig. Steroids 1985;45:119–34.

16. Lowry OH, Rosebrough NJ, Farr AL, Randall RJ. Protein measurement with the Folin phenol reagent. J Biol Chem 1951;193:265–75.

17. Luu-The V, Lachance Y, Labrie C, et al. Full length cDNA structure and deduced amino acid sequence of human 3β-hydroxy-5-ene steroid dehydrogenase. Mol Endocrinol 1989;3:1310–15.

18. Maniatis T, Fritsch EF, Sambrook J. Molecular cloning: a Laboratory manual. Cold Spring Harbor, NY: Cold Spring Harbor Laboratory, 1982.

19. Feinberg AP, Vogelstein B. A technique for radiolabelling DNA restriction endonuclease fragments to high specific activity. Anal Biochem 1983;132:6–13.

20. Sokal RR, Rohlf FJ. In: Sokal RR, Rohlf FJ, eds. Biometry. New York: WH Freeman, 1981:271–320.

21. Csapo AI. In: Knight J, O'Connor M, eds. The fetus and birth. Amsterdam: Elsevier, 1977;47:159–210.

22. Hoshina M, Boothby M, Boime I. Cytological localization of chorionic gonadotropin and placental lactogen mRNAs during the development of human placenta. J Cell Biol 1982;93:190–8.

23. Wide M, Persson H, Lunkvist O, Wide L. Localization of mRNA for the β-subunit of placental hCG by in situ hybridization. Acta Endocrinol (Copenh) 1988;119:69–74.

24. Petraglia F, Volpe A, Genazzani AR, Rivier J, Sawchenko PE, Vale W. Neuroendocrinology of the human placenta. Front Neuroendocrinol 1990;11:6–37.

25. Nestler JE. Insulin and insulin-like-growth factor-1 stimulate the 3β-hydroxysteroid dehydrogenase activity of human placental cytotrophoblasts. Endocrinology 1989;125:2127–33.

Author Index

Subject Index